GEORGE GAYLORD SIMPSON
PALEONTOLOGIST AND EVOLUTIONIST

GEORGE GAYLORD SIMPSON
PALEONTOLOGIST AND EVOLUTIONIST

Léo F. Laporte

COLUMBIA UNIVERSITY PRESS
NEW YORK

Columbia University Press
Publishers Since 1893
New York   Chichester, West Sussex

Copyright © 2000 Columbia University Press
All rights reserved

Library of Congress Cataloging-in-Publication Data
Laporte, Léo F.
George Gaylord Simpson : paleontologist and evolutionist / Léo F. Laporte.
p. cm.
Includes bibliographical references and index.
ISBN 0–231–12064–8 (cloth : alk. paper) — ISBN 0–231–12065–6 (pbk. : alk. paper)
1. Simpson, George Gaylord, 1902– 2. Paleontologists — United States — Biography. I. Title.

QE707.S55 L36 2000
560'.92 — dc21
[B]   00–029030

Casebound editions of Columbia University Press books are printed on permanent and durable acid-free paper.
Printed in the United States of America
c 10 9 8 7 6 5 4 3 2 1
p 10 9 8 7 6 5 4 3 2 1

For the children of the new millennium
Noel, Abby, Henry, William
A Posse Ad Esse

CONTENTS

Preface xi

Acknowledgments xv

1. Biographical Introduction   1

2. Paleontology and the Expansion of Biology   17
   Biology and Fossils at Yale   19
   Biology and Fossils at the American Museum   21
   Simpson's Biological Insights of the 1930s   25
   Tempo and Mode in Evolution   29
   Response to *Tempo and Mode*   31
   Adoption of the Population Concept of Species   34
   Disciplinary Independence for Paleontology   37

3. The Summer of 1924   41
   "A Very Promising Graduate Student"   41
   "Digressing All Over the Landscape"   45

"Some Discovery to Rival the Boss's"   47
"A Considerable Piece of Work"   50
"Bone is Abundant"   56
"Busy Collecting Other Men's Prospects"   57
"First Paper I Ever Wrote"   61
"Beginning a Series of Raids"   62

4. Darwin's World   67
Prologue   67
Presbyterianism and Apostasy   70
Evolution and Darwin   72
Six Darwinian Themes   73
Selected Simpson Commentaries on Darwin   78

5. Paleocene Mammals of Montana   80
The Invitation   81
First Field Season — 1932   83
South American Interlude   84
Second Field Season — 1935   86
"New Ways of Going at Things"   87

6. On Species   96
From Types to Statistical Inferences   98
From Statistics to Ecology   105
From Species to Higher Categories   107
Species as an Evolutionary Concept   109

7. *Tempo and Mode in Evolution*   112
Two Door-Openers   113
The New Genetics   115

Genetics and the Origin of Species   117
The Contemporary Paleontology   120
Tempo and Mode in Evolution   126
Micro-, Macro-, and Mega-Evolution   139
Low-Rate and High-Rate Lines   144
Inertia, Trend, and Momentum   147
Organism and Environment   151
Modes of Evolution   155
Reception and Impact   162
Final Perspective   171

8. Mentor for Paleoanthropology   174
The Evolutionary Synthesis   175
Simpson's Credentials   176
Simpson as Mentor   178
Simpson as Critic   182
Simpson as Apologist   189

9. Wrong for the Right Reasons   195
Background   196
Wegener and Fossils   197
Simpson's Response to Drift   198
Structure of Simpson's Rebuttal   200
The Plate Tectonics Revolution   213
Simpson's Conversion   215
Summing Up   217

10. The Mind's Eye   221
Showing What Is Observed   223
Visual Induction   223

Visual Deduction 227
The Visual as Effect and Affect 236

11. The Awkward Embrace 240
Prelude to the American Museum Appointment 241
Simpson Joins the American Museum 241
American Museum Research and Expeditions 243
New Directions with a New Director 245
Simpson as Administrator 247
Accident in Brazil 249
"Plus Ça Change . . ." 251

12. Concession to the Ineluctable 256
Magruder's Misadventure 257
The Nature of Time 259
Past Life's Lessons 261
Melancholia 263
Simpson as Sam Magruder 264

Notes 267
Bibliography 305
Index 319

PREFACE

He who understands everything about his subject cannot write it. I write as much to discover as explain.
— Arthur Miller, playwright

David McCullough, the historian, has said that he chooses the topics for his books he does because those are the topics he himself would like to read. I share that sentiment, because some two decades ago I wanted to know more about George Gaylord Simpson (1902–1984). As a paleontologist I had read most of his books, many of his articles, and a few of his monographs. My curiosity was thus piqued to learn more about how his theories developed, how his arguments were constructed, and from what specific knowledge they proceeded. At the time, there was very little literature analyzing Simpson's work and, as I came to discover, most was superficial and concerned with just a few aspects of his voluminous research. Occasionally too, authors couldn't resist the temptation to trivialize his accomplishments, if only to aggrandize their own efforts.

Initially I thought of writing a biography, but several circumstances soon dissuaded me. In 1978, Simpson published his autobiography, *Concession to the Improbable*, which provided most of the descriptive facts of his life. Also, as background for the book still forming in my mind, I interviewed some three dozen persons who knew Simpson — members of his family, former students, and professional colleagues. Despite their varying views about him, they led me to the conclusion that a biography would be of much less interest than a closer examination of Simpson's major areas of research. To be sure, such discussion would need relevant biographical background, but conventional full-scale biographical treatment was less urgent.

Another event was the mention to me by his sister Martha when I interviewed her that there was in the archives of the American Philosophical Society a batch of letters written by Simpson mostly to her, some to his parents, and a few to his future second wife that spanned a half century from his teenage years to late middle age. I went to Philadelphia, and reading the letters (156 in all) made me realize that these had precedent over anything I was ready to write, and deserved priority for publication. After Simpson's death, another fifty-four family letters turned up among his personal papers. I spent much of the next few years transcribing the letters, including decoding a number of them written in Simpson's own personal code to his future second wife. Eventually, all the letters were published, providing a rich biographical context to complement his autobiography (Simpson 1987).

The concept of the book thus evolved from conventional biography to a case of "reverse engineering," where one dismantles an object of interest into its constituent parts to understand how it is constructed and the way it works. The book now before you, therefore, is not biographical in the ordinary and strict sense. However, it does progress more or less chronologically through Simpson's life, focusing on major aspects of his research (fossil mammals, evolutionary theory, biogeography, taxonomy and systematics), his ways of approaching scientific problems (continental drift, visual way of thinking), and his impact on related disciplines (biology and physical anthropology). When particularly relevant, biographical material is included, such as his early education as a paleontologist (especially the influence of Darwin and W. D. Matthew), his often difficult relations with his museum colleagues, and the apparent melancholia toward the end of his life. There is also a brief biographical introduction that provides the essential facts of Simpson's life within which his work can be placed.

The chapters here have been published previously but adapted to create a single work. The justification for bringing them together in one volume arises from the broad range of their original places of publication: disciplinary journals (history of biology, history of earth sciences, physical anthropology, paleontology) and separate volumes published by the American Philosophical Society, American Society of Zoologists, California Academy of Sciences, and the Geological Society of America. Much of this variety reflects the diversity and scope of subject matter that Simpson addressed in his research. Some is simply the serendipity of my being invited to participate in a symposium or conference whose proceedings were then published by the sponsors. Bringing these articles together reinforces several times over the importance and impact that Simpson had on American science. I believe the whole, as presented here in one volume, is greater than the sum of its parts scattered in time and place.

Chapter 1 summarizes Simpson's life, thereby providing a biographic context within which his scientific work can be viewed. Chapter 2 gives a chronological overview of Simpson's scientific accomplishments that indicates their significance and how it is that Simpson became to be the most important paleontologist of the middle part of the twentieth century as well as one of its leading evolutionary theorists. The third chapter describes how Simpson, a young graduate student at Yale, got jump-started on his career and the special importance of the mentorship provided by William Diller Matthew, vertebrate paleontologist at the American Museum of Natural History in New York.

Chapter 4 reaches back into Simpson's late teenage years, showing the impact of his reading of Charles Darwin's *On the Origin of Species* and how it reinforced his turn away from formal religion. Darwin's influence was so strong that Simpson, over his long career, regularly celebrated what Darwin has to teach us, philosophically as well as biologically. Chapter 5 follows the various threads of Simpson's scientific development, especially as brought together in his several-years-long study of early mammals, that led to his most important single scientific contribution, which was the validation of the new genetics and population biology of the 1930s by their corroboration and further elucidation from the fossil record. In the sixth chapter we see how Simpson's concept of species moved from species-as-types, embodying some invariable ideal abstract quality, to species-as-populations, composed of variable organisms upon which Darwinian natural selection operated. Chapter 7 discusses in detail Simpson's *Tempo and Mode in Evolution*, which represents the culmination of his ideas on evolutionary rates and patterns and was his contribution to the evolutionary synthesis that was consolidating between 1936–1947.

Chapter 8 describes how the evolutionary synthesis was incorporated into the understanding of human evolution and the special role Simpson played in first educating the younger generation of physical anthropologists returning from World War II, and then subsequently supporting their interpretation of the human fossil record as another instance of Darwinian evolution. Chapter 9 examines the reasons for Simpson's strong opposition to continental drift, based as it was upon his independently developed principles of historical biogeography. Until the development of plate tectonics theory in the 1960s, Simpson's explanation for similar fossils on widely separated continents by the natural dispersal of organisms during their lifetime was scientifically more credible than appeal to Wegener's wandering continents carrying their postmortem remains by as yet unknown mechanisms.

The tenth chapter considers Simpson's use of visual imagery in formulating and expressing theoretical concepts, a style of reasoning that he particularly found most suitable. Chapter 11 describes Simpson's relations with his institutional employers — the American Museum and Harvard's Museum of Comparative Zoology — that were often made more difficult by hypersensitivity regarding what he thought was due him as one of the world's leading scientists of his day. Chapter 12 concludes the book by reviewing a posthumously published novella that Simpson wrote toward the end of his life. It reflects a melancholy view of life that contrasts with the more optimistic attitudes found in his earlier work as, for example, in his Darwin writings.

The span of years over which the research in this book has occurred — some two decades — is the time it has taken me to educate myself in each of the different spheres of Simpson's work and life. Moreover, until recently, academic and family responsibilities precluded my devoting full attention to this project.

This book is merely a first attempt at examining, interpreting, and evaluating exactly what George Gaylord Simpson accomplished professionally in his life as paleontologist and evolutionist. I hope — indeed expect — others will be stimulated by this initial effort to take up from where I have left off, whether by addition, expansion, or correction.

Léo F. Laporte
Redwood City, California

ACKNOWLEDGMENTS

The writing of history is notoriously slow.
— Mott T. Greene, science historian

A work as protracted as this in coming together in its present form has necessarily depended on the support and assistance of many friends and colleagues. First and foremost I greatly appreciate the initial encouragement of the late Anne Roe Simpson, who persuaded her husband to allow me access to his personal papers and soon after to himself for several long interviews at their home in Tucson. And never thereafter did she attempt to influence in any way my approach to this research, even though I know that more than once she felt her privacy and that of her family was being intruded upon. I have the same high regard for Simpson's three daughters — Helen, Joan, and Elizabeth — who generously gave me so much of their time in interviews. The latter two have continued to provide much helpful information and insight about their father, and like their stepmother, Anne, never tried to dissuade me of my own opinion in those few instances where we differed.

I am grateful to all the more than three dozen family members, colleagues, and former students who allowed me to interview them and offered their candid views about Simpson, both as a person and as a scientist. Although I have quoted here only a small fraction of what was told me, their various "takes" on Simpson provided a rich context in which to frame my research. They include: Anne Roe Simpson, Helen Simpson Vishniac, Joan Simpson Burns, Elizabeth Simpson Wurr, Margaret Simpson Peck (sister), Martha Simpson Eastlake (sister), Robert

Roe (brother-in-law); Francisco Ayala, David Bardack, Frank Beach, Joseph Birdsell, Preston Cloud, Edwin Colbert, Kenneth Cooper, Joseph Gregory, John Imbrie, Richard Kay, David Kitts, Richard Lewontin, Ernst Mayr, Paul McGrew, Malcolm McKenna, Louis Monaco, John Moore, Norman Newell, Rachel Nichols, Everett Olson, John Ostrum, Colin Pittendrigh, Rodolfo Ruibal, Bobb Schaeffer, G. Ledyard Stebbins, Richard Tedford, James Valentine, Sherwood Washburn, George Whitaker, Ernest Williams, Albert Wood, Florence Wood, and Nelda Wright.

I want to thank specifically the following colleagues who reviewed drafts of one or more papers that later became the basis for the chapters here — some of these reviews occurred long enough ago that they may be surprised to see their names here: Michele Aldrich, David Archibald, Joseph Cain, Richard Cifelli, William Clemens, Sheila Dean, Jon Endler, Phillip Gingerich, Elihu Gerson, Bentley Glass, Mott Greene, Alan Leviton, Bruce MacFadden, Jonathan Marks, Ernst Mayr, Kevin Padian, Donald Prothero, Ronald Rainger, Marc Swetlitz, Richard Tedford, Sherwood Washburn, David Webb, and Adrienne Zihlman.

Kennard Bork, Denison University, reviewed the entire manuscript and provided invaluable help in clarifying and correcting a number of grammatical, typographical, and logical infelicities and mistakes. His generous support came at a crucial moment when I needed all the help I could get.

I must also acknowledge and thank Kenneth Taylor, University of Oklahoma, and Hugh Torrens, University of Keele, who by their example and enthusiasm for the history of geology aided me in raising my own standards and staying the course.

I thank the American Philosophical Society, the American Museum of Natural History, and the National Museum of Natural History for access to their archives filled with Simpsoniana that I have barely touched.

The Faculty Research Committee, University of California at Santa Cruz, provided funds from time to time, over the years, to help defray some of the costs of this research.

I'm grateful, too, to publisher Holly Hodder, manuscript editor Roy Thomas, and assistant editor Jonathan Slutsky of Columbia University Press, whose experience and skill enabled this project to reach its desired completion. And, finally, I thank my wife Margaret who, in so many ways, constantly offered encouragement and support not only for this effort but also for all else worthwhile in my life.

GEORGE GAYLORD SIMPSON
PALEONTOLOGIST AND EVOLUTIONIST

FIGURE 1.1 George Gaylord Simpson at his desk at the American Museum of Natural History in the late 1940s. (American Museum of Natural History)

CHAPTER 1

BIOGRAPHICAL INTRODUCTION

The past is not the present: pretending it is corrupts . . . and thus . . . shrivels the imagination and conscience.
— Paul Fussell, literary critic

By the late 1930s and early 1940s Simpson, though relatively young, was already a distinguished paleontologist at the American Museum of Natural History in New York City. His achievements included a Yale doctorate in geology and paleontology, a position as a visiting research scholar at the British Museum, leader of two year-long fossil-collecting expeditions to Patagonia, author of two books and more than one hundred scientific articles and monographs, and newly elected fellow of two of the most distinguished honorary and scholarly societies, the American Philosophical Society and the National Academy of Sciences. As new discoveries, ideas, and theories in genetics were being published, Simpson understood their importance and kept himself informed. Before leaving for military service in late 1942, Simpson completed a major revolutionary text titled *Tempo and Mode in Evolution*, published two years later.

Simpson's book applied the concepts and conclusions of the new discoveries in genetics to the large body of fossil evidence of life's long history, and claimed that the "microevolution" of the geneticist could indeed be extrapolated to explain adequately the "macroevolution" of the paleontologist. That is, the mechanisms of generating and accumulating inherited variation, as described by laboratory geneticists and field naturalists, were sufficient to provide a parsimonious explanation of the adaptations, specializations, and evolutionary trends of the paleontologists, as measured in their fossils over long intervals of geologic time. In this respect *Tempo and Mode* became one of a half-dozen books that formed the basis for what came to be called the modern evolutionary synthesis. "Synthesis" because the new body of theory came from a variety of fields — genetics, ecology, anatomy, field biology, paleontology, botany as well as zoology, and biogeography — which were integrated into a unified whole.

Simpson thereby single-handedly brought the discipline of paleontology into the mainstream of biological research by validating the use of fossil evidence for solving evolutionary questions. Before Simpson, what fossils had to say to biologists as articulated by paleontologists was at best confusing, at worst contradictory, when compared to their own observations and theories. Simpson debunked, once and for all, many previous paleontological explanations of evolutionary phenomena that depended upon inherent or internally directed forces, like "inertia and momentum," which argued that once organisms began to evolve they continued to do so because of the impetus of past evolution; "racial senescence," the notion that organisms in a given line may exhaust their evolutionary reserves and become extinct; "orthogenesis," that organisms evolve toward some future goal and thus intermediate stages exist only as steps toward that goal rather than as viable ends in themselves; and "aristogenesis," that organisms are driven forward in their striving for perfection. Simpson demonstrated that such explanations were not consistent with modern genetic theory. He also provided the coup de grace to lingering claims for the inheritance of acquired characteristics through use and disuse. Thereafter, if paleontologists were to carry conviction, they had to ground their interpretations of macroevolution in terms of microevolution.

Another equally important contribution of *Tempo and Mode* was Simpson's identification of significantly varying rates of evolution — very fast, average, and very slow — and his explanation of how such differing rates would yield characteristic patterns of evolution within the fossil record (hence the title "Tempo and Mode in Evolution"). Simpson invoked a special importance for the environment, with all its physical, chemical, and biologic manifestations, in influencing evolutionary patterns and rate.

Simpson was thus preeminent among paleontologists and evolutionary biologists for the two decades following World War II; indeed he was a household word for these scientists. And as often happens when fame in a field is so great — even a field as arcane as paleontology — it spills over into public consciousness. Thus, in the 1950s and 1960s, Simpson appeared on the cover of the *Saturday Review of Literature*, was the subject of a full-page cartoon in the *New Yorker*, was featured in a radio broadcast by Lowell Thomas, and was periodically mentioned in the national newspapers, especially his hometown papers, the *New York Times, Herald-Tribune, Sun,* and *World-Telegram*. He even appeared on early television, guiding an anchorman through the fossil displays at the American Museum.

Despite this public exposure and despite his many writings — both professional and popular — George Gaylord Simpson was a difficult man to know. To most people, even those colleagues with whom he worked closely, he seemed reserved, often aloof, extremely guarded about his private life, and capable of sharp critical comment. He did not make friends easily, and those whom he referred to as good friends in his autobiography were surprised to be so considered. Because Simpson made no special efforts to cultivate friendship among his many acquaintances, he put most people off. Clearly more brilliant and more renowned than most around him, he accentuated the distance between them by his lack of warmth. By the time he died, some of his colleagues of his own generation had long given up knowing him; others, through hearsay, wrote him off as cranky, difficult, even embittered.

Toward the end of his life, Simpson's reputation waned somewhat, in part because his contributions became so thoroughly assimilated into current theory and practice that the identity of the originator was forgotten. Moreover, the next generation of evolutionists, especially among paleontologists, began to question some of Simpson's conclusions. In particular, some paleontologists challenged the idea that macroevolutionary events portrayed by fossils are merely long-term extrapolation of the short-term microevolutionary processes seen by the experimentalists and naturalists. Whatever history's final evaluation of Simpson's contribution may be, during much of his lifetime he was judged by many of his peers to be *the* leading paleontologist and one of the key founders of modern evolutionary theory.

* * *

George Gaylord Simpson was born in Chicago on June 16, 1902. He was the third and last child of Helen J. (Kinney) and Joseph A. Simpson, having been preceded in the world by his sisters, Margaret (1895–1991) and Martha (1898–1984). His father was an attorney who handled railroad claims, but he soon became involved in land speculation and mining in the West, which resulted in the family's resettlement in Denver while Simpson was still an infant. The family's Scots ancestry and missionary background led to a strict fundamental Presbyterian upbringing for the young Simpson, but he turned his back on formal religion by his early teens.

As a boy Simpson was curious about everything. He talked his parents into subsidizing his purchase of the now classic eleventh edition of the *Encyclopedia Britannica*, which he then read straight through. It became the foundation of what was to become a huge personal research library, and he was still using it at the end of his life. He also kept a notebook in which he recorded random entries, including such dubiously useful information — to a young boy at least — as descriptions of the densities of various materials.

Simpson had only a few close friends in childhood, chiefly a neighborhood chum, Bob Roe and his sister, Anne, whom Simpson would marry years later. In old age Simpson reminisced about his childhood and noted that being more intelligent, shorter, and redheaded had guaranteed antagonism from his peers. He was also afflicted with an eye condition that made it difficult to follow the flight of a ball — a serious handicap that prevented his participation in virtually all team sports. His father, sister Martha, and Bob Roe all enjoyed the outdoors, so Simpson spent much of his youth exploring the Rocky Mountain landscape, which undoubtedly fed his life-long interest in natural history.

Simpson attended Denver grammar and high schools. Although missing much school owing to eye ailments and appendicitis, he managed to skip grades and graduate from high school close to his sixteenth birthday. In the fall of 1918 he entered the University of Colorado at Boulder. He was vague about what to major in, although he thought he wanted to be a creative writer. In his second year he enrolled in a course in geology and was quickly converted, in part because of the enthusiasm of his instructor, Arthur Tieje, and the mutual respect each developed for the other. It was in this geology class that Simpson learned the importance of trusting his own observations, even those contrary to perceived truth. On one particular field trip, when he found a terrestrial dinosaur bone in a deposit that was clearly marine in origin, he was told that he was mistaken about the bone's location. Simpson quietly stuck to his guns, reasoning that the bone must have been washed in from a nearby river. And, of course, he was right.

FIGURE 1.2 Simpson with his future brother-in-law Bob Roe on homemade raft in Colorado. (Courtesy Robert Roe)

In his senior year Simpson transferred to Yale, because he was advised that if he wanted to be a geologist and paleontologist, Yale was the best place to study. Other factors may have played a role in his decision: he had lost his girl to his best friend and his position on the literary magazine he helped found was in dispute. Yale's requirements for graduation were more stringent than those at Colorado, so Simpson had to make up several general education courses, among them a foreign language. As a result, Simpson spent the summer of 1923 in France, quickly boning up on French so he could pass a test when he returned to Yale, enabling him to graduate retroactively with the class of 1923.

In February of his senior year, Simpson married Lydia Pedroja, whom he had met at the University of Colorado and who was now attending Barnard College in New York City. Yale at that time did not allow its undergraduates to marry, so Simpson and Lydia were married secretly. His parents did not know of the marriage right away either. This betrayal of his parents seems to have plagued Simpson all his life and may explain his overly deferential attitude toward them well into his middle-age.

The marriage thus got off to a troubled start. Simpson continued at Yale as a graduate student in paleontology, but Lydia's mental instability soon became apparent and compounded the normally difficult circumstances of graduate students, especially those burdened with young children and meager incomes. Despite their problems, Simpson and Lydia were sufficiently congenial to produce four daughters in their first six years of marriage: Helen, Patricia Gaylord, Joan, and Elizabeth.

In the basement of Yale's Peabody Museum, Simpson discovered a large collection in storage of primitive mammals from Mesozoic-age rocks of the American West. Despite the initial lack of encouragement from his doctoral adviser, the distinguished paleontologist Richard Swann Lull (1867–1957), Simpson decided to study these mammals for his dissertation. Lull's hesitation arose from his uncertainty that Simpson was qualified for the task. Worse, Simpson had earlier annoyed Lull by his elaborate preparation for Lull's lectures on vertebrate paleontology. In class Simpson would visibly tick off in his own notes various points as Lull covered them. Lull found this unnerving and asked Simpson to stop. Simpson continued the practice, but unobtrusively without Lull's noticing.

Simpson received his Ph.D. from Yale in 1926 and went on to the British Museum of Natural History in London to continue his study of primitive mammals by examining the British and European specimens held there. His wife Lydia decided that London was too unpleasant, so she took their two daughters and spent the year in southern France, which further stretched family finances. Yet the separation gave Simpson the freedom of a bachelor in an interesting foreign capital. He thoroughly scouted London, the English countryside, and made occasional side trips to the continent. He was able to meet the leading scientists in his field as well as indulge his growing interests in art, sculpture, and architecture, foreign languages and customs, and he imbibed a cosmopolitanism hitherto denied the provincial youth from a Western cowtown.

On his return from England in the fall of 1927 Simpson joined the scientific staff of the American Museum of Natural History as assistant curator of fossil vertebrates. He accepted the museum's offer after rejecting an offer made by Yale University, which had vacillated because of malicious gossip about Simpson's treatment of his wife. At the American Museum Simpson replaced William Diller Matthew, who had left to assume the directorship of the museum of paleontology and chairmanship of the reestablished department of paleontology at the University of California, Berkeley. Matthew was impressed with Simpson's studies of Mesozoic mammals and with the endurance and diligence that the young scientist displayed as his field assistant in the baking summer heat of Texas when Simpson was a first-year graduate student at Yale.

Having established himself as a paleomammalogist, Simpson wanted to continue his fossil mammal studies by examining the South American strata in Patagonia, which had yielded an unusual and important fauna that had evolved in isolation from the rest of the world during the long time that South America was an island-continent. Simpson befriended a museum patron, who put up the money — so critically short during the Depression years — for two expeditions to Patagonia, one in 1930–31 and the other in 1933–34. In later life Simpson remarked that he had spent many hours drinking with that patron, in persuading him that the expeditions were necessary, and quipped that he only regretted that he had "but one liver to lose for my museum." Simpson's first book, *Attending Marvels* (1934), is a travel journal of the first expedition. It was quite successful (it is still in print) and brought Simpson notice from the world outside of paleontology, including a radio interview in New York City and front-page coverage in the Sunday *New York Times Book Review*.

FIGURE 1.3 Simpson during the 1933–34 fossil-collecting expedition to Patagonia. (Courtesy Anne Roe Simpson Trust)

In the late 1930s and early 1940s, Simpson's work turned more theoretical, as he shifted some of his attention to more general problems of evolution, rather than focusing solely on fossil mammals. Just before American entry into World War II, Simpson published *Quantitative Zoology*, coauthored with his second wife Anne Roe (1939), and he completed two book-length manuscripts, which were published as *Tempo and Mode in Evolution* (1944) and "The Principles of Classification and a Classification of Mammals" (1945).

When Simpson left for South America in 1930, he had broken with Lydia, although they would not be legally separated until two years later and not finally divorced until 1938. Lydia's history of mental problems had begun even before she married Simpson, and some were serious enough to require hospitalization. The daughters had been moved about, at times with Lydia, at other times with their grandparents, and occasionally with their father. Domestic life was by no means tranquil, and one wonders if part of Simpson's motivation for his two South American expeditions wasn't the prospect of escape from an increasingly difficult situation at home.

During his last year of graduate work at Yale, Simpson chanced to meet Anne Roe (1904–1991), his childhood friend from Denver, who was working for her doctorate in psychology at Columbia University. Great friends as children and teenagers, they had gradually lost touch. She had gone to Denver University while he went back East to Yale. Moreover, Anne had had a brief adolescent conversion to fundamentalist evangelicism, which had rather discouraged Simpson. Thus, while staying friends, they had drifted apart. When Simpson saw Anne again in New York, he was much taken with her, and she with him. Both were now mature adults with fully developed, interesting personalities. Their renewed friendship quickly developed into a love that, obviously, could not be freely acknowledged. Within a few months Simpson left for his postdoctoral research in London, but he and Anne managed to correspond surreptitiously through his ever-reliable and supportive sister, Martha. Anne came to realize the impossibility of their situation and impetuously married while Simpson was in London. Rather than making things simpler, her marriage only created further complications, for Anne and Simpson began seeing each other again when he returned to New York in fall of 1927. By 1930, when Simpson left for a year's fossil-hunting in Patagonia, he was no longer living with Lydia, and Anne was questioning her own marriage.

When Simpson returned in 1931, he immediately filed for a legal separation from Lydia, which he obtained in early 1932. For the next six years Anne (by now separated from her husband) and Simpson lived together whenever their respective jobs permitted them to do so. For a while, Martha even joined the ménage, and all three happily lived together in a New York City apartment. Simpson eventually moved to Connecticut, commuting to the museum in New York, so he could file for divorce from Lydia on the grounds of mental cruelty, which was not permissible in New York State. His eldest daughter, Helen, was living with him, attending a nearby private school as a day student; Patricia Gaylord (called "Gay" in the family) was living with her maternal grandmother in Kansas, and the two youngest daughters, Joan and Elizabeth, were in temporary custody of their mother, pending the outcome of the divorce and final custody arrangements.

In April 1938, after a rather long court battle, Simpson was awarded a divorce from Lydia and custody of the two older children. (Not long afterward, he received custody of the other two girls as well.) A month later Simpson and Anne were married. Later that year they went to Venezuela on a fossil-collecting expedition; Anne put aside her work in clinical psychology and collected various mammals for the museum. The following summer the happily married couple set up house in New York City and brought together the far-flung family for the first time in years, except for Gay, who remained in Kansas where she'd lived practically all her life. The next three years were tranquil: Anne was caring for the family and working part-time as editor and researcher, the girls were in junior and senior high school, and Simpson was working on *Tempo and Mode* and "Principles of Classification," among many other projects.

In 1942 the new director of the American Museum was contemplating reorganization of the various departments, which included combining the paleontologists with the zoologists — that is, putting scientists working with living groups with those studying their fossils. For various reasons — some personal, others disciplinary — Simpson resisted this plan and became somewhat depressed about his future prospects. Before he was forced to make a decision, he enlisted in the armed services and in December 1942 he joined military intelligence as a captain in the U.S. Army. He must have flustered his superiors for he completed a six-week course in intelligence methods in a single week. By spring of the following year he was in North Africa, as part of the Allied Forces

FIGURE 1.4 Simpson in the uniform of the U.S. Army, 1944. (Courtesy Anne Roe Simpson Trust)

Command led by General Dwight Eisenhower. He moved on to Sicily and Italy during the invasions in the summer of 1943 and returned later to North Africa, where he remained until the fall of 1944, when he was shipped home with a severe case of hepatitis. By then, he held the rank of major and had been awarded two Bronze Stars.

When Simpson returned to the museum, the earlier reorganization plan had been scrapped and a different plan was imposed, which included a department of geology and paleontology of which Simpson was now named chairman. Simpson also accepted appointment as professor of vertebrate paleontology at Columbia University. In 1949 Simpson published a popular account of modern evolutionary theory from the point of view of the fossil evidence, *The Meaning of Evolution*, which was subsequently translated into a dozen languages and sold some half-million copies. This book further familiarized the educated public with Simpson the scientist. A few years later, Simpson completely reworked *Tempo and Mode* into *Major Features of Evolution* (1953), which was considered by some a more ponderous, detailed version of what had been a succinct and brilliant monograph.

In 1953 Simpson published a small volume, *Evolution and Geography*, that climaxed a series of writings published over more than a decade, all of which addressed the principles for explaining the past distributions of land animals, especially mammals of the Cenozoic Era, representing the last 65 million years of Earth's history. Strongly contrary to the ideas of Alfred Wegener, the German scientist who argued that drifting continents carried fossils of land animals far and wide from their original distribution, Simpson made a cogent case for the natural dispersal mechanisms of organisms over long geologic time-intervals to accomplish the same results on continents that did not move. So great was Simpson's authority and so persuasive his reasoning, that the theory of continental drift slipped still further in its credibility with North American geologists. Wrong for the right reasons, Simpson only converted to the new theory of plate tectonics when the geophysical evidence from the oceans provided compelling proof of seafloor spreading from mid-ocean ridges that carried the continents — and their resident floras and faunas — with them. Simpson, like most other scientists who live their allotted three-score-and-ten, was forced to acknowledge that what had been learned earlier in one's professional career was now either wrong, obsolete, or irrelevant. He did not relish the opportunity for recantation the way a number of his colleagues did, born-again enthusiasts for the new global tectonics.

The special place that Simpson enjoyed in the earlier revolution, that of the modern evolutionary synthesis, was visible beyond his contribution of learned treatises. Just before the war he helped found the new Society of Vertebrate Paleontology and became its first elected president. After the war, he helped organize the Society for the Study of Evolution, which brought together a whole range of life-scientists, including paleontologists, working within the new paradigm of the modern synthetic theory of evolution. He was the society's first president and, later, when the society wanted to begin publishing its own journal, *Evolution*, Simpson managed to arrange initial funding from the American Philosophical Society.

In the summer of 1956 Simpson undertook yet another South American expedition, this time to the upper headwaters of the Amazon River in Brazil to collect fossils to fill in gaps between Argentina to the south and Colombia to the north. The area was so covered with vegetation that the best prospects for developing outcrops were along the riverbanks exposed during seasonal low-water. Simpson headed a small party of North and South American scientists, assisted by several Brazilian workmen. Near the end of their river trip, camp was being cleared along the river's edge. The field party did not like to camp on the boats in the river itself, but preferred making a campsite on dry land, even though it usually entailed hacking through thick vegetation and even chopping trees down. In late August, while one such campsite was being prepared, a newly cut tree fell on Simpson, causing multiple injuries, the most serious of which was a bad break of his lower right leg. After a painful and dangerous trip back down the river and a series of transfers from boats to planes, he arrived a week later in New York City. He spent the next two years in and out of hospitals, undergoing twelve separate operations — one of which was necessary to remove a surgical sponge that had been left behind from the preceding operation. Simpson resisted all advice to have the leg amputated, and finally recovered the use of the leg, although it always pained him and he was somewhat lame.

During this difficult time Simpson's second daughter died. Gay had been born with a congenital heart defect and had been shielded from most of her parents' marital problems by living with her maternal grandmother in Kansas. Recently married herself and working as a librarian, Gay died of a brain abscess at age thirty-one.

In 1958 Simpson gave up the chairmanship of the department of geology and paleontology at the American Museum, under urging to do so from the museum's director. Shortly thereafter Simpson resigned from the museum altogether and was appointed to an Alexander Agassiz professorship at the Museum of Comparative Zoology, Harvard University. There had been some resentment among Simpson's colleagues of his periodic absences from the museum during his two-year convalescence and the delegation of his administrative chores to others. The director, intending to lessen Simpson's administrative burden, suggested he resign the chairmanship, but Simpson, feeling that under the circumstances he had been carrying out his responsibilities quite satisfactorily, was so insulted by the idea that he made up his mind to seek a position elsewhere. Despite the turmoil, Simpson managed to revise a major biology textbook, *Life: An Introduction to Biology* (1965), in collaboration with another Harvard biologist, see into print a number of articles started before his accident, and write a number of book reviews.

The centennial of Darwin's *On the Origin of Species* was celebrated in 1959, a year that not only signaled a fresh start for Simpson at Harvard but also brought him once again into the spotlight as a leading evolutionist. Conferences, symposia, and special events marked the centennial, and Simpson was often present either as a contributor — as in Chicago where he gave a keynote public lecture at the annual meeting of the American Association for the Advancement of Science entitled "The World into which Darwin Led Us" — or as an honoree — as in London where he received the Darwin-Wallace Commemorative Medal from the Linnean Society in whose meeting rooms Charles Darwin and Alfred Russel Wallace first announced their theory of natural selection as the mechanism for evolution a century earlier.

Darwin always held special interest for Simpson, not only for the obvious reasons but also because Darwin represented for Simpson the quintessential liberator of the human spirit, seeking to find answers to the puzzle of human existence without recourse to supernatural explanations. Simpson's own penchant for a naturalistic, positivistic view of the world was well reflected in his reading of Darwin. Simpson regularly took the opportunity throughout his career to speak and write about Darwin for nonscientific or nonspecialist audiences. Hence he was in his glory during the Darwin revival of the late 1950s.

Simpson and his wife Anne settled in at Harvard. He was busily writing once again, but because of his accident serious fieldwork was over. She was eventually granted a full professorship in the Graduate School of Education, and they were thus the first married couple to receive professorial appointments at Harvard. They spent winters mostly in Cambridge, summers at their simple home in the New Mexico mountains, and managed a number of trips abroad, including to England, France, and Spain, as well as Africa and Australia. These trips were as much opportunities to renew scientific friendships and receive further honors and awards as they were to examine fossil collections in museums and universities. The Simpsons had conversations with the Duke of Edinburgh, the granddaughter of Darwin, the Huxleys, and the

Leakeys, among others. Nor was Simpson ignored at home: in February 1966 he received the National Medal of Science from Lyndon B. Johnson, then president of the United States.

Yet all was not unmitigated satisfaction. Simpson and Anne both suffered from heart attacks and were hospitalized. In fact, some of their traveling was in part intended as recuperation. And although, strictly speaking, Simpson's Agassiz Professorship at Harvard did not explicitly require residence at the museum, his frequent absences once again made for some difficulty with the administration.

In 1967 the Simpsons decided to move to Tucson where they had earlier bought a house in preparation for retirement. He was given an appointment at the University of Arizona, which involved some teaching at first and later evolved to weekly luncheon meetings with interested students and faculty. By 1970 all formal ties with Harvard were severed. Simpson continued to publish books — on South American mammals, penguins, Darwin, fossils and the history of life, collections of essays, and his autobiography — as well as the usual monographs and articles on Cenozoic mammals, obituaries of deceased colleagues, and book reviews. He worked in a small building built next to his house, surrounded by his research files and extensive personal library, walls and surfaces scattered with honorary degrees, photographs from the past, and replicas of the many medals he had been awarded.

Despite the change in locale and greatly increased domestic tranquility, Simpson's life did not alter much. Rather, his Arizona years reveal an intensification of everything that he really loved: close, daily contact with Anne; small, informal get-togethers with a few intimate friends; occasional visits from children and grandchildren; trips of varying length in the United States and abroad; association with the academy without much formal responsibility; and long stretches of time to devote to his research and writing with minimal interruption or distraction. At times there were periods when his or Anne's health was poor, which then required a slow building up of energy and reserves to reestablish the routine.

In the last decade of his life, as his friends and colleagues passed on, Simpson became more a memory from the past than a prime mover in the present. Most thoughtful students were aware of what he had contributed but now took it for granted. They looked to the writings of younger paleontologists and evolutionary biologists for new ideas. In a

FIGURE 1.5 Simpson in his office annex in Tucson, Arizona (1984), examining *Fossils and the History of Life*, his next to last book, published the year before his death. (Courtesy Anne Roe Simpson Trust)

way, Simpson had outlived his fame and had become a living, mostly ignored monument of what had come before.

In the summer of 1984 Simpson contracted pneumonia during a South Pacific cruise. After returning home his condition worsened and he was in and out of the hospital several times over the next few months. He finally succumbed on the evening of Saturday, October 6, 1984, at the age of eighty-two. His remains were privately cremated and later dispersed in the Arizona desert. Anne survived him for several more years, eventually dying in 1991.[1]

CHAPTER 2

PALEONTOLOGY AND THE EXPANSION OF BIOLOGY

The geneticists tended to consider that paleontology was incapable of rising above pure description and they did not even take the trouble to study descriptive paleontology for its bearing on genetics. It was easier to conclude that it had no such bearing. The paleontologists were, as a rule, quite willing to accept this stultifying conclusion, which also spared them the trouble of learning genetics.
— G. G. Simpson[1]

Paleontology at the beginning of the twentieth century in the United States was something of an orphan within the biological sciences. Because of its emphasis on the older tradition of comparative morphology and the use of fossils by geologists in determining relative ages of rocks, paleontology had neither a home in the new institutional centers established for biology, nor was it considered a formal discipline within biology. Worse yet, it did not demonstrate "much likelihood of becoming a foundation for serious programs of research in biology."[2] The American Museum of Natural History provided the single exception to this otherwise bleak situation for paleontology. Henry Fairfield Osborn (1857–1935), William King Gregory (1876–1970), and William Diller Matthew (1871–1930) of the Department of Paleontology were among the few paleontologists of this era who consistently addressed biological questions in their study of fossils.

During the subsequent twentieth-century expansion of biology, paleontology continued to exclude itself from biology's ever-widening theoretical and observational base, especially the genetics and ecology of living populations of organisms, until the development of the "evolutionary synthesis" in the late 1930s and early 1940s. Given the American Museum's earlier history as an outpost of biologically oriented paleontology, it is not surprising that another, younger American Museum scientist, George Gaylord Simpson, catalyzed the reconciliation of paleontology with contemporary biology. As the opening epigraph suggests, until this reconciliation came about, paleontology and biology were at cross-purposes in their search for evolutionary mechanisms. Paleontologists had argued for internally directed evolutionary forces inferred from their fossils that biologists could not corroborate in the laboratory or the field, whereas biologists had focused on problems that seemed irrelevant to the paleontologist. Bones in stones were a world apart from fruit flies in bottles.

This chapter will examine Simpson's role as a prime mover in bringing paleontology and biology closer together during the evolutionary synthesis, thereby resulting in a mutual expansion of the two disciplines — genetics and ecology providing a theoretical context for the actual historical patterns documented by paleontology. As described in chapter 1, Simpson dominated American paleontology for some five decades spanning the middle of the twentieth century. This dominance was both quantitative and qualitative, for Simpson not only published hundreds of articles, monographs, and books (his bibliography includes more than 750 entries), but his work had major impact on contemporary views of the origin, evolution, and classification of mammals; concepts of historical biogeography; principles of taxonomy and systematics; and biostatistical methods, as well as the formulation of the modern evolutionary synthesis.

Because his research data ultimately derived from fossils in layered sequences of sedimentary rocks, Simpson necessarily straddled the two disciplines of geology and biology. Consequently, Simpson had to master both disciplines in order to enter the field of paleontology. His mastery of geology was accomplished by undergraduate and graduate matriculation in geology at Yale University. His education in biology was somewhat adventitious, yet it turned out to be crucial in his intellectual development, especially with respect to his subsequent contributions to evolutionary theory. Both by formal graduate study at Yale and professional association at the American Museum, Simpson realized early in his career the special value of the biological sciences in studying and interpreting vertebrate fossil remains. This deep appreciation and thorough understanding of the subject enabled Simpson to bring paleontology more fully into the mainstream of biology during the decade of the development of the modern evolutionary synthesis. However, for all

Simpson's realization of the importance of biology for paleontology, he was nevertheless equally convinced of the uniqueness of paleontology, particularly in terms of its raw data of fossils-in-rocks with their much longer temporal context, thereby justifying paleontology's own intellectual autonomy.

Biology and Fossils at Yale

Biology did interest me, and I began making up [at Yale] the serious gap on this side of my knowledge requisite for paleontological work.
— G. G. Simpson[3]

In 1922, Simpson transferred from the University of Colorado to Yale University for his senior year of college work. Reasons for his transfer included getting what he perceived as the best education in paleontology then available.[4] At Colorado Simpson had a year-long course in paleontology, but no biology.[5] Yale's liberal arts curriculum required a year of biology, so Simpson enrolled in the two-semester class taught by Lorande L. Woodruff, who founded and taught the basic biology course and whose textbook *Foundations of Biology* had just been published. "Woodruff gave a good introduction to Mendelian genetics as of the 1920s. As to evolution, he adhered to what he called 'clarified Darwinism,' accepting natural selection as the only natural explanation of adaptation, but believing that many 'variations are neutral to selection' . . . and hinting . . . at . . . some vitalistic factor."[6] Further insight about Woodruff's views on evolution at this time are reflected in an essay on biology that Woodruff wrote for his edited volume *The Development of the Sciences*: "Today no representative biologist questions the fact of evolution . . . though in regard to the factors there is much difference of opinion. It may well be that we shall have reason to depart widely from Darwin's interpretation of the effective principles at work in the origin of species."[7]

Simpson continued at Yale as a graduate student in the department of geology studying vertebrate paleontology, and because biological form and function were considered important, as much as chemistry and physics was for other geologists, the department required its paleontologists to take courses in zoology.[8] Thus, Simpson took classes in vertebrate morphology with Ross Harrison, embryology with G. A. Baitsell, and comparative anatomy of vertebrates with W. W. Swingle.[9] Simpson has said about Harrison that he "taught morphology in a rather nineteenth century Teutonic way, very thorough and wonderful basic knowledge, with no emphasis on evolutionary theory."[10]

Simpson's dissertation adviser at Yale, Richard Swann Lull, was himself trained as a biologist. He had bachelor's and master's degrees in zoology from Rutgers before taking his Ph.D. in vertebrate paleontology at Columbia University under Henry Fairfield Osborn, who was also trained as a biologist. Lull's biological background was clearly evident in his textbook *Organic Evolution* (1917), which Simpson read over the summer of 1923 before entering graduate school.[11]

The 1929 revision of that book was issued after Simpson left Yale, but it indicates the approach Lull took to the subject when Simpson was at Yale just a few years before. Throughout, Lull integrated contemporary biology with the vertebrate fossil record. In Part II, entitled "The Mechanism of Evolution," Lull had seven separate chapters — on genetic variation; mutation; heredity; artificial, sexual, and natural selection; as well as on inheritance of acquired characters; orthogenesis; and kinetogenesis. In the rest of the book dealing specifically with particular fossil groups, Lull also included as much biology of living organisms as was appropriate to understand the fossils. For each fossil group he discussed its place in nature, form and function, habitat and habit, particular specializations, and ontogenetic as well as phylogenetic history (see chapter 7 for a more complete discussion of Lull's text).

Although the book had obvious Lamarckian and vitalistic overtones, Simpson was certainly taken with Lull's paleobiological approach. According to Simpson, Lull infused "life into the bones . . . the whole fauna, flora, and landscape of the distant past."[12] Simpson later remarked that "Lull was considered a great authority on evolutionary theory — and he was, in the sense of being well informed on the ideas, old and new, in this field. He never had an original idea of his own about theories of evolution, but just expounded everyone's views as if all were equal. That was very useful to me only in telling me what the established alternatives were and what to read."[13] So both by formal education and the example of his dissertation adviser, Simpson learned early in his career the importance of biology in thinking about fossils.

Particular biological influences with respect to the biology of living organisms or evolutionary theory were not especially apparent in Simpson's dissertation and postdoctoral work on Mesozoic mammals.[14] Like most paleontological research of the era, these latter studies emphasized descriptive morphology, taxonomy and systematics, and phylogenetic interpretation. However, there were two predissertation articles by Simpson — both written before age twenty-four — on Mesozoic mammals in which he did explicitly consider fossils as once-living. He referred to the first article as "a study in paleobiology, an attempt to consider a very ancient and long extinct group of animals not as bits of broken bone but as flesh and blood beings." The fossils in question were multituberculates, the longest-lived mammalian order with worldwide distribution and considerable diversity, although now no longer extant. Simpson described and interpreted these fossils as "living animals," using the skull and dentition of the marsupial rat-kangaroo as a living analogue to infer a herbivorous diet of "cycads, cones and nuts, and angiosperm fruits and seeds." Morphological analysis of the limbs of one multituberculate genus in light of the locomotor function of living animals led Simpson to the further conclusion that the extinct form was "probably a swift moving and agile quadruped."[15]

In the second article, Simpson reconstructed the paleoecological relationships of the terrestrial and aquatic biota of the late Jurassic Morrison Formation of Wyoming, showing the inferred ecosystem of the fossils that included invertebrates, fishes, and reptiles as well as mammals.[16] As Simpson himself remarked years later, this was among the first, if not the first, such application of the ecological concepts of trophic levels and food webs to a fossil association, which was not to become commonplace for several more decades.[17] While the specific content of both these studies was original for the time, their spirit certainly is anticipated in Lull's approach to fossils as developed in his text *Organic Evolution*.

Biology and Fossils at the American Museum

I joined the . . . American Museum . . . [as] a raw recruit but rapidly gain[ed] experience and a measure of expertise of various sorts.
— G. G. Simpson[18]

After finishing his Yale doctorate and a year of postdoctoral study at the British Museum, Simpson returned from England in the fall of 1927 to take up a position as assistant curator in the department of vertebrate paleontology at the American Museum. This move was to have enormous implications for the rest of Simpson's career. The museum's collections were outstanding, providing Simpson with new fossil mammals to work on and extensive specimens of living species for reference, as well as a superb research library close at hand. His prescribed duties too — at the beginning at least — were sufficiently minimal that he had most of his time for his own research. The museum was even more important intellectually for Simpson in that his senior colleagues included W. K. Gregory and H. F. Osborn, both of whom always approached fossils as once-living, complex biological systems. W. D. Matthew, another distinguished paleontologist, whose position at the American Museum Simpson filled in 1927, played an even earlier role in Simpson's life, when Simpson was Matthew's young field assistant on a museum expedition to Texas in 1924 (discussed more fully in chapter 3). In his first year of graduate school at Yale, Simpson had applied to the American Museum for summer employment, willing to work either as a preparator of fossils in the laboratory or as a field assistant.

In his autobiography Simpson said that he was "trained by Matthew at least as much as by my major formal professor, Lull."[19] In a posthumously published personal memoir first written shortly after Matthew's death in 1930, Simpson gives some details about Matthew's influence on him that hot Texas summer. Simpson recalled that Matthew revealed "his life's dream (one never gratified) to build up an exhibition of vertebrate paleontology . . . to show not so much the zoologic classification as the grand history of life, the succession of fauna & the associations of animals at one time & in one place."[20]

One might suppose that Matthew was confiding to Simpson in the field what he would subsequently tell his Berkeley students in the classroom: "You cannot understand or appreciate the past history of life without knowing something about animals and plants, about their structure and mechanism and habits and how they are classified and arranged — what they are and where they live and how they live."[21] Matthew also had strong thoughts about what paleontologists could contribute to evolutionary biology. In his 1922 presidential address to the Paleontological Society he remarked, "I do not agree with a distinguished Columbia professor [the geneticist T. H. Morgan] who declared not long ago [in 1916] that paleontologists had no business to reason or draw conclusions from their specimens, but should content themselves with describing and illustrating them. . . . [On the contrary,] paleontologists, with the facts before them as to what actually did take place in the evolution of a race of animals, may claim the right to reason and draw conclusions from these data as to the methods and causes of the transmutation of species."[22] Likewise, Matthew's interests in historical biogeography, as reflected in his classic "Climate and Evolution," also influenced Simpson's own extensive subsequent research in the past geographic distribution of terrestrial mammals.[23]

W. K. Gregory had become a formal member of the scientific staff at the American Museum in 1911, after taking his Ph.D. at Columbia under H. F. Osborn, although he had been Osborn's assistant for a dozen years before that. Gregory's influence on Simpson is evident from Simpson's memorial tribute to Gregory.[24] Gregory's scientific interests were very broad and several overlapped with those of Simpson's, including mammalian classification, Mesozoic mammals, evolution of mammalian dentition, and functional anatomy. Some flavor of Gregory's scientific philosophy that particularly impressed Simpson is captured in the latter's words: "Gregory was a pioneer in the study of the anatomy of parts of animals, both recent and fossil, not merely in a descriptive way but with primary consideration for their functions in the lives of whole, living organisms." Simpson did not formally acknowledge this debt to Gregory in any of his publications, but he did point out in his tribute to Gregory the significance of Gregory's "palimpsest theory" of differing rates of evolution of individual characters and character complexes — Gavin de Beer's "mosaic evolution" in present-day terms.[25] This theory may have contributed to Simpson's own views about differential evolutionary rates within a given lineage that he developed in *Tempo and Mode in Evolution*.

Perhaps the single, most specific, catalytic effect Gregory had on Simpson's career (that we can document) was Simpson's preparation of a paper "following a suggestion by Professor Gregory" for a symposium that the latter was organizing for the American Society of Naturalists in 1936 on "supra-specific variation," that is, differentiation and evolution above the species level.[26] At the time, however, when he sent the manuscript to Gregory, Simpson noted that "it has never been entirely clear to me exactly what sort of paper you wanted, and I may have missed the point altogether. This is in any case an unusually labored effort and for some reason I have had great difficulty in selecting and organizing, from the embarrassingly vast amount of data at hand."[27] Whatever ambivalence Simpson may have had at the time, the ideas introduced in the paper about "tempo" — or rates — and "mode" — or styles — of evolution were to assume central importance in his 1944 classic.[28]

Not the least important influence on the young Simpson was H. F. Osborn, who was virtually synonymous with the American Museum in the early decades of this century. Osborn's training in various biological disciplines — including histology under William H. Welch in New York, embryology under Francis M. Balfour at Cambridge, comparative anatomy under Thomas H. Huxley in London, and neuroanatomy at Princeton, as well as his having collected Eocene mammals in Wyoming and studied Mesozoic mammals in England — resulted, according to Simpson, in his communicating "a soundness of training and a passion for research that have had a profound effect on vertebrate paleontology."[29] Owing to Osborn's intellectual and material resources as well as his energetic and dynamic leadership, he created an institution of singular distinction. Within vertebrate paleontology more specifically, Osborn developed a department in which "problems of biogeography, the process and pattern of evolution, or the relationship between inheritance and development, more than the traditional issues in systematics and stratigraphy, were the heart and soul."[30]

Osborn thus gave warrant to paleontologists to pursue theoretical issues, which Simpson did not hesitate to acknowledge. Shortly after publication of *Tempo and Mode*, Simpson reviewed "some of the historical background of modern evolutionary theories," including especially his own. Because "none of the mechanistic schools . . . seemed fully adequate to interpret paleontological observations . . . [and] paleontologists themselves failed to find a satisfying alternative[,] . . . some paleontologists began to support a non-mechanistic, that is, in a broad sense, a vitalistic explanation." Simpson went on to note that an outstanding example of such a nonmechanistic explanation was Osborn's theory of aristogenesis, and that "it was almost inevitable that such an escape from the dilemma would be sought by a deeply philosophical paleontologist of his period."[31] Osborn provided a forceful example to Simpson of what the important questions were in paleontology. In his biographical sketch of Osborn, Simpson repeated Osborn's own words to the effect that he "always found the mere assemblage of facts an extremely painful and self-denying process . . . [whereas] the discovery of new principles is the chief end of research."[32] Thus it was into this professional ambience that Simpson stepped, at age twenty-five, when he joined the American Museum.

Simpson's Biological Insights of the 1930s

Now age thirty-five and more than ten years out of graduate school I dared to take modest personal steps in the direction of principles and theory.
— G. G. Simpson[33]

In the mid-1930s, Simpson worked on several projects that were seminal for virtually all his subsequent work. In spring of 1936 Simpson completed a monograph describing Paleocene mammals of Montana, which included brief discussions of the paleoecology of the faunas, their postmortem preservation (today's "taphonomy"), and the value of univariate statistics in aiding taxonomic decisions (discussed in more detail in chapter 5).[34] Later that year Simpson delivered two papers, one that addressed general problems of the evolutionary interpretation of fossils, and a second that specifically illustrated how such interpretations can be made using the extinct mammalian order Notoungulata.[35] During this same time, Simpson was also collaborating with the psychologist Anne Roe (whom he married in 1938) on a book for zoologists and paleontologists explaining how statistics were the "best means of describing and interpreting what animals are and do."[36] More generally throughout the 1930s, Simpson was working on a new classification of mammals that took a unified approach to the analysis and interpretation of living and extinct animals, as well as informing himself of the new developments in population genetics.[37] It was his reading in 1937 of Dobzhansky's *Genetics and the Origin of Species* that crystallized all his previous training and thinking about biological paleontology and encouraged him to take modest steps in the direction of principles and theory. "When I read [Dobzhansky] it opened a whole new vista to me of really explaining the things that one could see going on in the fossil record and also by study of recent animals. I began pulling things together into this framework but also with a good many points that were not involved in the work of the geneticist, [I] began thinking of what my own by this time rather long studies in the history of life might mean within this context."[38]

The bulk of the monograph on Paleocene mammals contained standard morphological description and taxonomic discussion of fossils that Simpson and others had collected over the years in the Fort Union Group of Montana. However, the point of view taken by Simpson was rather original for such a paleontological monograph. Simpson used elementary statistics for describing mammalian teeth: their individual lengths and widths, means and ranges of observations, standard deviations, and coefficients of variability as well as the t-test for comparing samples. The purpose of these calculations was to assist him in making objective and reproducible inferences about the probable limits of variation acceptable in defining species. As Simpson noted, "Most other writers on the question of species making in paleontology have insisted on making due allowance for variation, but they have adduced no real, objective criteria as to what 'due allowance' should be." Rather than just paying lip service to the new concept of species as populations of breeding individuals rather than as idealized types, Simpson was advocating that "the methods of statistics provide the desired means of measuring variation accurately and the necessary criterion as to whether this variation is . . . of the sort normal in a species."[39]

Simpson used these statistical results, for example, to justify his inference that there were indeed two similar but statistically distinct species of carnivore that lived in geographically different areas of the region, thereby confirming the principle of "ecological incompatibility." Simpson observed that the principle had been qualitatively applied previously by W. D. Matthew.[40] But here Simpson was invoking the principle using a quantitative statistical test to support what biologists would later refer to as G. F. Gause's hypothesis of "competitive exclusion."[41]

In the paper "Patterns of Phyletic Evolution," Simpson further recommended the use of the population concept of species, noting that "paleontological psychology has been to direct attention, to a disabling extent, to the individual specimen, and the number of publications on fossil mammals in which a suite of specimens has been correctly studied as representative of a group can be almost counted on the fingers of one hand."[42] Using a small hypothetical set of data, comprising a range of values for a given morphologic character that increases in dimension through a sequence of rock strata, Simpson indicated the three sorts of qualitative interpretation that might result. The data might record the decline over time of one smaller species and the rise of another larger one, or possibly just one species in which the character was increasing through time, or perhaps even the gradual divergence of one larger species from the smaller ancestral one. Simpson then explained the basis for discriminating among the three alternative interpretations by simple statistical analysis of the data, stratum-by-stratum, through the rock sequence. Simpson declared years later that this paper "marked his abandonment of the typological thinking of my college teachers and started aiming me toward statistical biometry and the deeper investigation of evolutionary theory and taxonomic stance."[43]

A third seminal paper of this period by Simpson was the one on supraspecific variation that leads to evolutionary diversification above the species level (as noted above, this was a topic suggested to him by W. K. Gregory). In the first half of the paper, Simpson reviewed the history of an extinct group of hoofed herbivores, the notoungulates, almost entirely confined to South America during the Tertiary Period. Simpson demonstrated that rates of evolution can vary markedly within different characters of an otherwise homogeneous taxon, leading to distinct higher categories (e.g., different suborders within the order Notoungulata). He showed that close parallel evolution is often manifest between otherwise distinctly different taxa (e.g., similar limb elongation, toe reduction, and increase in molar-tooth height both in notoungulates and true horses). Finally, Simpson inferred that there is inherent variability within interbreeding populations whose subsequent segregation can eventually produce supraspecific taxa (e.g., evolution of distinct lineages having characteristic types of molar teeth from a highly variable ancestral species whose variability spans that of the separate, descendant lineages). The second half of the paper critiqued an earlier paper by Alfred C. Kinsey (1894–1956) with Simpson "adding what seems to be the best or most generally held paleontological opinion on each concept" regarding various principles for the recognition of higher taxa that Kinsey had himself debated.[44] What was significant about this discussion was that a paleontologist, using fossils as evidence, was prepared to argue about biological principles and concepts on equal terms with a biologist.

Simpson made the same point in *Quantitative Zoology* by recommending that paleontological samples be given the same statistical analysis one should use for biological samples. As implied by the title and illustrated by examples throughout the whole text, Simpson intentionally blurred the distinction between biological data and paleontological data, and the conclusions one could draw from them. For him there was no important qualitative difference between living and fossil: "Paleontological mensuration differs little from that of the hard parts of recent animals."[45] Therefore, theoretical interpretations based on fossils were, in principle at least, as sound and cogent as those relying on living materials. Whatever else one might be concerned with about the size and quality of a given sample — fossil or living — both kinds of samples had similar validity, all other things being equal.

Simpson further conjoined biology and paleontology in his work as curator at the American Museum. There he had to identify, label, sort, and catalogue fossil mammalian specimens. Simpson soon developed a set of concepts and principles to relate them to living taxa so he could physically organize the museum's collections. Like any workable filing system, Simpson had to set up a catalogue that had some rhyme and reason to it.[46] Progressively elaborated over the next two decades, Simpson formulated a comprehensive approach to taxonomy and systematics. This resulted in an outline classification of mammals (1931), followed by a monograph that expanded on the previous outline and included a discussion of the principles of classification (1945). Still later he produced a full-blown exposition of the principles of animal taxonomy (1961). In all three works, Simpson observed no formal distinctions between living and fossil.[47] On the contrary, every piece of information, whatever its source and disciplinary jurisdiction, was useful: "The data of neozoology are highly pertinent to the problems of phylogeny and major classification, but this work has fallen more and more into the field of the palaeozoologist who should, for this purpose, be a competent general zoologist as well as a palaeontologist. . . . In so difficult a study [as taxonomic classification] it is inexcusable to reject offhand any line of evidence that might give light."[48]

Simpson thus threw down the gauntlet in 1937 when he asserted "that there is no natural barrier between genetic and paleontological research and that both must eventually unite in any final synthesis of modes of evolution."[49] Now fully prepared to undertake a more comprehensive treatment of the subjects he had been exploring these several years, Simpson picked up the gauntlet himself in the spring of 1938 when he began work on *Tempo and Mode in Evolution*.

Tempo and Mode in Evolution

*The present purpose is to discuss the 'how' and . . . not the 'what.'
. . . [The how] is more immediately interesting to the nonpaleontological evolutionist.*
— G. G. Simpson[50]

Simpson's *Tempo and Mode in Evolution* is one of the half-dozen or so books that are the pillars of the modern evolutionary synthesis; it also helped bring paleontology back into the mainstream of biological science.[51] Hence, it is tempting for paleontologists today to see this book and its author sui generis, like Venus rising from the sea, without historical roots or intellectual genealogy. But as several articles by Ronald Rainger cogently argue and the previous sections suggest, that view is historically inaccurate.[52] The genesis of *Tempo and Mode* was itself evolutionary, not revolutionary. By his previous training and the examples of his mentors, Simpson recognized the importance of the larger biological issues raised by fossils, and he had learned the skills to address those issues. He was no mere describer and namer of fossil bones and teeth. In fact, as a paleontological treatise, what was left out of *Tempo and Mode* was as unexpected as what was included. There was very little discussion of specific fossil taxa per se — whether their detailed morphological description; their temporal progression, geologic period by geologic period; or their phylogenetic sequence, from ancestor to descendant.[53]

Simpson's task in *Tempo and Mode in Evolution* was to demonstrate that what happens on the individual level of populations of living organisms — genetic and phenotypic variation, natural selection, differential survival and reproduction, acclimatization and adaptation, migration, and so on — is necessary and sufficient to explain the much longer-term, morphological transformations of skeletal hard parts seen in the fossil record. Thus Simpson proposed to show how the microevolution of population genetics of organisms, viewed within an ecological context of those organisms interacting with their diversified environments, could explain the macroevolution documented by paleontology. That is, he was projecting the three-dimensional concepts of biology into the fourth, temporal, dimension of paleontology. As he phrased it a decade later, "I am trying to pursue a science that . . . has no name: the science of four-dimensional biology or of time and life. Fossils are pertinent . . . but [the fruit fly] *Drosophila* is equally pertinent."[54]

As a way of encapsulating his argument, Simpson developed the concept of the "adaptive grid," a field of discrete, noncontinuous ecological zones of increasingly finer grain. When considered on a timescale ranging from ecological (microevolutionary) to geological (macroevolutionary) time, populations might become adapted early on within a narrow zone with little subsequent change (very slow evolution, or Simpson's "bradytely"). If the prospective adaptive zone is wide enough, populations might diversify and become increasingly specialized within various subzones (slow to fast evolution, or "horotely"). Rarely, populations might jump the gap between major adaptive zones and radiate rapidly into a new adaptive zone altogether (very fast evolution, or "tachytely"). What was especially original about *Tempo and Mode* as a paleontological treatise was that it expressly addressed the "how" of the evolutionary transformations familiar to the paleontologist in terms that fit well with experimental genetics and field biology of living organisms.

Of course, to get to the point where he could discuss the "how," Simpson had first to demonstrate that fossils do record highly varying rates of evolution (his chapter 1); that population genetics explained the mechanisms of evolution on the ecological timescale (ch. 2); that large, systematic gaps in the fossil record were due to particularly high rates of evolution (ch. 3) whereas very low rates of evolution of so-called living fossils could be attributed to "survival of the unspecialized" in unchanging, long-lived habitats (ch. 4); and therefore, all intrinsic evolutionary mechanisms, such as orthogenesis, inertia and momentum, and racial senescence, hypothesized by earlier paleontologists, were incorrect (ch. 5).

These chapters addressed two different audiences. On the one hand, they made accessible to paleontologists the recent discoveries of population genetics as well as introducing important ecological concepts. For example, how the rapid shift might have occurred in the Oligocene epoch when some browsing horses "jumped the gap" to a grazing way of life as certain favorable genetic variants underwent strong selection. On the other hand, these chapters were meant to convince the biologists that the fossil record, indeed, had useful content that could make genuine contributions to evolutionary theory. For instance, one could measure relative and absolute rates of evolution, either in terms of changes in morphology within lineages (like fossil horse molars) or in terms of taxonomic categories between fossil lineages (like horses and marine cephalopods)[55] Simpson thus began the process of winning over the two factions who had had a history of talking at cross-purposes, when they talked to each other at all.

Having brought the two camps together, Simpson could next develop how interactions between "organism and environment" (ch. 6), expressed generally but not always through natural selection, led to adaptation. Depending upon the temporal and ecologic scales used, various outcomes ensued, whether new species ("microevolution"), new genera and families ("macroevolution"), or new orders or higher taxa ("megaevolution").[56] Thus, one moved along a continuum from the realm of the population geneticist and field biologist to that of the paleontologist. The final chapter of *Tempo and Mode* (ch. 7) summarized the full argument of the book by relating specific patterns of evolution ("mode") to their complementary rates of evolution ("tempo") against the backdrop of environmental context.

Response to *Tempo and Mode*

I am consoled by the conviction that [Tempo and Mode] had some historic value, that it was a success at least to the extent that it did bring a new field of study and a new thesis into the development of the synthetic theory, and that its thesis has stood up well.
— G. G. Simpson[57]

Simpson devoted three of the seven chapters of *Tempo and Mode* to explicitly biological subjects: the "determinants of evolution," "organism and environment," and how they are integrated in varying "modes of evolution." Thus, almost one half of *Tempo and Mode* was "evolutionary biology," which illuminated what Simpson called in his introduction the "how" of evolution, and in the other half of the book "phylogenetic examples [from fossils] are introduced as evidence and to give reality to the theoretical discussion [based on genetics]" (1944b:xviii). Intentionally or not, Simpson thus gave equal time to the two disciplines he was trying to integrate, biology and paleontology. But the initial response to *Tempo and Mode* was by no means equally enthusiastic in both camps.

Reviews by biologists were quite lavish in their praise: "from now on no competent discussion of the mechanisms of macroevolution can be made outside the ambit of Simpson's analysis" (Dobzhansky); "hereafter no treatise on the causes and modes of evolution can ignore *Tempo and Mode*" (Glass); the "most important contribution yet to come from paleontology on the methods by which evolution takes place" (Hutchinson); Simpson "has performed the double task of reminding neozoologists of many facts of paleontology which they have tended to overlook as unfamiliar or inconvenient, and at the same time showing the possibility of accounting for them in genetic terms" (Huxley); and "the reconciliation [of genetics] with paleontology" (Wright).[58] Of course, almost every reviewer had the usual, minor reservations about one thing or another, but the overall impact was definitely one of unanimous approval of Simpson's book by the biological community.

No doubt part of this approval from biologists was self-congratulatory because Simpson fully accepted and integrated the data and interpretations of the population geneticists into his reading of the fossil record. But surely it was more than self-interest, because by extending the explanatory power of the new genetics not only to a different set of data, but data viewed at completely different temporal and ecological scales of reference, *Tempo and Mode* made the genetic argument more consilient. Simpson by this time had been elected to the American Philosophical Society and the National Academy of Sciences, so the biologists could only be pleased that a leading student of paleontology validated their results and expanded their realm of explanation.

The paleontological community, however, paid little attention at first to *Tempo and Mode*. For example, it was not reviewed by the leading North American paleontological journal. The only paleontologist to take note of it was Glenn Jepsen of Princeton University in a review, ironically, for the *American Midland Naturalist*. Jepsen praised "its service as a debunking device for some traditional and stereotyped opinions" of paleontologists, and he noted the enthusiasm of the biologists for the book, but he also pointed out aspects of it that they ignored: namely, Simpson's "most original contributions to interpretation of fact and epistemology . . . [that is,] the expansion and correlation of scientific reasoning in paleontology."[59]

Why the lack of immediate response by the paleontologists? It wasn't because they were too obtuse to understand the argument, but rather that the argument did not directly involve their own day-to-day research. As Rainger has noted, there was a strong tradition in paleontology in the late nineteenth and early twentieth centuries that concerned itself chiefly with empirical morphological description, and its change and sequence.[60] Although by the 1930s there was some theoretical interpretation occurring within the discipline by people like Gregory, Matthew, and Osborn (as noted above), the great majority of its practitioners were still not so theoretically inclined. So while paleontologists accepted the overall thesis of *Tempo and Mode*, there wasn't an obvious display of it in their published writings. Simpson himself remarked upon this lack of response several years after the publication of *Tempo and Mode*. He tabulated the subject matter of "principal papers" in the *Journal of Paleontology* for the years 1939 and 1949, that is, before and after the publication of *Tempo and Mode*. On the basis of fifty-eight papers published in 1939 and sixty papers in 1949, he noted virtually no difference in either the absolute numbers or relative proportions of articles devoted to "descriptive morphology and systematics" — 71 percent for both years. "Mainly geological" papers accounted for 17 percent versus 11 percent, whereas those "mainly biological" in content stood at 8 percent versus 7 percent. (The remaining categories were "methodology and bibliography," 4 percent versus 9 percent, and "general orientation of paleontology," 0 percent versus 2 percent.) Most surprising of all, within the "mainly biological" category there was just a single paper on "evolutionary theory" published in 1939 and none in 1949, and just a handful in either year concerned with "principles of systematics, phylogeny, and functional morphology."[61]

Simpson introduced this analysis by noting that it is "the opinion of some paleontologists that there has been some change in the scope and emphasis of paleontological research" (presumably owing to the formulation of the evolutionary synthesis) such that "descriptive work would tend to broaden, to emphasize increasingly the bearing of the fossils described on their geological and biological settings . . . and perhaps give more attention to such subjects as ecology, functional and broadly comparative morphology, [and] evolutionary processes." After tabulating his statistics, Simpson concluded that the "figures certainly reveal no trend away from straight descriptive studies," and he asked rhetorically, "does this [lack of change] betoken desirable stability and maturity of research programs or does it indicate lack of progress and undesirably narrow, routine, and unimaginative approaches to research?"[62]

Because scientific journals obviously are supported by people who have a vested interest in what the journal publishes, both as subscribers and, more importantly, as prospective authors, they tend to resist change in editorial policy. In fact, Simpson already had sensed this resistance and therefore was instrumental in the founding in 1947 of the journal *Evolution*, published by the fledgling Society for the Study of Evolution, whose members spanned the diverse spectrum of specialties unified by the evolutionary synthesis. As the first president of that society in 1946, Simpson helped raise the money to get the journal started.[63] In the early decades of *Evolution*'s existence, a number of paleontologists, both invertebrate and vertebrate — including Simpson — contributed articles dealing with "evolutionary theory." Their contribution proportionally waned somewhat in later years, if only because the number of such journals interested in "theory" had since increased.

There were a few contemporaries of Simpson who paid more than lip service to the synthesis, but full-scale incorporation of the results of the evolutionary synthesis into a "paleontological tradition" required the subsequent education of a new generation of paleontologists in microevolutionary theory — especially genetics and ecology — as well as the new macroevolutionary theory, rather than a making-over of those already established in the discipline.[64] E. C. Olson has recalled that, as editor of *Evolution* in the 1950s, he invited Alfred S. Romer (1894–1973), one of America's leading vertebrate paleontologists, to "please write us some articles," but Romer demurred, saying, "But I don't write that kind of evolution."[65] Nevertheless, by the mid-1960s, two decades after the consolidation of the evolutionary synthesis (and the publication of *Tempo and Mode*), paleontology had indeed broadened its scope and emphasis, as symbolized by common usage of the word *paleobiology* to distinguish the new practitioners from the old.

Adoption of the Population Concept of Species

A change was then [1939] in the air, especially as regards systematics which among all the ramifications of zoology necessarily remains its basic discipline. . . . The population approach has now [1960] become usual in systematics and has spread into all branches of zoology.
— G. G. Simpson, A. Roe, and R. Lewontin[66]

Although *Tempo and Mode*, and more generally the evolutionary synthesis, were slow in having an impact on the ongoing, day-to-day practice of paleontology, there was at least the passive effect among paleontologists of abandoning the older, now outmoded concepts. As Mayr has noted, the evolutionary synthesis refuted a "number of misconceptions that had the greatest impact on evolutionary biology. This includes soft inheritance, saltationism, evolutionary essentialism, and autogenetic theories."[67] Simpson's *Tempo and Mode* endorsed hard inheritance; claimed that evolutionary transformation was gradual (in terms of the degree of genetic changes from one generation to the next, although the rate of change could be quite variable, from very slow to very fast) and thus that any apparent gaps among fossils were artifacts of the record; debunked inherent factors in driving evolution (like racial senescence, inertia, momentum, and orthogenesis); and accepted an anti-essentialist, populational view of species. *Tempo and Mode* thus was crucially instrumental in catalyzing the end of one conceptual era in paleontology and in ushering in another, even if the transition did not occur overnight.

However, there was one specific area where there was fairly immediate, active response, and that was the shift from the typological to the population concept of species. As quoted above (at note 42), Simpson had remarked about the disabling effects of the attention paid to the individual fossil specimen by paleontologists. Prior to the evolutionary synthesis, species were commonly defined on the basis of a single tooth or bone or shell that provided the "type," exemplifying the sort of neo-Platonic essentialism noted by Mayr. During the writing of *Tempo and Mode*, Simpson addressed this particular issue in a separate article, by proposing the concept of the "hypodigm" to replace the type: "All the specimens used by the author of a [new] species as his basis for inference, and this should mean all the specimens that he referred to the species, constitute his hypodigm of that species. . . . The hypodigm is a sample from which the characters of a population are to be inferred."[68]

As two of his distinguished paleontological contemporaries observed some time later: "In earlier years the all important specimen of the species was the type: a sort of enthroned little god, in the image of which all other individuals of the species were supposed to have been made. We are coming to quite a different position now . . . what we must consider is the population as a whole — the hypodigm to use Simpson's term."[69] "This whole scheme [of taxonomic nomenclature] collapsed like a house of cards when George Simpson published his short but epoch-making paper on 'Types in Modern Taxonomy.' Simpson emphasized that in an interbreeding population each individual is as much as any other a part of the species. Therefore a true concept of the species — which he called the hypodigm — must encompass the extremes as well as the means of variation within such a population."[70]

This conversion to populational thinking was, of course, more than cosmetic because any serious thinking about the natural populations of fossil organisms led to broader considerations of genetic variation, selection, interactions of organism and environment — in short, many of those elements included within the evolutionary synthesis. So even if the practical significance of *Tempo and Mode* might initially have seemed somewhat remote to paleontologists, in fact this new perspective on more ordinary descriptive morphology and taxonomic work continuously reinforced *Tempo and Mode*'s biological concepts. It emphasized for paleontologists that the theoretical contributions that biology was making to paleontology could be reciprocated because fossils had to validate the biological theory. Simpson observed that whereas "early paleontologists regularly cited the fossil record as evidence *against* the reality of evolution . . . [now] no theory of evolution can long be satisfactory, even to the geneticists and systematists, unless it is explicitly shown to be harmonious with the factual record of evolution as revealed by paleontology. Moreover, there are very essential parts of a general theory of evolution that cannot be based on the study of recent animals and plants, alone."[71]

The biologist Ernst Mayr, too, acknowledged this role for paleontology: "The study of long-term evolutionary phenomena is the domain of the paleontologist. He investigates rates and trends of evolution in time and is interested in the origin of new classes, phyla, and other higher categories. Evolution means change and yet it is only the paleontologist among all biologists who can properly study the time dimension. If the fossil record were not available, many evolutionary problems could not be solved; indeed, many of them would not even be apparent."[72]

In one important aspect of the populational species concept, Simpson did indeed distance himself somewhat from the biologists. As an evolutionist interested in Darwinian "descent with modification," yet constrained as a paleontologist in observing that modification always in terms of morphological transformation of hard parts (bones, teeth, and shells) through a sequence of sedimentary rocks, Simpson recognized a distinction between evolutionary species and biological species. "An evolutionary species is a lineage (an ancestral-descendant sequence of populations) evolving separately from others and with its own unitary evolutionary role and tendencies." The biological concept of species as given by Simpson did not contradict Mayr's definition: "groups of actually or potentially interbreeding natural populations, which are reproductively isolated from other such groups."[73] Rather, Simpson the paleontologist was preoccupied with which species gave rise (over time) to the next species, whereas Mayr the biologist would be preoccupied with physical isolation (across geographic area) giving rise to new species. Thus, Simpson's species concept emphasized the vertical, or temporal, aspect of the species while Mayr's definition emphasized the horizontal, or geographic, aspect.

Although both concepts of species are consistent with each other, the differences in emphasis reflected the very different sorts of data available to the field biologist and the paleontologist — the former working within a virtually zero-time dimension of the present moment across fine-grained geographic space, the latter working within a greatly expanded temporal dimension across much coarser-grained geography. Given the nature of the fossil record, it is therefore extremely difficult for paleontologists to resolve time and space with sufficient precision to document Mayrian speciation the way biologists can infer it. Clearly, then, it is the longer-term, larger-scale changes and trends in hard-part morphology that will more often occupy paleontologists. However much they might be interested in observing speciation frozen in time, the inherent nature of the fossil record usually makes it inaccessible or impracticable.

Disciplinary Independence for Paleontology

Paleontology is characterized, but not fully defined, by having its own objective subject matter: fossils. Fossils occur in rocks, and they are organisms. Their extended study necessarily overlaps widely into both of the broader (or more miscellaneous) sciences of geology and biology.
— G. G. Simpson[74]

For all of Simpson's awareness and acknowledgment of the importance that biology had for paleontology, he was at the same time equally cognizant that the study of fossils was a unique and separate discipline.

The source of this disciplinary identity was the nature of the raw, basic data: fossils in rocks. It is "the flow of new discoveries and data from the field, laboratory preparation of specimens, and study of their morphology and taxonomy" that provide the "most basic essentials for continued progress in vertebrate paleontology." But however important such data and study were — what Simpson was fond of calling the "what" of paleontology — he also believed it important to "consider problems of broader and more theoretical biological interpretation that arise after the basic data, taxonomic and geologic, are in hand" — what he called the "how" of the science.[75]

Simpson was so adamant about maintaining disciplinary identity for paleontology that he considered resigning from the American Museum when the new director, Albert Parr, disbanded the department of vertebrate paleontology in June 1942 and transferred Simpson and his two paleontological colleagues to other biological departments (see chapter 11 for a more complete discussion of this incident). A more positive example of Simpson's belief in paleontology's mission was his role in the founding of the Society of Vertebrate Paleontology (SVP) in the early 1940s. Early in this century vertebrate paleontologists had their own professional society, but in 1907 decided to join with the somewhat more numerous invertebrate paleontologists to create the Paleontological Society.[76] Almost immediately, however, the Paleontological Society became a section of the still larger Geological Society of America. Soon the vertebrate paleontologists, always a small coterie of scientists, were greatly outnumbered by increasing numbers of invertebrate paleontologists and geologists. Gradually, the vertebrate paleontologists found that their interests at national meetings were given low priority. Initially, the vertebrate paleontologists tried to remedy their loss of identity by becoming a formal section of the Paleontological Society, but eventually this too did not work out to their full satisfaction. After several years of discussion, the SVP became formally organized in 1941, with a membership of some 150 vertebrate paleontologists.

Simpson's role in the founding of the SVP was central. He was the secretary-treasurer during its provisional first year in 1940, and its elected president in its first year of formal existence. An archival photograph, taken of the organizational meeting of the SVP at Harvard, shows Simpson (secretary-treasurer) sharing the dais with Alfred Sherwood Romer (president), another doyen of twentieth-century American vertebrate paleontology, both looking out over two dozen vertebrate paleontologists gathered below. This photograph is a synecdoche for the two dominant strains within paleontology: the "how" of the theoretician (Simpson) juxtaposed with the "what" of the comparative anatomist (Romer).

During 1941, the organizational year of the SVP, Simpson as secretary-treasurer published several small issues of a mimeographed "news bulletin," which not only contained the usual who-what-when-where but also an essay by him on "Some Recent Trends and Problems of Vertebrate Paleontological Research." In 1941, Simpson was also in the middle of writing *Tempo and Mode* and his monograph on the principles of classification. So we might expect a plea for broader, more theoretical consideration of the meaning of the basic raw data of vertebrate paleontology in such an essay. On the contrary, the essay is entirely concerned with the increasing decentralization of vertebrate fossil collections, such that it was more and more difficult for researchers to have easy access to them. Moreover, lack of adequate preparation, curation, and library resources aggravated the problems. "This society is a proper forum for their discussion and could perhaps be a means toward their solution. . . . It is certainly practical to improve the conditions for research without retarding, indeed while facilitating, the broader distribution of collecting, exhibition, and public education."[77] Obviously, however important Simpson viewed the construction of theory, he recognized that theory depended in the first instance upon the robust physical evidence of the fossils.

Given Simpson's fundamental contributions to the evolutionary synthesis and to other theoretical subjects, one might easily overlook his extensive writings on descriptive morphology, classical taxonomy, and systematics. Tabulation of Simpson's oeuvre counts 109 titles and 6675 pages of his work as theoretical and synthetic, and 224 titles and 5785 pages as empirical collection-oriented.[78] Clearly, in word and deed, Simpson was as committed to the "what" of vertebrate paleontology as to the "how." Accordingly, while he saw the importance of biology for illuminating the "how," he also was firmly convinced that the "what" deserved full recognition on its own terms as a separate discipline of paleontology. It therefore could not be subsumed under biology — or geology for that matter. Consequently, while he strongly believed that paleontology should draw upon observations, principles, and concepts within the biological sciences, he was equally convinced that all such theorizing must be firmly grounded in the fossils themselves and their geologic context.

*   *   *

> *I have collected a great many fossils, described even more, and named a good number of them. . . . Beyond that, I have taken a broader stance and a more theoretical and subjective one always in part in geology but increasingly also in organismal and evolutionary biology.*
> — G. G. Simpson[79]

When Simpson entered the field of paleontology as a neophyte in the 1920s, the discipline was estranged from the rest of biology because of the apparent contradiction between the putative mechanisms of evolution endorsed by biologists and the evolutionary history of organisms as interpreted by paleontologists. This estrangement, in turn, was a continuation of the more general, ongoing dichotomy between the older morphological tradition of the nineteenth century represented by paleontology and the newer experimental fields, like genetics and ecology, that were expanding within biology in the twentieth century.

During the 1930s and 1940s, Simpson demonstrated that the contradiction between paleontology and biology was only apparent and that in fact the two disciplines were mutually supportive of each other's conclusions. Simpson was thus instrumental in expanding contemporary biology's theory into the deep time represented by fossil history. This accomplishment resulted from his training in biology by distinguished teachers as well as from day-to-day institutional contact with a few paleontologists already sympathetic to what biology had to offer their discipline. The mutual relationship of the two disciplines, each expanded by the other, in the latter half of the twentieth century thus reflects in no small measure the contributions of George Gaylord Simpson. His election to the presidencies of the American Society of Mammalogists (1962), the Society for Systematic Zoology (1962), and the American Society of Zoologists (1964) was obvious acknowledgment of his contributions to biology, as was the award of more than a dozen international medals and prizes by scientific societies and organizations.[80]

CHAPTER 3

THE SUMMER OF 1924

The professional relations among scientists are important in the ongoing process of science. Unless one pays attention to them, one cannot begin to understand science.
— David Hull, philosopher

On the face of it, then — whatever importance history may ultimately assign him — during his lifetime Simpson was supremely successful as a scientist. But even the most preeminent must modestly begin somewhere, and so it interests us to know something of the first, early steps taken that eventually led to such a long and distinguished career. In this chapter, therefore, I focus on what retrospectively turns out to have been a critical period in Simpson's life — one that catapulted him onto the paleontological scene at the tender age of twenty-two (fig. 3.1).

"A Very Promising Graduate Student"

In the middle of his first year of graduate work in vertebrate paleontology at Yale in 1924, George Gaylord Simpson began looking about for employment for the coming summer. He needed a job that would not only further his paleontological education but also, with a wife and infant daughter to support, one that would pay him a salary, however modest. He eventually obtained a position prospecting for Tertiary mammals in Texas and New Mexico as a field assistant to William Diller Matthew of the American Museum of Natural History (fig. 3.2).

By the end of the summer, Simpson established himself as an energetic and highly successful field man, having made two major fossil dis-

FIGURE 3.1 G. G. Simpson upon his graduation from Yale College, 1923. (Courtesy Peabody Museum of Natural History, Yale University)

FIGURE 3.2 William Diller Matthew, 1922, two years before Simpson was his field assistant in Texas. (Courtesy Margaret Matthew Colbert family)

coveries, thereby impressing both R. S. Lull, his major adviser at Yale, and Matthew, whom he would eventually succeed at the American Museum as curator of fossil mammals. When Simpson returned to Yale in the fall, Lull, despite his earlier refusal, permitted him to study the Marsh Collection of Mesozoic mammals for his dissertation. Matthew, too, was enthusiastic about Simpson's demonstrated abilities, for he became Simpson's mentor, acting as informal off-campus adviser for his dissertation and eventually an advocate for Simpson's appointment at the American Museum.

Simpson also learned the hard way about scientific protocol and professional territoriality when a short paper he wrote describing the geologic results of his work in New Mexico was suppressed by Childs Frick, honorary curator of the department of vertebrate paleontology who had supported the New Mexico (and Texas) excursion with his own funds. Frick's very substantial financial support of the American Museum apparently gave him greater influence than Matthew who, as chairman of Vertebrate Paleontology, had initially approved Simpson's paper for publication in the museum's *Bulletin*.

Simpson's academic abilities were demonstrated early on. He had accumulated a grade point average of 93.7 at the University of Colorado in Boulder where he spent his first three years of college, before transferring to Yale. In his senior year at Yale his grade point average was 91.6, which enabled him to graduate with "High Honors" and be elected into Phi Beta Kappa. Simpson continued performing at this high level of achievement as a graduate student: in his first year of graduate work at Yale in 1923–24, he earned seven "Honors" and one "Good" in eight courses.[1] The only untoward event during his first year of graduate work was Lull's annoyance at Simpson's ticking off so visibly in his notebook the points Lull was making in his lectures.[2]

Perhaps as a result of this, Lull initially refused to let Simpson study the Marsh Collection of Mesozoic mammals that Simpson had discovered in the basement of the Peabody Museum. Lull, who was museum director, claimed that "those are much too important . . . very delicate and highly significant . . . for a young graduate student."[3] However, Lull still thought highly enough of Simpson to inquire about a possible summer position for him. He wrote H. F. Osborn, president of the American Museum of Natural History in New York, on Simpson's behalf in January 1924 to see if there was a planned field expedition that he might join. Lull told Osborn that while Simpson was "a very promising graduate student . . . what he greatly needs is a chance to do field work." Lull assured Osborn that Simpson "would give loyal and efficient service, as he is a hard worker, quick to learn, and can use his hands."[4] Three days later Osborn dictated a departmental memorandum to W. D. Matthew, who was chairman of the department of vertebrate paleontology, in which he suggested "that Mr. Gaylord Simpson [sic] be given a trial during the coming summer at manual work in the laboratory, this to be followed by manual work in the field if he proves to be promising and helpful in this way."[5]

Shortly after, Matthew wrote Lull explaining that he was planning "some reconnaissance work in Texas" and wanted to know "what sort of hand is [Simpson] with a car — both as to driving and as to tinkering

... [and] would $100 per month and expenses satisfy him?" Matthew also made a point of saying that it would be "incumbent on [Simpson] to devote himself not to having an interesting trip, nor to acquiring paleontological training, but to aiding me to get the results I want for the Museum." Finally, Matthew added that he "would want to size [Simpson] up personally" before making any definite arrangement.[6] Apparently, when Lull led his vertebrate paleontology class on a field trip to the American Museum, Lull did introduce Simpson to Matthew, and Simpson made a sufficiently good impression that Matthew ended up hiring him.[7]

W. D. Matthew, Canadian-born vertebrate paleontologist, was one of the leading paleomammalogists of the day. He was particularly known for his research on Cenozoic mammals of the American West and for his monograph "Climate and Evolution" (1915). He was also a leading student of horse evolution.[8] Besides giving Simpson the opportunity to work closely in the field with a distinguished vertebrate paleontologist who was also an accomplished stratigrapher, Simpson was earning a good salary for the four months when the university was not in session. During the academic year, Simpson had been able to support himself, his wife, and infant daughter on his university fellowship, supplemented by tutoring in the evenings and on weekends. But he had to look elsewhere for support during the summer. So the prospect of a monthly salary of $100 for doing fieldwork in vertebrate paleontology was undoubtedly irresistible.[9]

"Digressing All Over the Landscape"

By April, Matthew had decided on Simpson as his field assistant and was now writing him notes about arrangements. Simpson was to round up the necessary field gear — picks, chisels, geologic maps, relevant field reports — and be in Amarillo, Texas, by mid-May, where he was to make preliminary inquiries about buying a car and any local knowledge of places where fossil bones had been reported. Simpson was to do whatever advance scouting that he could conveniently accomplish on foot or by train until Matthew arrived.[10]

Once Matthew arrived, Simpson was assigned the driving of the new Model T Ford, although he had had no prior driving experience. Matthew was searching for late Tertiary fossils in the panhandle of Texas to fill a gap in the evolution between *Equus*, the modern horse, and *Plio-*

hippus, its Pliocene ancestor, a few million years older. Matthew, as the major authority at the time on horse evolution, was anxious to explore the Texas strata because they had yielded a few teeth recognized several years earlier by J. C. Merriam as suggestive of such an ancestor.[11] In the beginning the search was hardly promising. As Simpson wrote to his sister Martha, "I've been digressing all over the landscape. . . . Now & then we find a fossil — every third day or so, if small fragments count. . . . Poor Dr. Matthew gets madder & madder[:] 'First Tertiary formation in which I couldn't find mammals.'"[12] (fig. 3.3). Part of Matthew's chagrin no doubt arose from Simpson's own clumsiness. On one of the first few days in the field, Matthew found some fairly important horse teeth and, after treating them with shellac, laid one of them on the ground to dry. Simpson promptly stepped on the tooth, breaking it into several pieces. Matthew glowered, telling Simpson, "Go stand over there," some twenty-five feet out of the way, and didn't talk to him for several hours. But by the next day the incident was forgotten.[13]

For most of the six weeks they were in the field together, Simpson

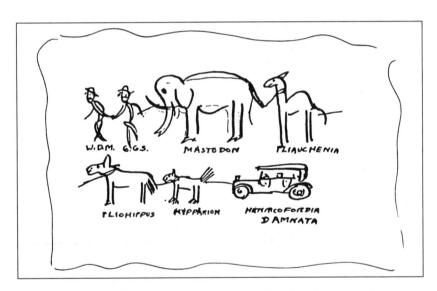

FIGURE 3.3 Sketch by Simpson in a letter to his sister Martha written from Texas in 1924, showing Matthew ("W.D.M."), Simpson ("G.G.S."), and reconstructions of several fossil mammals (whose bones they were discovering) pulling their new Ford field-vehicle nicknamed "Henricofordia damnata" by Simpson. (Simpson Papers, American Philosophical Society Archives, GGS/APS. Courtesy American Philosophical Society)

and Matthew "shared a room in a village boarding house . . . and became as well acquainted as possible for a man of fifty-three at the top of his profession and a neophyte of twenty-two who hoped to enter it."[14] When not preoccupied with finding fossils, Matthew "expounded a great deal of his scientific and personal creed" to Simpson, discoursing on what he considered the proper ethics and desirable motivations for scientific research, the nature of scientific truth, the role of museums, and even his political views. Simpson learned, too, about Matthew's taste for cigars, late afternoon soda pop, and pulp westerns. The fieldwork was not only intellectually heady but also physically demanding: "Most of the time we were there the thermometer was above 100° every day, & the gullies in these blazing rocks were a veritable hell. I was a mass of huge water blisters. Every movement was agony, & at the slightest exertion sweat rolled off in a steady trickle."[15]

Eventually, their combined efforts paid off. In mid-June, one day after Simpson's twenty-second birthday, Matthew found an almost complete skeleton of the animal he was looking for, although it lacked the skull. A week later Simpson made his own first important fossil find, the complete skull and partial skeleton of the same species (fig. 3.4).[16] Combined together, the fossils produced precisely what Matthew had hoped to find: the missing piece from the latter part of the evolutionary lineage of the horse (fig. 3.5). Matthew reported that the "size and proportions of the skeleton . . . are those of a rather small *Equus*, the pattern of the teeth is intermediate but nearer to *Pliohippus*, [and] a tiny nodule of bone remains to represent the fifth digit of the fore foot."[17] The scrappy material found earlier by others had formed the basis for a new species, *Equus simplicidens*, named by E. D. Cope in the late nineteenth century. Matthew retained Cope's species designation but placed the species in a new genus, *Plesippus*, intermediate between *Pliohippus* and *Equus*.[18] A composite skeleton reconstructed from the two major fossil finds was placed on exhibition, soon after, at the American Museum as part of an imposing display on Darwinian evolution as exemplified by horses.

"Some Discovery to Rival the Boss's"

Simpson's field notes for Texas and New Mexico, as now preserved in the archives of the department of vertebrate paleontology at the Amer-

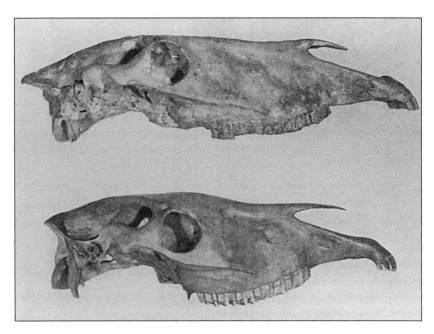

FIGURE 3.4 Simpson's first fossil discovery was a horse skull (*top*) compared with modern horse (*bottom*). (W. D. Matthew 1926, "The Evolution of the Horse," 162; © University of Chicago Press)

ican Museum, contain descriptions of twenty-two localities in Texas, thirteen of which produced some vertebrate bone. The notebook — 5 inches wide by 7 ½ inches high with a stitched binding across the top and "Yale University" on the cover — seems to be a "fair copy" that was rewritten after return from the field, as judged from its neat handwriting and clean condition as well as occasional post hoc commentary within the running text, such as "This is the area especially stressed in this summer's work [in Texas]."[19] Simpson also seems to have included descriptions provided by Matthew, which are indicated by the latter's initials following Simpson's handwritten descriptions.

Simpson's notes for his own important discovery of a fossil horse state:

> [Quarry] # 18 G. G. S. Up left hand fork of central one of the three larger tributaries of this big draw. (See sketch map.) At the very bottom of the draw, in a vertical cliff 7′–8′ high. Practically at the base of the formation, clearly at a lower horizon (by 10′–20′) than the mastodon jaw (#13 [quarry]) from the right hand fork of the

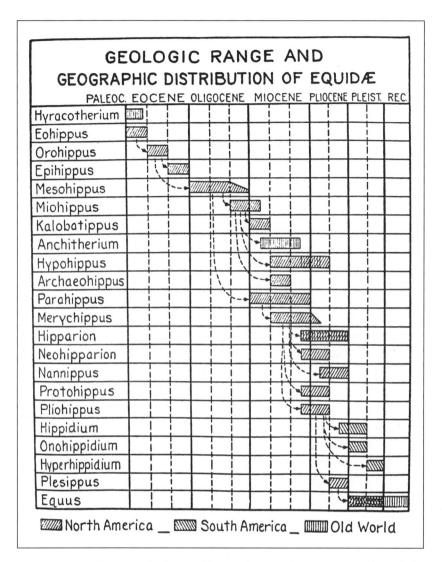

FIGURE 3.5 Horse evolutionary history showing important position of the Texas fossil horse, *Plesippus*, as ancestral to modern living horse, *Equus*. (W. D. Matthew 1926, "The Evolution of the Horse," 167; © University of Chicago Press)

same ["small," crossed out] branch. Matrix — A fine white slightly gritty clay, fairly soft in places, but generally very hard about the bone. Many small cavities. The bone is in part good & firm, but with a tendency to flake. Much, however[,] has suffered solution until it is spongy & the surface checked, or even until it is entirely gone, leaving only the cavity. Lead — Well exposed — sections of broken fore & hind limbs, pelvis, cervicals, & the anterior lower part of skull. Mold of lower jaw and part of a cervical afterwards found in the wash. The bone occurs in a lens about 10′–15′ long & a foot or so thick. No fragments observed outside this lens. Above & below are alternating layers 3″–6″ thick of banded soft clays and hard limy platy layers. The outcrop is covered by a peculiar fine chocolate colored dust. Bones belonged to a single individual of the rare species of *Pliohippus* as in #14 [the quarry where Matthew found his headless horse a week earlier]. Mostly removed [by Simpson and fieldwork assistant Charles Falkenbach] in blocks. Apparently several cervicals, an articulated fore-foot & parts of the other, a few ribs, part of the pelvis, both articulated hind limbs. Also a skull probably complete save for the extreme anterior part, and the mold of the lower jaw found in the wash.

These particular notes are accompanied by a map showing section-line roads and washes where several quarries, besides those mentioned above, are located as well as a stratigraphic section in which quarries 14 and 18 were located (fig. 3.6). There is no scale on the map, but by comparison to a current U.S. Geological Survey topographic map seems to be about one-tenth mile to the inch. The stratigraphic section spans some sixty-five feet of the Pliocene Blanco Formation exposed in the embankment of the wash (fig. 3.7).

Simpson was, of course, pleased that he was able to prove himself as a fossil prospector. Years later in his book *Horses*, Simpson reminisced about his discovery: "hoping to make some discovery to rival the boss's, [I] found a lead to a partly buried fossil and breathlessly worked in to see what [I] had. Soon [I] was running back to Matthew, wild with excitement. [I] had a skeleton, too, a much less complete skeleton in all but one respect: [it] had the head still there!"[20]

"A Considerable Piece of Work"

Soon after, Matthew returned to New York, leaving Simpson with Charles Falkenbach, a young preparator who had also come out from New York to help with the fieldwork. As the Texas work was concluding, Matthew wrote to Simpson telling him of another area the museum wanted explored, this time in north central New Mexico, before he and

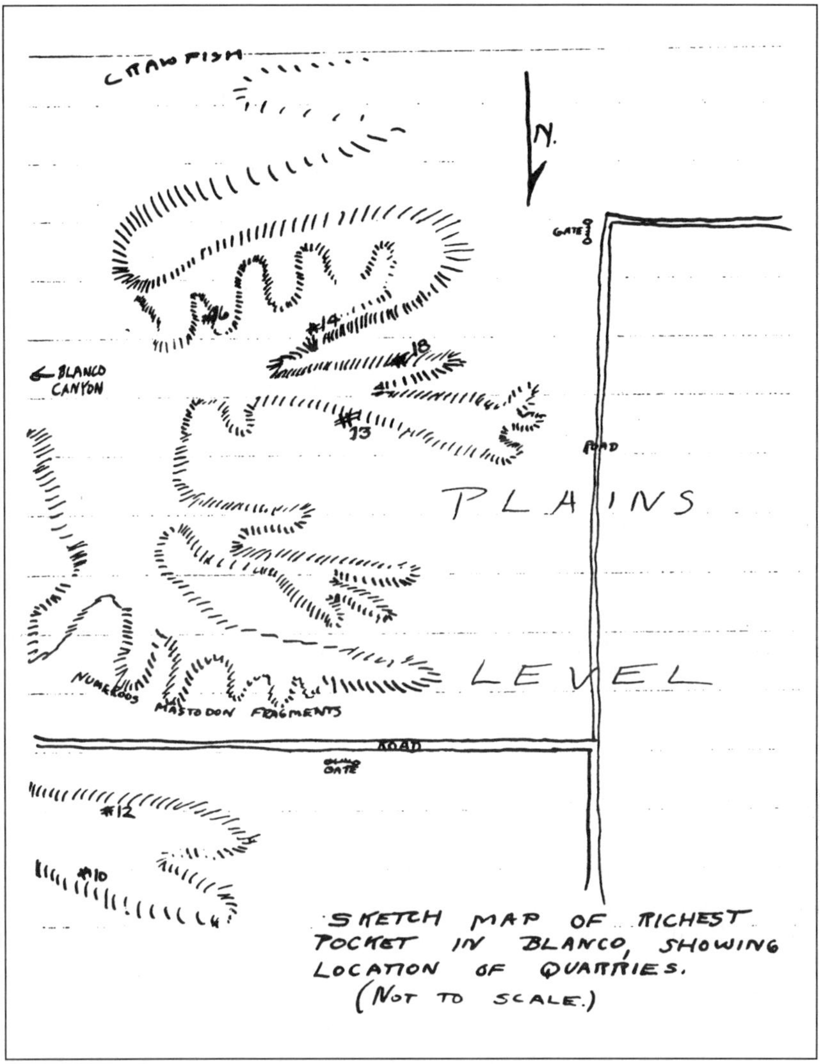

FIGURE 3.6 Simpson's hand-drawn field map of localities in Texas showing where his (#18) and Matthew's (#14) horse bones were found. (Courtesy Henry Fairfield Osborn Library, American Museum of Natural History)

Falkenbach returned home: "The Santa Fé beds you will find probably a considerable piece of work to examine, and calling for pretty careful stratigraphic work and faunal records."[21]

The New Mexico work was being specifically undertaken for Childs Frick, honorary curator at the museum. Childs Frick (1883–1965), the in-

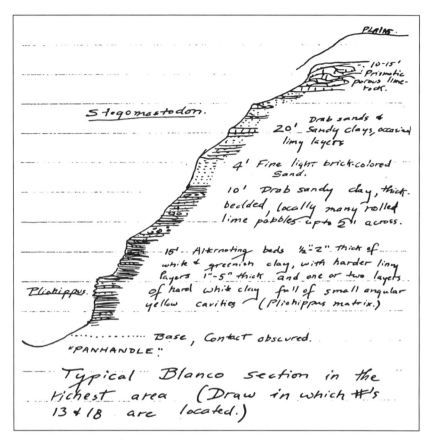

FIGURE 3.7 Simpson's field sketch of rock strata where his fossil horse skull was found. (Courtesy Henry Fairfield Osborn Library, American Museum of Natural History)

dependently wealthy son of Henry Clay Frick, had developed an interest in paleontology at Princeton University and established an informal association with the American Museum upon graduation. After his father's death in 1919, the younger Frick was elected a museum trustee in 1921 to take his father's place and immediately put on the Finance Committee.[22] Frick established early on a Paleontology Field Fund "after consultation with Professor Osborn . . . for field work in extinct mammals. The thought was to provide the curator of fossil mammals with a nucleus to build on in order that he might eventually have an annual income that would be sufficient to guarantee seasonal expeditions into the field."[23] Money from this fund had already supported Matthew's Texas work and was now going to be specifically used for Frick's own research. Frick's ongoing generosity to the museum allowed his having laboratory facilities within the museum and hiring preparators to work on his collections of late Cenozoic mammals from the western United States. Eventually Frick set up his own research facilities at his home in Roslyn, Long Island, that included a private staff of field collectors and laboratory preparators.

As Frick observed at the time, his interest in the Santa Fe formation of Miocene and early Pliocene epochs was stimulated one day when he was

> surprised to discover among other interesting material forwarded . . . by my industrious field assistant, Mr. Joseph Rak, some teeth of . . . a much discussed but little-known carnivorous genus [*Hemicyon*] of the French Miocene, related perhaps to the peculiar carnivore of Monte Bamboli [Italy] . . . and to certain of the huge and inaptly named "hyena-bears" of the later Pliocene, but heretofore unrecognized in America. Subsequently I happened to note the strong resemblance of a fragmentary lower jaw in the National Museum collection to certain of this . . . material, and later was able definitely to determine it, too, as *Hemicyon*. This specimen had been found by Cope, then vertebrate paleontologist of the U.S. Geological Survey, in the vicinity of Santa Fé, a locality that had lain untouched by fossil hunters since his visit in 1874. Was material still to be found there? Might the old locality yield more of Hemicyon?[24]

Simpson went on to Santa Fe, as requested by Matthew, but he had in the meantime checked back with Lull at Yale. Simpson then wrote Matthew saying that Lull agreed that he could stay on there as long as the museum wished. Simpson added a cryptic short sentence in his letter to Matthew: "My plans for the winter are still entirely unsettled."[25] As noted above, Simpson's earlier request to work on Yale's Marsh Collection of Mesozoic mammals had been rebuffed by Lull, so perhaps Simpson felt uncertain of his standing with Lull and might have been considering leaving Yale for New York, where he could pursue a dissertation under Matthew at the museum and matriculate at Columbia for his doctorate. There must have been some face-to-face conversation between Simpson and Matthew about this possibility, because Matthew next wrote Simpson, saying, "Professor Osborn has had under consideration the possibility of making an offer to you here this winter, in case you do not find it practicable to continue your Yale p.g. [post-graduate] work. . . . He is very much pleased with the results of your work so far."[26]

Simpson and Falkenbach left Texas on July 19, arrived in New Mexico the next day, and, as Falkenbach informed Frick, "spent the afternoon in the hills, saw quite a few pieces of bone, feel quite certain we are going to make a good fine [sic], from the looks of this country it will take a long time to look it all over."[27] Simpson meanwhile was concerned with more ordinary considerations, writing Matthew, "It seems that if we are to navigate in peace here we must have license plates. To get these we must have written permission from you to operate the car. . . . Could you please send us a note to that effect?"[28] Simpson sent Matthew a second letter that same day outlining what he thought ought to be their prospecting strategy: "The Santa Fé formation is exposed between here [in Espanola] and Santa Fé — a strip several miles wide and thirty long. It also continues, says Darton in a bulletin, for an 'unknown distance south' of Santa Fé, & from here splendid exposures can be seen extending for at least a number of miles north. I think for the present we shall confine ourselves to the region covered by Gilmore — the area directly south & east of here, mainly on the Santa Clara, Pojoaque, & Nambé Pueblo lands. Even this area is so large & rough that we shall need to use horses for one or two days to locate the richer areas if we are to get much good out of a very limited remaining time."[29]

Matthew was obviously satisfied with what his young field assistant had accomplished, for he wrote Simpson the following week that, "If, as seems very likely, you are not able to finish up the basin this season, you will want to do some preliminary scouting for next season, so that we can plan the expedition for next year. . . . I would hope to take charge of it, and to have you and Charlie join if you are free."[30]

Even as Matthew was writing this, Simpson and Falkenbach had already discovered exactly what they had been sent for. As Falkenbach announced to Matthew: "[I am] writ[ing] you of our good luck. We found the end of a skull . . . it is a large dog . . . complete back bone, pelvis, lower jaws, ribs, several limb bones. . . . It is going to take quite some time to take this material out."[31] Matthew replied immediately, "I congratulate you both most cordially. . . . A large dog skeleton or indeed any large carnivore is a great desideratum for the Miocene and Pliocene faunas. If it should prove to be the Santa Fé species described by Cope from a lower jaw as *Canis ursinus*, it will be of special interest to Mr. Frick as he is of the opinion that this species belongs to the genus *Hemicyon* of the Miocene of France. . . . I will want to have a letter from Mr. Simpson with recommendations as to how long it will take to clean up the Santa Fé beds. . . . This I should have to take up with Mr. Frick. . . . Subject to his approval I would be inclined to authorize you to stay with it this season as long as the fossils hold out."[32]

The next day, Matthew wrote American Museum president Osborn telling him of the important fossil discovery and remarked that the "success of the Santa Fé work may serve as further ground for considering making an offer to Simpson to join our staff this winter if he cannot continue his post-graduate course at Yale. Lull probably will find a place for him in his laboratory or field staff if we do not make the offer, but I think he [Simpson] would give us the preference. Only we ought not to delay action too long." At the bottom of the letter Osborn scrawled "Delighted" with his initials "HFO."[33] But he had second thoughts about Simpson's joining the museum staff, for a few days later he wrote to Matthew, "Would it not be well to see Doctor Lull and talk over the Simpson assistantship? If possible, I would like to have a man more representative in appearance, so that in your absence he could more adequately represent the Department."[34] Perhaps Simpson's slight build (5 ft. 7 in., 130 lbs.) as contrasted with Osborn's own large frame (over 6 feet tall and more than 200 lbs.) had something to do with Osborn's opinion about Simpson's physical bearing.

Meanwhile, in New Mexico, Simpson and Falkenbach were battling the weather to recover their fossil. Simpson wrote Matthew, "Our work has been much delayed. The rainy season is setting in and it rains hard every afternoon. Our only hope is to get the block [with the fossil in it] firm enough to place on a bed of hay & move it to shelter where it can dry thoroly [sic]."[35] Simpson was also finding time to do some basic geology as he and Falkenbach scouted around for additional fossil-collecting localities. In this same letter, he devotes several pages to the composition, thickness, structure, and geomorphology of the region, concluding with, "There is much geological & stratigraphical work. . . . I do not think we can possibly get done this year. At present I feel that five years would be required." He ends his letter with the remark that he is still "hoping to return to Yale, altho still uncertain as to the possibility."

"Bone is Abundant"

Simpson's field diary for the New Mexico work indicates seventeen localities where vertebrate fossils were found, most of them within less than a five-mile radius southeast of Espanola, some twenty-five miles north of Santa Fe (fig. 3.8). As Simpson noted, "This is the area especially stressed in this summer's work. . . . Bone is abundant . . . excellent exposures, very favorable for prospecting." But Simpson also observed: "Owing to the entire separateness of these blocks [of exposures], the impossibility of tracing any stratum laterally even within a block, the impossibility of referring a horizon to the top of the formation, seen only in the cliffs W[est] of the Rio Grande (if indeed this surely represents a level as high as or higher than any E[ast] of the Rio), the absence of basal contact, the varying dip & strike & probable presence of undetected minor faults, it would be impossible to determine

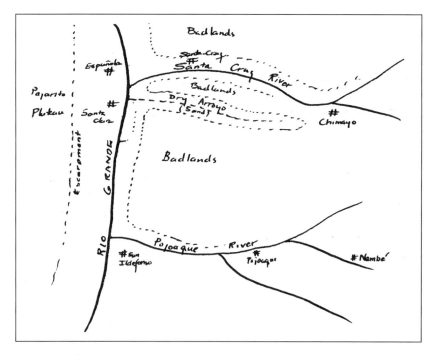

FIGURE 3.8 Simpson's hand-drawn field map of the area in New Mexico where the fossil of the "dog bear," *Hemicyon*, was discovered. (Courtesy Henry Fairfield Osborn Library, American Museum of Natural History)

the levels of finds accurately in most cases save by detailed E-W traverses from the Rio Grande fifteen miles or more to the base of the S[an] F[rancisco] Range."[36]

With respect to *Hemicyon* (fig. 3.9), Simpson described the fossil occurrence as follows:

> [Quarry] #5. Skeleton of carnivore. 7/25/24 Large block — Skull complete. Tip of nose exposed but only tip of right c[canine?] broken off, & this recovered & pasted to top of skull. Jaws separate from skull, somewhat crushed, hind end exposed & broken but all of left ramus recovered. No teeth exposed. Vertebrae complete and articulated from axis to sacrum, atlas & tail probably in block. Pelvis broken but nearly complete. Femur. Other limb bones not exposed. Many ribs. Small block — Scapula (left) etc. Separate — Several ribs, humerus complete. From wash — lower part of right scapula. [H]umerus nearly complete. Parts of other limb bones.
>
> The bone is hard and firm with a splendid surface. It breaks clean from the matrix. The matrix is a fine sand with thin (1" or less) lenses of greenish clay. In places it is hard next to the bone. Rather micaceous.
>
> Locality — About 1 mi. S. of Dry Arroyo 2 mi. E. of Rio Grande.
>
> Horizon — Ca 200' Below top of beds exposed in this block (E. of Rio Grande, bet. Dry Arroyo & Pojoaque River).

By the end of the month, Falkenbach reported to Matthew that, "We have our dog out and packed," but "we are having lots of trouble with our car, the roads are terrible . . . the sand is to[o] heavy for a Ford. Most of the time [one] of us is driving while the other is pushing. . . . Between motor trouble and tire trouble it takes most of our time."[37]

"Busy Collecting Other Men's Prospects"

Frick was so encouraged by what Simpson and Falkenbach had accomplished that he sent Joseph Rak, a collector in his full-time employ, out from New York to expand the prospecting effort, starting what would become a five-decade project in the Santa Fe formation, ending only when Frick himself died in 1965. Simpson was perhaps getting wind of the importance of what he had been doing, given the flurry of interest

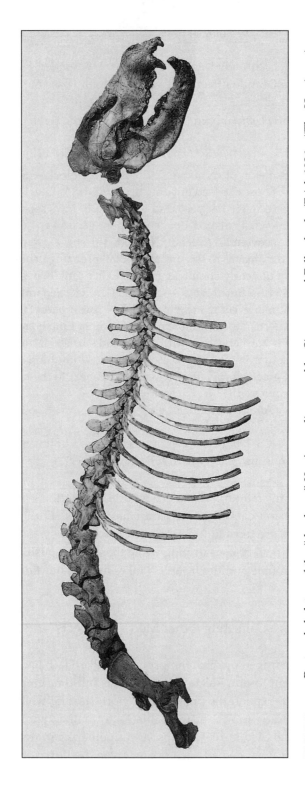

FIGURE 3.9 Restored skeleton of the "dog bear," *Hemicyon*, discovered by Simpson and Falkenbach. (Frick 1926a, "The Hemicyoninae and an American Tertiary Bear," *American Museum of Natural History Bulletin* 56: frontispiece/Neg. 311116; Photo: E. M. Fulda. Courtesy Department of Library Services, American Museum of Natural History)

and communications between the museum and the field. Matthew was becoming sensitive to what Simpson might perhaps be feeling, so in late September, as Simpson's work was winding down — he had by then spent four and a half months in the field — Matthew reassured him. "You are situated a little as [Walter] Granger was during his first year in Mongolia [on the museum's Central Asiatic Expedition], when he was kept busy collecting other men's prospects. Of course you will understand that the success of an expedition is based on the finds and on the way they are collected, and that expert collecting is valued a great deal higher than prospecting. I state this because I know how you and Charlie may feel about spending your time on taking up another man's prospects, in an area where you had done all the preliminary reconnaissance, and I want you to know that you don't lose anything on it, if that is the situation." Having reassured Simpson about what he had accomplished, Matthew let him down easy about his future with the museum in the next paragraph. "I haven't been able to get anything definite as to prospects on taking you on here in case you cannot continue at Yale. Probably that means that the President [Osborn] would like well to take you on in such a case, but sees too many other prospective calls for the money to be willing to commit himself. However, you had better let me know as soon as you can be sure one way or another [about Yale], as I think Mr. Frick may want some work done that you could handle to advantage."[38]

Simpson undoubtedly didn't lose anything working other men's prospects because Lull and Yale did come through after all, for Simpson was not only offered a fellowship for the coming academic year but was also given permission to undertake his original idea for a dissertation. Writing to Matthew in the fall from New Haven, Simpson said that, "Rather to my surprise Professor Lull has given me the Mesozoic mammals to work over."[39] Matthew replied the following month: "As to the Mesozoic mammals, that is a most important assignment. I think you'll run up against rather serious difficulties. . . . I am sending you herewith a transcript of my notes on British Mesozoic mammals, made in 1920. . . . These notes are quite at your service."[40] And thus began a new correspondence about Mesozoic mammals between Simpson-the-student and Matthew-the-mentor that was to last several more years.

Meanwhile, the New Mexico fossils had arrived in New York. As Frick recalled it two years later, "Months later this block was received and opened up at the Museum. Before us lay a nearly complete skeleton of *Hemicyon*, revealing a beast with somewhat tiger-like proportions and dog-like teeth, till lately quite unknown excepting for the few partial jaws and palates from the Mid-Miocene of France. Since then we have worked for many months at Santa Fé, and have secured a splendid series of remains representative of this ancient fauna, but to date, this first great trophy is our most complete specimen of any of those ancient animals, and, save for two fragmentary teeth, our only example of *Hemicyon*."[41]

In his technical report on *Hemicyon*, Frick noted that "on the fifth day of their reconnaissance Messrs. Charles Falkenbach and G. G. Simpson, of our party, had the good fortune to come upon the magnificent skull and skeleton. . . . Curiously enough this was the one and only specimen of *Hemicyon* found during some nine weeks in the field."[42] While explicitly acknowledging Simpson's role in this discovery, Frick was also clearly indicating that Simpson was "of our party," thereby staking out his own claim for both the fossil and all other associated results. As the one who suggested and funded the project in the first place, he obviously felt justified in being proprietary about its outcome.

Years afterward, Simpson claimed that he did not learn until later that Frick had provided the financial support for the work; he had thought he was working strictly for the museum and Matthew. He certainly would have known had he seen the letter from Matthew to Falkenbach, July 31, 1924, where Matthew discusses taking up with Frick the latter's continued support for fieldwork (see above). One would suppose, too, that Falkenbach must have told Simpson sometime during their days together that he was an employee of Frick's and not the museum's.[43] However, on the face of it, knowledge of Frick's support would not necessarily have precluded Simpson's publishing on some of the geology on the Santa Fe beds because Matthew, who did know, encouraged Simpson's efforts in this regard.

In late November, Simpson sent Matthew an abstract and brief article on his reconnaissance of the Santa Fe formation, with the note that "previous publications have been very sketchy (even more so than this) and quite inaccurate."[44] About a week later Matthew wrote back saying, "I . . . congratulate you upon an excellent piece of work. I have turned [the paper] over to Mr. Frick for study and sent in the abstract [to the Paleontological Society]." Matthew backed up his opinion by volunteering to deliver the paper orally at the upcoming meetings of the Paleontological Society in upstate New York if Simpson himself couldn't be there, and then he would "submit it for publication in the American Museum Bulletin."[45] Given that Matthew was not one to mince words, his enthusiastic support of Simpson's work in New Mexico indicates that Simpson had indeed proved his mettle not only as a prospector and collector of fossils but also as an able field geologist. Matthew's letter also shows that, while he knew of Frick's interest and stake in the results, he did not see any obvious objection to Simpson's publishing the geological results of the New Mexican work.

"First Paper I Ever Wrote"

The manuscript sent to Matthew comprised thirteen pages of text, four pages of footnotes (about one-third page of actual text for each of the four footnotes), a page each for bibliography and for figure captions, and eight 3.5 × 8.5 inch, black-and-white contact prints of field photographs.[46] There is no field map, although Simpson had indicated to Matthew that he would eventually provide a sketch for one to be finished by a museum draftsman.[47]

The paper starts with a brief review of previous work in the region, including pointing out that the name, Santa Fe marls, originally applied by F. V. Hayden in 1869, is inappropriate in that "true marl is, in fact, by no means typical of the formation." Instead Simpson recommended calling the rocks in question the "Santa Fé formation" (p. 1). The body of the rest of the manuscript describes the geology of the area, indicating where fossils are especially abundant: "the [Rio Grande] river has cut a rugged and narrow gorge thru a series of basalt flows . . . slopes are rounded, covered with pebbles of granitic and metamorphic rocks and arkosic debris . . . no really good exposures and no fossils were found . . . badlands are cut into isolated blocks or massifs by tributaries of the Rio Grande . . . one rich in fossils, lies between Tesuque creek and the upper reaches of the Pojoaque river" (pp. 2, 3, 4). Simpson noted that the varied attitude of the strata and the possible presence of faulting not only made it difficult to estimate accurately the overall thickness of the strata but also made "it impossible for the present to refer any fossils to any very definite horizon within the formation" (p. 6). He recorded color, bedding characteristics, composition, and local unconformities of the various units, and emphasized the rapid changes in lithology both vertically and horizontally. Simpson also described his "impression of the general character of the fauna, gained while collecting, [that] may . . . have some bearing on the mode of origin of the formation": proboscidians, rhinoceroses, camels, cervids and antilocarpids, carnivores, beavers, rodents, tortoises, and even a vulture (p. 10).

Simpson ended the paper with his interpretation of the environments of deposition, noting that while earlier writers "spoke of the Santa Fé beds as lacustrine in origin . . . all recent writers agree in calling them subaerial, fluviatile, or fluviatile and eolian. . . . The present writer can only add evidence and detail in support of this view" (p. 11). In particular, Simpson interpreted the finer-grained sediments, where the fossils were mostly found, as floodplain deposits "in which burial was most quick and efficient. Contrary to the popular belief skeletal remains seem to have very little chance of becoming fossilized once they are washed into a main stream channel. The scattered nature of the remains and their comparative scarcity seems [sic] to indicate that death was not usually due to bogging or drowning but to other causes, and that there were few semi-permanent ponds or marshes on the flood plain. It is a suggestive fact that not infrequently bones are found which bear the marks of the teeth of contemporary carnivores" (p. 11).

Simpson recommended that future work "from the standpoint of geology" include petrographic and grain-size analysis of the sediments, measurements of direction and angle of cross-bedding, and relative frequency of the different rock types (p. 13). At the bottom of the manuscript there is a handwritten note: "This is the first paper I ever wrote, and inexperience both in the field and with a pen is obvious. The observations, however, seem to be correct and include a number of facts which have not been published even now. GGS June, 1933" (fig. 3.10).

"Beginning a Series of Raids"

Despite Matthew's initial favorable response to Simpson's paper, several months later he wrote Simpson that "Mr. Frick has decided to take charge of the Santa Fe work himself and it does not appear that you would fit into his plans for the development of the work."[48] Although Matthew didn't specifically indicate the future fate of Simpson's paper, apparently Simpson got the message either indirectly by this letter, or else orally soon after, that the paper wasn't going to be published after

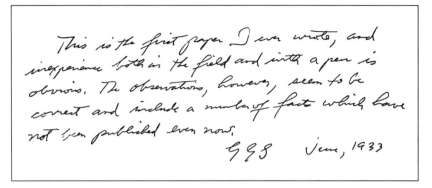

FIGURE 3.10 Simpson's 1933 note at the end of his unpublished 1924 paper. (Simpson Papers, American Philosophical Society Archives, GGS/APS. Courtesy American Philosophical Society)

all. Matthew went on in his letter to squelch any other hopes that Simpson might have entertained about working for the museum: "As matters stand at the present, I do not see any very good prospect of our putting you in the field elsewhere. We have two young men on the department staff, to whom I should have to give precedence." Matthew then reassured Simpson: "I would like to make it clear that the above is my real reason, and not any dissatisfaction with your record last year, which I consider was very good." Matthew then changed the subject by saying that he was much interested in Simpson's Mesozoic mammal work and reflected on the presumed differences between late Cretaceous and early Paleocene mammals, and wondered whether these differences were more a function of facies changes than any great time break.

Simpson wrote back immediately the next day, saying, "Thank you for letting me know Mr. Frick's plans so promptly after his decision," and then going on to describe his financial plight, once more asking Matthew if he happens "to know of any other expeditions which are going into the field next summer & which might still have room for me?" Then, like Matthew, turning to a mutually more pleasant subject, Simpson went on for two pages about his dissertation.[49] A week later Simpson turned down a subsequent offer from Matthew to work as a laboratory preparator because, by then, Lull had made arrangements for him to be paid while doing his thesis research in New Haven.[50]

At the end of the unpublished manuscript on the Santa Fe formation, there is another note, this one undated, in Simpson's handwriting: "This paper was passed for publication by W. D. Matthew, then Chairman of the Dep't. in A.M.N.H., but Frick requested that this information on a collecting field where he intended to (& did) continue collecting be suppressed, & it was" (fig. 3.11). As Simpson recalled years later: "I didn't realize it at the time, but it later turned out that I had been actually working not so much for the American Museum as I thought, but for Childs Frick, who was then beginning a series of raids, one might almost call it, on the fossil beds."[51]

Considerable animosity between Frick and Simpson remained over the years, presumably because of this episode. Simpson claimed in his autobiography that "his dislike of me was so great that Frick threatened to break off financially beneficial relationships with the museum unless I was fired. To my surprise the director refused to fire me, and to my somewhat lesser surprise Frick did not carry out his threat. His magnificent [fossil] collection, which I was not allowed to see, did go to the American Museum after his death."[52]

In 1946, as chairman of the department of geology and paleontology, Simpson wrote to Frick in the "usual practice to tell you of [our] plans for using the income from the Paleontology Field Fund, which you generously donated to the Museum for its field work on fossil vertebrates" for expeditions in the Triassic of Arizona under E. H. Colbert, curator of fossil reptiles, amphibians, and fishes, and in the Paleocene and Eocene of New Mexico under his own direction.[53] Frick shot back that the fund was set up for "field work in extinct mammals," and not for Mesozoic reptiles and amphibians, and therefore he was "sure that you [Simpson]

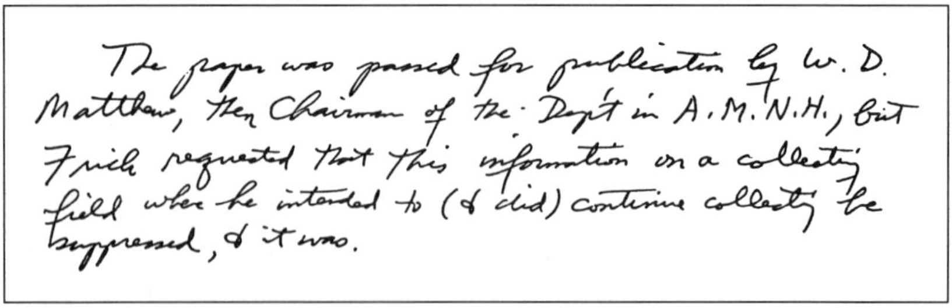
The paper was passed for publication by W. D. Matthew, then Chairman of the Dep't in A.M.N.H., but Frick requested that this information on a collecting field where he intended to (& did) continue collecting be suppressed, & it was.

FIGURE 3.11 Another later, undated note by Simpson at the end of the unpublished 1924 paper. (Simpson Papers, American Philosophical Society Archives, GGS/APS. Courtesy American Philosophical Society) will wish to see that terms of the gift are adhered to."[54] Simpson quickly replied that upon "checking the written record I found only that [the Paleontology Field Fund] was for field work in vertebrate paleontology. This accounts for any misunderstanding . . . to work in the Triassic. Your comments have clarified the record of the terms of the gift and of course we are willing and anxious to adhere to those terms."[55]

In 1949 Simpson wrote Frick about a field conference of the Society of Vertebrate Paleontology being planned for the following year in northwestern New Mexico that would include visits to areas being studied by Colbert and Simpson, among others. "In view of your long interest in the Santa Fé formation and your collectors' intensive knowledge of it, we would all be pleased if the Frick Laboratories would join us" in selecting the most favorable places to see typical fossiliferous exposures, logging the route, and preparing a portion of the guidebook.[56] Frick declined the invitation, saying, "I have consulted with my associates involved" and "this is to advise you that when our long continued explorations in the New Mexican Late Tertiary are developed to the expected point and reports are published, the Frick Laboratory will then gladly join with the University of New Mexico in sponsoring what we hope may be an instructive tour of the particular area. . . . I am sure that on further consideration you will appreciate that your proposal is premature."[57] In a letter to Colbert the following month Simpson fumed, "By the way, Frick turned down any cooperation in a way not only final but also literally insulting. I am so angry that I can't trust myself to write an acknowledgment to him. His constructive suggestion is that we drop the whole thing until an unspecified future date when the Frick Laboratory & the Univ. of N. M. (without A.M.N.H.) can put on a good conference! Needless to say, I am going right ahead with the original plan, including the Santa Fé beds, & Frick can do as he pleases. I am through trying to be polite to him or trying to cooperate as I have honestly tried to do hitherto. My extreme anger is not just from his turning us down or his silly alternative suggestion but from his coldly & openly insulting not only me but also all those connected with the plan & accusing us of unethical conduct because we want to look at his private collecting grounds."[58]

In his part of the guidebook for the 1950 field conference, Simpson published at last a short, somewhat updated version of the suppressed 1924 paper on the Santa Fe formation, adding the remark that, "Pending completion of the Frick Collection, and revelation of its stratigraphic positions and associations, no satisfactory review of the rich Santa Fé mammalian faunas can be provided."[59]

*   *   *

The summer of 1924 clearly was significant in launching Simpson on what was to become an eminent career in paleontology. We have no way of knowing, of course, if his career might have taken a different direction if he had spent the summer in some other pursuit. But certainly his future path was at least eased by demonstrating to both Lull and Matthew that he could find and recover valuable fossils, that he was a dedicated and industrious paleontologist, and that he was interested in placing his fossil materials within a broader geologic context. The upshot was that Lull did permit Simpson to study Yale's fine Marsh Collection of Mesozoic mammals for his dissertation which, when published, immediately became a classic monograph.[60] Matthew was also sufficiently impressed by Simpson that he maintained a cordial and supportive relationship with him for the next half-dozen years, until Matthew's premature death in 1930. Matthew generously shared his own ideas about the Mesozoic mammals both in person and in personal correspondence as well as promoted Simpson as his replacement at the American Museum when he, Matthew, left for the chairmanship of the newly revived department of paleontology at the University of California, Berkeley, in 1927.[61]

The backing of men as distinguished and influential as Lull and Matthew had to be an enormous asset for Simpson in a discipline as small as vertebrate paleontology was in the 1920s.[62] Although Simpson had inadvertently stepped on Frick's toes, Frick's subsequent negative opinion of Simpson had no real impact on his career, no doubt because Frick was an amateur and outsider. While certainly neither necessary nor sufficient for ultimate success, this episode in Simpson's life shows how a young scientist's career can be catapulted forward by influential supporters — as was true for Darwin a century earlier and still is true today once that young person shows the "right stuff."

CHAPTER 4

DARWIN'S WORLD

Scientists . . . do not believe one episteme is superior to another because they are in it; rather, they are in it because they believe it is superior, and this belief is a product of their experience.
— Neal C. Gillespie, psychologist

Prologue

I got over the effects of Presbyterian theology before I was far into my teens. . . . While still in graduate school I had of course read The Origin of Species . . . [and was] already anti-vitalistic [and] a firm Neo-Darwinian. . . . By the 1930s . . . I had reached a philosophical and theoretical viewpoint that seemed to have some originality and some importance.
— G. G. Simpson[1]

This quotation comes from Simpson's response to a questionnaire prepared by Ernst Mayr dealing with the development of the modern evolutionary synthesis. At a follow-up conference on the history of the development of the modern evolutionary synthesis at Harvard University in October 1974, several historians of science commented on these remarks of Simpson.[2]

> JOHN GREENE: What is this philosophical and theoretical point of view that [Simpson] is talking about?
> ERNST MAYR: My personal impression is that Simpson had developed a rather consistent interpretation of the world prior to 1930. . . . Most scientists, when speaking of their "philosophy," mean their interpretation of the world.

WILLIAM PROVINE: By and large scientists don't really talk about their larger metaphysical view of things, no matter how important it might be in their personal approaches to their science.

STEPHEN GOULD: Among working scientists this is the predominant view — but it's all wrong — philosophy cannot be epiphenomenal.

JOHN GREENE: I don't think you can eliminate this [philosophical] dimension and I strongly suspect — I can't prove it — that George Gaylord Simpson quite early in life . . . had some very strong feelings about nature and about religion and about science. . . . In other words he was very much concerned about the meaning of evolution from the beginning.

GARLAND ALLEN: Too bad Simpson isn't here; we can't ask him what he read, or whom he was influenced by.

After reading these comments, Simpson wrote, "I had in childhood very strong feelings about nature, or about the world. I wanted to know things; even as a child too young to know this was impossible I wanted to know everything, not just nature in the narrow sense of the word. Religion I was exposed to, but could take it or leave it, and I did leave formal, dogmatic religion as soon as I decided there was nothing real to know about it. . . . I was not concerned about the meaning of evolution 'from the beginning' — beginning of what? When I learned, late teens, that evolution has occurred, of course I then wanted to puzzle at its meaning, if any. It took me quite a while still to decide that it has a meaning — i.e., mainly an explanation and a significance for our lives, and to puzzle out just bits of that. I'm still at it."[3]

In late middle age Simpson asserted that, "I do not think evolution is supremely important because it is my specialty. On the contrary, it is my specialty because I think it is supremely important."[4] This autobiographical insight was volunteered in an essay on Darwin written in a professional journal for educators. The subject (Darwin), the audience (nonspecialists), and the personal revelation about philosophical motivation were by no means an atypical combination for Simpson.

From private letters and personal papers one can trace Simpson's childhood apostasy from the fundamental Presbyterianism of his parents to a temporary and partial conversion to the natural theology of

Rev. Charles Kingsley. Maturation of a naturalistic, positivistic worldview occurred in college with his exposure to geology and paleontology, and especially with his reading of *On the Origin of Species*. Thereafter throughout his adult life, Simpson repeatedly celebrated several Darwinian themes that reflected his own deeply held views about the philosophical content and methodological possibility residing in evolutionary phenomena. These Darwinian testimonials, incorporated in five decades of public lectures and popular writings, were directed either at educated nonscientists or scientists outside his own discipline of paleontology and evolution.

In this chapter, I suggest that interesting perspectives on the questions raised in the prologue emerge if we focus on that part of Simpson's work dealing with Darwin. From 1947 until his death in 1984, Simpson regularly asserted what it was that Darwin did, why it was important then, and why it remains so today. These assertions are either major asides on Darwin in works not specifically dealing with him, or they are full-length discourses treating one or another aspect of Darwinian science.

Simpson prided himself that he had read all that Darwin had written[5] — and yet, given the volume of what both Darwin and Simpson published, a rather remarkable situation occurs. First, Simpson chose relatively few Darwinian themes for his commentaries. Second, his views on what these themes signify never wavered over five decades of discussing them. Third, Simpson selected themes that served to articulate his own interpretation of animate nature. Finally, Simpson almost always wrote for the instruction and edification of either scientists outside his own discipline or else for the thoughtful nonscientist, and thus he played an important role in educating this particular audience. He did not write expressly for historians of science or for philosophers of science.[6]

It would appear, then, that Simpson found in Darwin a point of view, a philosophical stance that resonated with his own. As indicated in the opening quotation of the prologue, Simpson discovered Darwin by his reading of the *Origin* at Yale by his early twenties. But what of his intellectual and philosophical development before his learning of evolution and the reading of Darwin? Obviously, Simpson was not a postadolescent tabula rasa, a clean slate waiting for suitable content. From personal letters, private papers, and scattered clues in his published writings, we can, in fact, trace in broad outline the stages of belief through which Simpson passed that brought him to the state of mind ready and willing to convert to a Darwinian worldview.

Presbyterianism and Apostasy

After her own mother died when she was quite young, Simpson's mother was raised by her grandparents who were lay missionaries in Hawaii. She was brought up in a strongly religious setting that carried over into her domestic adult life, for Simpson claimed that as a child he "commonly attended three services on Sunday [at the Capitol Heights Presbyterian Church in Denver] as well as the midweek prayer meeting, and in addition had family prayers and psalm recitations." Simpson's father had a religious upbringing as well, for his maternal grandfather was a Welsh Presbyterian minister. Simpson was made a formal member of the church at the age of nine, but soon after deconverted when he decided in a fit of childish peevishness that he did not want to forsake forever being "naughty." Whatever else this meant, it indicates that Simpson well remembered the first stirrings of personal willfulness against external authority. He recalled, "I then started the long, always difficult and sometimes desperate process of unlearning all that had to be unlearned and learning as much as I could of what had to be learned in order to find out what I was and am."[7]

What might have been involved in this unlearning and learning process? It was also at about the age of nine that Simpson cajoled his parents into guaranteeing half the cost of the new eleventh edition of the *Encyclopedia Britannica*, which he saw advertised in the local newspaper, if he could in turn come up with his half. Before long, to his parents' surprise and financial dismay, he had accumulated his share through a series of odd jobs in the neighborhood. He then proceeded to read all twenty-eight volumes straight through. "I think it gave me my first conception of the world of learning as a whole, my first definite feeling for organized facts, and my first inkling of how to go systematically about finding out such facts."[8]

Simpson has also recalled reading about this time a children's book on natural history by Rev. Charles Kingsley, entitled *Madam How and Lady Why*, first published in 1869. Simpson claimed that he could "trace very definitely and without doubt . . . my first understanding, however dim then, of the scientific method and, more distinctly but equally surely, the vague beginnings of a scientific philosophy."[9] Charles Kingsley (1819–1875) was an English cleric, poet, novelist, who was for a time Queen Victoria's chaplain, later professor of history at Cambridge, and still later canon at Chester, then Westminster. He was interested in natural history, Christian socialism, opposed to the Oxford movement, and communicated his enthusiasms and beliefs in a series of published works. His *Water Babies* (1863) and especially *Madam How and Lady Why* were written to encourage (male) children to use their own faculties of sight and reason to give meaning to the world around them. "Mere reading of wise books will not make you wise men: you must use for yourselves the tools with which books are made wise; and that is — your eyes, and ears, and common sense." The purpose of such thinking for oneself, of course, was to arrive at the Godhead behind the Universe. "God has ordained that you . . . should begin learning something of the world . . . by [your] senses and brain. . . . The more you try to understand things, the more you will be able hereafter to understand men, and That which is above men."[10]

That this book, subtitled "First Lessons in Earth Lore for Children," made a great impression on Simpson is evident from the fact that years later he published a collection of his articles with the title *Why and How*, which he said was "based on gleanings from my past efforts to think properly over a great many years."[11] It may well be that Kingsley gave clerical permission to the youthful Simpson to substitute his own individual observations and experiences, and conclusions therefrom, about the world in which he found himself for the predigested views of others handed down by sectarian authority. Simpson thus exchanged one authority for another, but one that put him on the irreversible path for ultimately thinking for himself.

In the ensuing years Simpson developed his own views about the nature of the world and what, if anything, it might mean. We see a reflection of these maturing views in a series of letters he wrote to his former childhood playmate and teenage companion, Anne Roe, whom he would marry years later. Anne had a brief adolescent interest in evangelical Protestantism, which rather annoyed Simpson, and he made no bones about expressing it in his letters to her from the University of Colorado in Boulder, at age eighteen.

> If I didn't fear I'd do you harm . . . I'd try to make you an atheist. I really do think that you are a deluded follower of mistaken and superstitious & cowardly theories. That's as far as I'll go.

> ... Everyone who worships a god worships a force back of all nature, no matter what they call him or it and even if they call his *aspects* by different names & have many "gods." If there really *is* such a force, then all people who worship *any* god or gods, worship the same god. I'd just as soon call him Ishtar or Baal or Jehovah. They're merely names for the *same* idea.[12]

Another letter to Anne from this same time:

> The destruction of illusions always brings sorrows — a child who has lost a cherished rag doll cannot be consoled by the substitution of a real baby, even; just so when we simians lose our false "truths" & cults we often find little consolation even in actual truth. I never received any spiritual good (so-*called*) from religion anyway, so welcome the truth without heartache. . . . Where do we get the idea that we're the lords of creation? Bunk, utter bunk. . . . Things weren't made for us. We're just here; that's all. . . . Infinity probably doesn't exist. Man merely made up the idea, then he turns around & says it is incomprehensible & had better be lumped. How humorous! It is to laugh! Infinity is where parallel lines meet. It is where a hyperbole turns & comes back the other way, it is the value of a tangent of a 90° angle, it's the boundary of "our" smug little, tight little collection of universes. Just let your mind flow at ease & it compasses infinity with ease. Infinity indeed!. . . . As for creation — I believe I gave you my views before. Briefly — creation implies time. Time is a figment of man's imagination. Only in time must things have a beginning or end. Actually *nothing* is in time. Time itself is in our *perception* of things, not *in* the things or *governing* them.[13]

Evolution and Darwin

It was at the University of Colorado that Simpson first encountered evolution in a formal way when he took introductory geology and paleontology courses. Although initially uncertain as to what subject to choose for his major at college, Simpson eventually decided on geology as it satisfied his native curiosity about the natural world. His decision was so determined that he transferred to Yale when he learned that Yale was superior in both these fields. The intensity of this determination can be measured by the greatly increased financial cost and inconvenience that resulted from this move. He left a nearby public institution (his family was living in Denver at the time) for a more expensive private school, thousands of miles away.

Although his autobiographical notes are not consistent on this point, it was either as a senior or first-year graduate student at Yale that Simpson first read Darwin's *On the Origin of Species*. We do not have a record of the precise impact his first reading of Darwin had on him, but in late middle age, in an essay entitled, "One Hundred Years Without Darwin Are Enough," we can reconstruct the successive steps in his conversion to a Darwinian worldview. "Nothing learned [in high school] had any bearing at all on the big and real questions. Who am I? What am I doing here? What is the world? What is my relationship to it?"[14] By the time he entered college, he had come to the conclusion that "life is the most important thing about the world, the most important thing about life is evolution. Thus, by consciously seeking what is most meaningful, I moved from poetry to mineralogy to paleontology to evolution."[15] And finally, this intellectual realization led him to the study of evolution because, as quoted near the beginning of this chapter, he found it "supremely important."

Given that Simpson wrote so much about Darwin for the educated nonspecialist audience and that he tended to repeat the same general themes, an examination of these themes should inform us as to what it was in Darwin's work that so captured Simpson's intellectual attention and lifelong commitment. Six of these Darwinian themes articulate Simpson's own personal ontological and epistemological views of the world — that is, his attitudes about the nature of being and existence as well as the nature and limits of knowledge itself.

Six Darwinian Themes

The first Darwinian theme that Simpson emphasized is how Darwin brought the study of biology into the mainstream of scientific inquiry that previously had only included the physical sciences of astronomy, chemistry, geology, and physics. In his 1960 essay, "The World into which Darwin Led Us," Simpson remarked that "by early Victorian times . . . living things belonged outside the realm of material principles and secular history. . . . Perception of the truth of evolution was an enormous stride from superstition to the rational universe.

[With] Darwin's theory . . . the essential point was the demonstration that material causes of evolution are possible and can be investigated scientifically."[16]

Elsewhere Simpson asserted that the study of biology is, after all, the study of evolution, and given that humans too had clearly evolved, then the study of human existence falls within the realm of what he called "material principles," thereby obviating for Simpson recourse to immaterial, mystical causes. Thus, said Simpson, Darwin "finally destroyed the last stronghold of the supernatural, the providential, and the miraculous."[17]

The second Darwinian theme that Simpson championed is how Darwin solved what, until then, was the "main problem of his era," namely that "the appearance of purposefulness is pervading in nature in the structure of animals and plants, in the mechanisms of their various organs, and in the give and take of their relationships with each other. Accounting for this purposefulness is a basic problem for any system of philosophy or science."[18] For Simpson, Darwin's great contribution was explaining the apparent purposefulness of organic adaptation by means of the natural selection of inherited random variation. Simpson exclaimed that this is another "obvious lesson . . . that in science one should never accept a metaphysical explanation if a physical explanation is possible or, indeed, conceivable."[19]

As an evolutionist Simpson confronted periodically the so-called arguments from design for a Creator — the "watchmaker argument" — both in serious philosophical debate and in popular religious tracts. Given that his parents were what he called "strict fundamentalist Presbyterians" through whose direction he regularly attended Sunday school, we can be sure that he also encountered this argument as a child. But, as he said, he "ceased to be a Christian at about 12. . . . I became more and more critical and increasingly realized that practically nothing the preachers throw at you has any likelihood of being true."[20] His reading soon after of *Origin of Species* undoubtedly further confirmed his apostasy.

The third Darwinian theme highlighted by Simpson derives from the second — namely, that "for adaptation to evolve there must have been some kind of feedback between organism and environment. . . . Darwin correctly placed the whole feedback in the population and discovered its basic mechanism [of natural selection]. That was his greatest accomplishment."[21] Simpson, of course, throughout all his own work repeatedly acknowledged this critical interaction of organisms with their environments in understanding evolutionary history. In fact, much of the impact of Simpson's *Tempo and Mode in Evolution* was precisely owing to his demonstration that evolutionary ecology explains apparent trends and direction in the history of life, without needing to resort to those orthogenetic forces usually called upon by his predecessors. Both *Tempo and Mode* and its successor *Major Features of Evolution* elaborated in detail how differing degrees of environmental opportunity determine the rates and patterns of evolution.[22]

The fourth Darwinian theme repeatedly highlighted by Simpson, already alluded to above, is how Darwin "gave an answer to the tremendous question that so deeply concerns us. . . . 'What is Man?' [Darwin] answered this question to the effect that Man is a natural product of the universe; . . . man is an animal, a vertebrate, a mammal, and a primate."[23] Simpson continued, "[B]y bringing man into the evolutionary picture, Darwin finally took the last step in our emancipation and finally made our world rational. [Yet] Darwin felt humility and awe that seem to me truly religious."[24]

Casual readers of Simpson usually find him without much religious sympathy. Simpson would deny this conclusion, as the last quote suggests. Elsewhere he claimed that he thought "any sensitive person must feel a basically religious awe in face of the mysteries of life and of the universe, but belief in an anthropomorphic god, in a savior, or in a prophet is nonsense."[25]

The fifth Darwinian theme that claimed Simpson's attention was Darwin's continual interest in not only describing the "what" in nature but equally his attempting to explain the "how." Said Simpson, "Darwin brought together an enormous body of solid pertinent fact. . . . He also produced a particular theory as to how [those facts] occurred."[26] Simpson further asserted that Darwin's handling of facts with respect to their articulation of a theoretical explanation "demonstrates that Darwin had a consistent methodology which is the so-called hypothetico-deductive procedure."[27] In his early thirties Simpson claimed for himself this same desire to explain the how behind the what: "I have endeavored to develop or to follow an ideal of obtaining good central data and then to relate this to the whole body of relevant knowledge . . . to use the new materials as a means and an occasion to renovate the whole structure."[28] Simpson believed he had achieved this ideal when much later in life he published the collection of writings that he entitled *Why and How*.[29]

A sixth Darwinian theme that Simpson recognized is related to this last, and that is the thoroughly eclectic methodology of Darwin, which Simpson viewed as a varying combination of inductive, deductive, and comparative logic. Simpson found this eclectic approach to problem-solving best exemplified in Darwin's evolutionary taxonomy where his theory of "descent with modification" predicts some morphological characters that will indicate taxonomic propinquity. Thus inferences can be made about common ancestry both from a temporal succession of changing fossil forms as well as from comparative data from contemporaneous organisms.[30] Simpson was careful to point out that Darwin was "well aware that characters in common do not necessarily indicate common ancestry," owing to convergent evolution.[31] Simpson also saw Darwin's eclectic methodology well displayed in his work on historical biogeography, wherein Darwin used all three present-day approaches of dispersal, vicariance, and ecology to explain the geographic distribution of life.[32]

Of all the Darwinian themes, this last is perhaps the one Simpson most directly invoked in his own research. He often claimed a similar eclectic methodology when discussing these same subjects of systematics and biogeography to demonstrate that he, Simpson, was neither ruled by theoretical apriorism nor by simple-minded empiricism, but rather that he considered the full range of available fact and theory to arrive at his conclusions. "It remains true . . . that all these methods and instruments are only aids to the one indispensable instrument for reconstruction of phylogeny: an instructed and experienced human brain."[33]

Using Samuel Butler as a foil for Darwin's — and by implication, his own — methodology, Simpson asserted that Butler (and presumably Simpson's critics) "never did comprehend [Darwin's] attitude, [his] whole approach to [a] problem, the laborious amassing of facts, the self-discipline of submitting the most cherished hypotheses to the cold judgment of facts, the admission of uncertainties, and the critical weighing of probabilities in light of objective evidence. In short, what [Butler] did not grasp was science, its methods and its whole philosophical orientation."[34]

This chapter has not addressed what Simpson had to say about the specific content and significance of Darwin's published work with respect to evolutionary biology per se. What Simpson said in this regard was fairly commonplace and already apparent to knowledgeable students: Darwin as a population biologist rather than as a typologist; Darwin's impact on Charles Lyell and vice versa; Darwin as a gradualist with respect to small incremental changes from generation to generation, but not necessarily in rate of change; Darwin's approach to taxonomy that included not only the "what" but also "how come"; Darwin's eschewing of plan, purpose, and direction in his theory of descent with modification; Darwin's insights on the ecological incompatibility of species, adaptive radiation, increasing diversification and specialization of newly evolved groups, as well as the inherent bias of the fossil record and sweepstakes dispersal; Darwin's recognition that the first forms of life were primary heterotrophs, or that the causes of extinction are difficult to establish, although they are probably related to interspecific competition.

Simpson provided no new insights or breakthroughs in the understanding of Darwin; hence, Simpson's writings on Darwin have been ignored by historians of science and philosophers of science. This is in sharp contrast to Simpson's writings in other areas of his scholarly research where what he said often had important impact on the field. Indeed, his points of view and interpretations at times became the benchmarks against which other research was measured. Not so for his writings on Darwin. In fact, some commentators, like Ernst Mayr, have taken Simpson to task for concentrating on "vertical" speciation and failing to recognize Darwin's ignorance of the true mechanism of "horizontal" speciation and resultant diversification.[35]

Instead, my purpose has been to uncover the reasons why Simpson was so fascinated by Darwin, why he was so preoccupied by him that he wrote hundreds of pages on him. That is what strikes one who reads Simpson on Darwin: more is revealed about the commentator than about the person being commented upon. What focuses one's interest is not whether or not Simpson's commentary is precisely correct or not, but rather his drumbeat of a few persistent Darwinian themes: more is revealed about Simpson than about Darwin.

* * *

It is impossible to know if Darwin converted the unformed, impressionable mind of an adolescent Simpson, or whether — as is more likely — Simpson was already temperamentally inclined toward a positivistic, naturalistic worldview that was reinforced by his youthful reading of Darwin. What is clear is that by the time he began to practice his profession in his early twenties, George Gaylord Simpson was thoroughly committed to a positivistic philosophy.[36]

Thus I interpret Simpson's continued preoccupation with selected Darwinian themes as, first, the result of his own conviction of what the really important questions are and where the answers are to be found. As Simpson asserted, "It is evolution that can provide answers, so far as answers can be reached rationally and from objective evidence, to some of the big and universal questions."[37] Second, I believe Simpson's many writings on Darwin, specifically for persons not expert on Darwin or evolution, were an effort to convert others to his philosophical viewpoint. The name and reputation of Darwin provided Simpson with an entrée into the whole subject. Third, I suspect that Simpson found his periodic commentaries on Darwin rejuvenated his own, earlier youthful excitement, energy, and enthusiasm for the continued study of evolutionary phenomena. Thus, it was in this role as explicator of Darwin for a broad, lay audience that Simpson's Darwin writings have had their greatest significance.

Selected Simpson Commentaries on Darwin (in whole or part)

There are aspects of the subject of such transcendent importance that they bear frequent repetition.
— G. G. Simpson[38]

1947   "The Problem of Plan and Purpose in Nature." *Scientific Monthly* 64: 481–95.
1949   *The Meaning of Evolution.* New Haven: Yale University Press; rev. ed., 1967.
1950   "The Meaning of Darwin." In Sir Francis Darwin, *Charles Darwin's Autobiography,* ed. Henry Schuman, 1–11.
1951   *Horses.* New York: Oxford University Press.
1953   *The Major Features of Evolution.* New York: Columbia University Press.
1957   *Life: An Introduction to Biology* (with C. S. Pittendrigh and L. H. Tiffany). New York: Harcourt, Brace (2d ed., 1965).
1958   "Charles Darwin in Search of Himself" (review of *The Autobiography of Charles Darwin,* ed. Lady Nora Barlow). *Scientific American* 199 (1958): 117–22.
1959   "Anatomy and Morphology: Classification and Evolution — 1859 and 1959." *American Philosophical Society Proceedings* 103 (1959): 286–306.

1959	"Darwin Led Us into This Modern World." *The Humanist* 5: 267–75.
1959	"Foreword." *The Life and Letters of Charles Darwin*, ed. Francis Darwin, v–xvi. New York: Basic Books.
1960	"The World into which Darwin Led Us." *Science* 131: 966–74.
1961	"Lamarck, Darwin, and Butler: Three Approaches to Evolution." *American Scholar* 30: 238–49.
1961	"One Hundred Years Without Darwin Are Enough." *Teachers College Record* 62.8 (May 1961): 617–26.
1961	*The Principles of Animal Taxonomy*. New York: Columbia University Press.
1962	"Foreword." In Charles Darwin, *On the Origin of Species*, 5–9. 6th ed. New York: Collier Books.
1962	"The Status of the Study of Organisms." *American Scientist* 50 (1962): 36–45.
1963	"Biology and the Nature of Science." *Science* 139 (1963): 81–88.
1966	"The Biological Nature of Man." *Science* 152 (1966): 472–78.
1969	"The Present Status of the Theory of Evolution." *Royal Society of Victoria Proceedings* 82 (1969): 149–60.
1970	"Darwin's Philosophy and Methods: A Review of *The Triumph of the Darwinian Method* by Michael T. Ghiselin." *Science* 167: 1362–63.
1974	"The Concept of Progress in Organic Evolution." *Social Research* 41 (1974): 28–51.
1977	"A New Heaven and a New Earth and a New Man." In D. W. Corson, ed., *Man's Place in the Universe*, 53–75. Tucson: University of Arizona Press.
1980	*Splendid Isolation*. New Haven: Yale University Press.
1982	*The Book of Darwin*. New York: Washington Square Press, Pocket Books.
1983	*Fossils and the History of Life*. San Francisco: W. H. Freeman.

CHAPTER 5

PALEOCENE MAMMALS OF MONTANA

I see I began to hit my real stride and to do work of more basic and broad importance around 1938.
— G. G. Simpson

Simpson's *Tempo and Mode in Evolution* (1944) is acknowledged as one of several pillars of the consolidation of the evolutionary synthesis.[1] Although *Tempo and Mode* has since been discussed by various commentators with respect to its content and its impact, not enough attention has been paid to the development of the concepts and approaches that Simpson introduced in that work — concepts and approaches that were then quite novel for a paleontological treatise (see chapter 7, this volume). *Tempo and Mode* may have seemed without pedigree to most of its readers, but one can find its origin in much of Simpson's earlier work.[2]

In this chapter I trace the roots of several of the major concepts and approaches that Simpson articulated so cogently in *Tempo and Mode*, including species-as-populations, organism-environment interactions, the bias of the fossil record, and biometrical analysis. In particular, Simpson's study of the mammals of the Fort Union Group (Paleocene) of south-central Montana — the "Crazy Mountain fossil field" — gave him the opportunity to explore these concepts in monograph form as applied to a rich fossil fauna, well-defined in time and space.[3] The Fort Union research inspired two additional papers in 1937 that Simpson called, years later, door-openers to "formalizing and publishing my ideas" on the principles of classification and to moving "in the direction of principles and theory" of evolution.[4] It also served as a précis for the textbook *Quantitative Zoology*, written with Anne Roe in 1937–38 shortly before they were married.[5] Soon after, in the spring of 1938, Simpson began writing what was to become his most important book, *Tempo and Mode in Evolution*.

Simpson himself seems to have realized, retrospectively, the value of the Fort Union study because, although he referred to *United States National Museum Bulletin* 169 as "a more or less routine monograph," he acknowledged that "it did introduce some new ways of going at things."[6] And in 1943, while overseas on military duty, Simpson had the opportunity to remark in a letter to his wife that, "As I look back, I see I began to hit my real stride and to do work of more basic and broad importance around 1938."[7]

The Invitation

In October 1931, upon returning to the American Museum of Natural History from his first Patagonian expedition, Simpson was asked by Charles W. Gilmore (1874–1945), curator in the department of vertebrate paleontology at the U.S. National Museum (USNM), if he would be "interested in finishing up the work" on Paleocene mammals started by James W. Gidley (1866–1931), who had recently died.[8] Shortly afterward, Simpson discussed the offer further with Gilmore during a visit to Washington, and a week later received a formal invitation from F. A. Wetmore (1886–1978), assistant secretary of the Smithsonian Institution. In his letter Wetmore noted that the collection "is one of the most valuable in existence . . . and I know of no one more competent than you yourself to undertake its study. I shall await your reply with interest."[9] Simpson replied almost immediately to say that he accepted the arrangement as discussed, and he ended his letter with the observation that Gidley had told him a few years earlier that his projected monograph on the Paleocene collection would be his monument and "the task of erecting this monument now falls to me, but I hope that it will [be] no less a tribute to Doctor Gidley's memory."[10] Simpson's acceptance letter was seen by R. P. Bassler, head curator at the Smithsonian, who attached a note ("Happy over this successful outcome").

The collection under consideration was a composite accumulation of fossil mammal remains from sedimentary rocks of Paleocene age in south-central Montana, east of the Crazy Mountains. Although others had explored the area in the early twentieth century, most of the best fossil localities and specimens were discovered by Albert C. Silberling, who started collecting fossils as an adolescent and continued thereafter for the rest of his life. In his Fort Union monograph, Simpson complimented this amateur collector for his "skill, persistence, and devotion . . . [who] established a system of field records of the greatest accuracy . . . [and] so carefully studied and correctly interpreted [the] geology."[11] Some of Silberling's smaller, earlier collections had gone to the Carnegie Museum in Pittsburgh and to the American Museum of Natural History in New York, but the largest collections were made by him during 1908 through 1911, by arrangement with the U.S. Geological Survey and the U.S. National Museum. In the years following, J. W. Gidley, assistant curator of fossil mammals at the National Museum, published several papers on some of the material, but much more study still needed to be done at the time of his death.[12]

Simpson's credentials for being offered this project were several. He had already demonstrated his competence as an expert on early mammals with the publication of his Yale dissertation and postdoctoral monograph, both of which dealt with Mesozoic mammalian fossils.[13] In addition, because of this expertise and his subsequent curatorial position in the American Museum, Simpson had by 1932 published a dozen papers on early Tertiary mammals, particularly multituberculates. He also had special familiarity with Fort Union Paleocene mammals previously collected by others from a coal mine in Bear Creek, Montana, south of the Crazy Mountains.[14]

Upon Simpson's agreement to finish the project, Wetmore sent him three boxes weighing a total of 264 pounds and containing 831 fossil specimens, along with Gidley's unpublished notes and drawings. It didn't take Simpson more than a few days' cursory examination of the fossils to realize that it was "almost impossible to get a really clear idea of the various [stratigraphic] levels & faunas involved. As a tentative suggestion . . . it might be advisable or necessary to spend two or three weeks going over the ground with Silberling before the monograph is completed."[15] In a later letter Simpson emphasized to Gilmore that further fieldwork was important so he could "become more familiar with the localities and stratigraphy than to collect much material," and he estimated that he could do that in four to six weeks at a cost of about $500.[16] Simpson also wanted the National Museum to pay an illustrator for the projected monograph. Before long, Simpson's request for the support of fieldwork was approved but, given the economic hard times of the Great Depression, Wetmore had to defer any definite decision about paying for an artist. "It is hardly required to inform you that the Smithsonian Institution like every other organization in the world at the present time, is struggling with its budget."[17]

First Field Season — 1932

Simpson arranged to meet Silberling, the local amateur collector, in late June 1932. Simpson had high regard for Al Silberling, despite his lack of professional credentials. "Al . . . was an ideal field worker and companion. He was crazy about fossils, tireless and ingenious in their pursuit, good-natured and good-humored whatever the circumstances. . . . [A]t this stage in his life he was spending winters as an engine wiper for the Milwaukee Railroad in Harlowton, Montana. Al's wife, a Seventh Day Adventist, was a violent anti-evolutionist. Al, who considered evolution an obvious fact, was not at all perturbed, but the home atmosphere was not always congenial for evolutionist visitors."[18]

Two days after arriving in the field Simpson wrote home to his parents: "Tomorrow we leave for a three day reconnaissance along the flank of the Crazy Mountains — good looking mountains, too, the highest in the state. We'll have to use horses & I'll be sore. The geology here is *very* confusing. I think I'll get it fairly straight, but it's lucky I came, as it would be quite impossible to do anything without studying it here in person. This is not the usual type of fossil country — no badlands to speak of, but vast stretches of rolling hills & valleys, relief of about 3000 feet but everything so spread out that really steep slopes are few, and almost all grass covered & very green now. . . . Rock exposures are small & widely scattered, which makes study hard, & the rocks I'm concerned with are about 6000 feet thick. You have to travel thirty or forty miles to see much of a [vertical] section — altogether the biggest sort of stratigraphic job, & it's no wonder it is so poorly understood."[19] To his sister Martha, he wrote that same day: "here I am . . . , once more enjoying sunburn, wind, & open plumbing openly arrived at. I'm . . . climbing rapidly over sandstones & andesitic shales of Fort Union age all day & sleeping like a baby all night. I eat several bowls of oatmeal & stacks of pancakes for breakfast — that will give you some idea. This is an interesting & difficult job I have here."[20]

After less than two weeks of such field reconnaissance, Simpson realized how sketchy the field location and stratigraphic position were for Gidley's fossils, and so he asked Wetmore to send him Gidley's earlier publications on the Fort Union.[21] Simpson's field diary indicates that by the end of the third week he had completed the "revision of maps, mapping of formation boundaries, [and] detailed field observations."[22]

After one month in the field, he returned to New York and sent Wetmore his final accounting for the field season, which was $2.28 less than his original estimate of $500.[23] Early the next year Simpson sent Wetmore the map he had prepared the previous summer "that sums up all the data on the areal geology of the mammal-bearing Fort Union, and shows the exact locality from which each of your [museum's] specimens came."[24] Wetmore obviously recognized the value of Simpson's effort for he replied, "Mr. Gilmore and I are greatly pleased to have this map. . . . It is a fine piece of work. . . . I can readily appreciate that with this information you now have it will be much easier to interpret the relations of the fossils you are studying from this region and to understand their correlations."[25]

During this first field season Simpson apparently realized that the Fort Union of the Crazy Mountains Basin had still more valuable fossils to be discovered because he wrote in March telling Wetmore that he wanted to visit the Montana localities again — not to supplement the previous fieldwork, which he believed finished, but to collect additional fossils for the American Museum. Simpson added, however, that he didn't think that he could get a second field season financed, so he would put further fieldwork off for the time being.[26]

Wetmore eventually came through with some funds to pay an artist, Sydney Prentice, to begin illustrating the Fort Union fossils. Prentice came to New York and worked for a month on the Fort Union materials under Simpson's supervision, which Prentice found anything but superficial. In a letter to Wetmore he remarked that "Doctor Simpson has an eye which quickly detects any inaccuracy in a drawing. . . . I will spare no pains in making them accurate in every respect."[27]

South American Interlude

Simpson had other major research projects in place at this time, too, particularly a second extended trip to South America that he was then planning. His willingness to put off further Montana fieldwork resulted possibly as much from pressure from these other demands as scarcity of funds. He told Wetmore that only about half of the necessary illustrations were completed for the monograph he would be preparing on the Gidley collection and that the rest would be ready sometime in late 1934. He thus promised Wetmore that "the manuscript will be ready for delivery to you as soon as the illustrations are."[28] The delay in illustrating, presumably because of the continuing lack of money, gave Simpson breathing room for more than a year, which allowed him to leave gracefully for South America in October 1933.

This trip was the second of two long field seasons in Patagonia with additional time in the museums of Buenos Aires and La Plata. Both South American trips (1930–31; 1933–34) were made possible by gifts to the American Museum from Horace Scarritt, a Wall Street banker and museum patron. As it later turned out, Simpson was able to use some of the Scarritt funds to support his second summer of later fieldwork in Montana in 1935, after his return from his second South American visit.

The second trip to South America also provided an interlude in Simpson's turbulent personal life. During the two years between the first and second South American expeditions, his domestic life was in turmoil: "1931 and 1932 were a period of almost unbearably intense unhappiness & emotional stress in my personal life."[29] When he returned from the first Scarritt expedition at the end of 1931, and shortly after receiving the offer to complete the study of the Gidley collection, Simpson filed for legal separation from his wife. In February 1932, while discussing with Wetmore his plans for summer fieldwork in Montana, Simpson was granted the separation as well as temporary custody of his four young daughters, three of whom were living with relatives and one of whom was in boarding school. During this time, his wife was committed for a while to a mental hospital, but when out, she harassed him at work and in public.

To complicate matters still more, Simpson had renewed his earlier friendship with his Denver childhood neighbor Anne Roe, then a doctoral candidate in clinical psychology at Columbia University. Before long, he and Anne were covertly living together, sharing a New York apartment with his sister Martha. In spite of — or perhaps because of — these domestic problems, Simpson maintained his scientific output. If anything, he increased it, for these years of the early and middle 1930s were among his most productive, both in terms of publications and in laying the theoretical foundations for his later work. One can speculate that the second Scarritt expedition, which lasted from October 1933 to May 1934, not only allowed Simpson to complete his South American mammal studies but also gave him the psychological relief to sort out his personal difficulties.

If Wetmore was in any way disappointed that Simpson was going to take an almost yearlong break from finishing up work on the Gidley collection, he did not show it. He ended a short note to Simpson about paying off Prentice, the illustrator, by commenting, "I understand that you are to go to South America again and I wish you all success in your work."[30]

Second Field Season — 1935

From early June to late September 1935, Simpson returned to the Fort Union, this time not only again with Al Silberling but also with colleagues in the museum's department of vertebrate paleontology, including Dr. Walter Granger, curator of fossil mammals, and Albert ("Bill") Thomson, a museum preparator. Besides prospecting surface localities, Simpson and Silberling reopened the Gidley and Silberling quarries that had been worked in prior years by others, including Gidley, and developed an entirely new quarry, the Scarritt Quarry, named after Horace Scarritt, who financed the effort.[31] This second field season yielded a bounty of fossil finds: 440 upper and lower jaws, about 200 partial skulls, and some 900 isolated teeth of mammals — in all, an amount about equal to that of the original Gidley collection at the National Museum.[32] Although the Scarritt Quarry yielded an abundant number of specimens, it only did so by much labor: "Our [field] party averaged about one jaw per day per collector."[33] This new collection was "still Paleocene in age but somewhat younger than the fossils previously known from that area."[34] The success of this second field season depended on the geologic map, accurate siting of localities, and the stratigraphic section that had been prepared during the first expedition in 1932.

The 1935 field season was, however, long and difficut. It rained and hailed more than usual, and Simpson was also inundated with visitors, including his benefactor Horace Scarritt; Mr. and Mrs. Fenley Hunter, friends and patrons of the American Museum; the local Seventh Day Adventist preacher; and even Simpson's mother, who stopped by on her way by train from Washington, D.C., to California.[35] And judging from his letters — in his own code — to Anne Roe, Simpson was longing to get back to New York City and to her: "I just wrote to [my lawyer] . . . and told him that I *must* have a divorce, at almost any cost. . . . I miss you so, darling. I love you like the devil. I never want to leave you again for more than five minutes, at most. . . . [T]ime will overcome this [separation] . . . and remember how very much I love you. Surely this must be the last bad time we have apart. . . . I love you."[36]

Following the 1935 field season, Simpson spent another six months completing the manuscript for the USNM monograph, which continued to compete for time and effort with Simpson's other major project of the moment, his study of the South American mammals. In the spring of 1936, Simpson was able to write Gilmore that he was "within gasping distance of having the Fort Union memoir completed. . . . It will certainly be a load off my mind when that goes to press. It has been a much greater job even than I anticipated."[37] In the eventual publication, Simpson placed a prominent notice: "Dedicated to the Memory of James Williams Gidley, Ph.D. 1866–1931."[38]

"New Ways of Going at Things"

Because of the separate institutional ownership of the Gidley collection (U.S. National Museum) and the 1935 field collection (American Museum), Simpson was obligated to publish the detailed descriptions of each collection separately. The fossils from the Scarritt Quarry and others collected under American Museum auspices were described in the *American Museum Novitates* series, and those collected for the National Museum were published in the *Proceedings of the United States National Museum*. These publications are just a few tens of pages each of systematic paleontology, with an occasional brief foray into more theoretical issues.

However, it is with the publication of *United States National Museum Bulletin* 169 that Simpson truly does "introduce some new ways of going at things."[39] These new ways included (1) considering individual fossil specimens as members of a variable biological population rather than as representatives of typological species; (2) use of statistical analysis to support inferences about the validity of apparent differences among specimens and to what extent, if any, such differences should be reflected in their taxonomic classification; (3) paleoecology of the sedimentary rocks and their included fossils, and the extent to which different environments influence the aspect of a given fossil assemblage; and (4) preservational bias in accounting for observed variations in the taxonomic composition of assemblages.

These approaches were not only "new ways of going at things" for Simpson, but for paleontologic practice more generally. They may all be subsumed under a broader point of view that fossils are the remains of living organisms interacting with their environment, the record of which can have significant postmortem bias. It was a view that was maturing all through the 1930s for Simpson, strongly influenced as he was by the new population genetics of the time as well as by his own growing experience with Cenozoic mammalian faunas from all over the world. For Simpson, the Fort Union monograph brought together all the elements of this new viewpoint into a coherent whole. It thus provided for him a platform from which he then began to launch his more theoretical work, starting with the "two door opener papers"[40] and culminating with his classic *Tempo and Mode in Evolution*.

Species as Populations, Not Types

In the Fort Union study, Simpson fully embraced the concept that the fossil specimens he was studying were from original local populations of organisms, and as such could be expected to exhibit individual degrees of variation, one from another. That is, rather than the "typological concept" of species, he applied the "population concept," which was to become one of the hallmarks of not only the evolutionary synthesis of the coming decade but also, more specifically, of all Simpson's own subsequent work (see chapters 2 and 6).

For example, in considering the taxonomy of the multituberculates, which "made up a large proportion of the collection and are the most important single element in the fauna," Simpson employed some "useful methods . . . in some parts unfamiliar to many paleontologists."[41] Rather than beginning "with a generic grouping of the specimens, as is the usual practice," Simpson found it "necessary to treat the whole collection as if it represented only one genus, to distinguish the species present, and then to attempt to place them in genera" (73). Simpson relied on simple statistical analysis (as discussed below) of dental characters to discriminate among the several dozen specimens of rodent-like, ptilodontoid multituberculates, ending up with eight species. "These statistical data, furthermore, when considered from a taxonomic biological viewpoint, suggested the degree of variation to be expected in species of this family and also gave a criterion for judging the greater or less usefulness of certain characters for taxonomic distinction. . . . [E]ach [species] represents a variable morphological unit . . . [and] the variation in each is not greater than commonly occurs in natural species" (76).

Simpson then addressed the question of whether perhaps eight species of a single family from a single horizon at a single locality were too many. He concluded that "there is really nothing extraordinary in this number" for "ptilodontids are analogous to small rodents, and there is, for instance, hardly any region of the United States today that does not have more than eight species of [the family of field mice] Cricetidae" (76). And by considering his fossil specimens as drawn from a morphologically variable interbreeding population — that is, a species — Simpson was formulating a null hypothesis: namely, that unless he could objectively determine significant differences in morphology, the specimens had to be judged as coming from the same species. By thus allowing for inherent variability within fossil populations, Simpson was in accord with the contemporary view of geneticists and field biologists — that sexually reproducing species universally demonstrate some degree of individual variation. Such an approach to the recognition of species — living or fossil — thereby replaced the "typological concept" of species with the "populational concept," a hallmark of the evolutionary synthesis.[42]

This made for an important turnabout in taxonomic practice, for now any observed variations in morphology were not to be taken as conclusive evidence for the presence of genuine differences between species; rather, the burden of proof was now on the classifier to show, objectively, that such variations were taxonomically significant (i.e., that more than one species was indeed represented by the specimens studied). Although Simpson had taken such a populational approach to species in some of his earlier, shorter papers in the 1930s, it wasn't until the Fort Union monograph that he applied it broadly and consistently (see chapter 6).

Statistical Analysis and Taxonomic Classification

Another "new way of going at things" in the Fort Union monograph was related to the one just discussed: how to determine objectively if observed morphological variations among fossil specimens were indeed significant in the sense that the specimens represented more than a single species. Obviously, if one was dealing with specimens drawn from species belonging to widely different genera or families, then mere visual inspection would suffice to recognize the taxonomic variety. However, if one were examining specimens drawn from closely allied species — for example, within the same genus — then any observed differences might be sufficiently subtle that one couldn't confidently attribute them to either intraspecific or interspecific variation. As Simpson noted: "the methods of statistics provide the desired means of measuring variation accurately and the necessary criterion as to whether this variation is or is not of the sort normal within a species. These tests and this logical background have been the basis for taxonomy in this study."[43]

Simpson's approach to this issue of species discrimination was to apply relatively simple statistical analysis to his materials. By measuring observed ranges, means, standard deviations and errors, coefficients of variability, and by applying R. A. Fisher's t-test, Simpson was able to provide objective criteria for judging whether a particular collection included one or more species.[44] Although he could still make erroneous taxonomic decisions (and apparently he did), nevertheless subsequent workers would know the objective basis for his judgment, correct or not.

To return to the example of the ptilodontoid multituberculates, just discussed, Simpson noted that the classification of this large collection into genera and species was "peculiarly difficult" and could only be resolved by statistical analysis (73). By measuring the length and width of premolars and molars, numbers of cusps and serrations, presence or absence of the third lower premolar, shape of lower fourth premolar, and character of incisor, Simpson was able to sort out several dozen specimens into three genera and eight species. Moreover, such an analysis had additional implications: "If we accept the specific groupings finally adopted as valid, some interesting conclusions regarding variability and the value and significance of various characters for taxonomy in these animals are possible. The length of P_4 [lower fourth premolar], the most useful single dimension as this is far the commonest tooth in multituberculate collections, has a coefficient of variability of 9.3 . . . in the *sinclairi* group. This is high, but comparably high coefficients have been recorded for linear dimensions of teeth of [fossil and living] mammalian species" (78). Here Simpson gave additional justification for statistical analysis in that one can make direct comparisons of extinct fossils with living organisms and thereby create additional, objective bases for species recognition. In a long footnote, Simpson also commented: "How misleading the best judgment may be when not aided by statistical treatment is shown by the fact that although Gidley clearly relied on size of P_4 chiefly for specific separation . . . he placed the small *sinclairi* specimens in one species but divided the large *montanus* into three species, although the variability of the former is nearly twice that of the latter. The misleading factor is that the absolute difference in the extremes is less for the small than for the large species. Although this is the striking character to the eye, it is not the essential factor either from a statistical or from a biological point of view" (78n33).

Similar statistical treatment among arctocyonid carnivores led Simpson to reclassify taxa, previously studied by E. D. Cope, Gidley, and W. D. Matthew, that had been assigned to two genera and six species. In 1883 Cope had recognized a new genus of arctocyonid carnivore, *Claenodon*, with two new species, *C. ferox* and *C. corrugatus*. In 1919 Gidley referred these species to a new genus, *Neoclaenodon*, and added three more new species, *montanensis*, *silberlingi*, and *latidens*. In the late 1930s, W. D. Matthew introduced yet another species, *procyonoides*, in unpublished notes and a posthumous monograph, but while accepting the reality of Gidley's new genus Matthew rejected all of Gidley's diagnostic characters for it. In Simpson's study of the Fort Union arctocyonids, he provided a more objective, quantifiable basis for the discrimination of all these fossils.

Using his statistical approach and arguing against the use of single characters to define taxa, Simpson threw out Gidley's new genus and collapsed five previously defined species into two species, after comparing them with equivalent-age type specimens from New Mexico. By doing so, Simpson showed that measurements of the length of the lower second molar, M_2, in eighteen specimens of *Claenodon* approximated a normal curve of distribution and that the slightly smaller specimens previously referred to as *C. corrugatus* ought to be included within *C. ferox* (178–79). "These data do not prove that two species are *not* present: Such proof of a negative is practically impossible. . . . They do show that in this sample it is impossible to distinguish two size groups . . . that the distribution is not inconsistent in modality, variability, etc., with a single species" (179; emphasis in original).

Although Simpson used such statistically based taxonomy in some earlier publications in the 1930s, the Fort Union monograph provided broader scope for an extended discussion of its methodology and rationale. Simpson developed this biometrical approach much more fully in *Quantitative Zoology*, which he was starting to write with Anne Roe, who in her own work in clinical psychology was using such small-sample statistics.[45]

Environments and Assemblages

In the first field season of 1932, Simpson worked out the stratigraphy of the Fort Union, dividing it into "three mappable lithologic units of very unequal thickness" following earlier field observations by Silberling.[46] Simpson designated the lower two stratigraphic units as lower and upper Lebo and correlated them with the middle Paleocene Torrejon of New Mexico, although he did not interpret the specific depositional environment of the Lebo, some 1,400 feet of andesitic sandstones and shales. Simpson named the third overlying lithologic unit the Melville, more than 5,000 feet of lenticular sandstone and shales, which he interpreted as fluviatile in origin with numerous channel and flood sandstones.[47]

Besides describing and classifying the mammalian fossils within these units, Simpson also addressed their probable ecology. His review of what was known of the fossil plants suggested to him "that the whole region was heavily forested, chiefly by deciduous trees. . . . It demonstrates the presence of a well-developed arboreal habitat and of abundant food for browsing and frugivorous animals and suggests (but by no means proves) that the more open type of plains habitat was here relatively restricted or absent" (58). The abundant remains of freshwater bivalves, snails, fishes, and turtles indicated the continued presence throughout the Fort Union deposition of plentiful sources of water.

For the mammalian fossils Simpson divided his Fort Union specimens into four "lots." The first two lots were those specimens collected from the Silberling and Gidley quarries; the third lot comprised fossils from the younger Scarritt Quarry; and the fourth lot included specimens collected as surface samples from four other localities of the same age as the Silberling and Gidley quarries. Comparing the relative proportions of taxa in each of these lots, Simpson concluded that the "surface fauna" appeared to be a "normal flood-plain facies," with smaller carnivores and ungulates accounting for 90 percent of the total fauna

(62). The quarry faunas of middle Paleocene age were notable for their "rarity of large animals . . . [with p]henacodonts and pantolambdids [being] relatively rare. . . . [This] warrants the tentative conclusion that [the] fauna is largely arboreal, which is well in accord with the evidence that the quarries were in a swampy and heavily forested area and would go far toward explaining the unusual facies of the quarry faunas" (62, 63).

Simpson's careful statistical discrimination of species among the large number of fossil remains, particularly from the Gidley Quarry, allowed him to draw further conclusions about the ecological relationships of the faunas. Referring to the somewhat earlier work by Matthew and Argentine paleontologist Ángel Cabrera, Simpson applied the concept of "ecological incompatibility," namely that closely related species would not occur together in the same time and place.[48] Simpson noted that in the Gidley Quarry there were thirty genera of fossil mammals, each represented by a single species filling its own particular ecologic niche. Other closely related, yet distinct, species occurred within the Fort Union, "coming from different horizons and localities within the Lebo [stratigraphic unit]."[49] For Simpson, then, species were to be conceived not simply as abstractions to be defined and identified, but instead as biological entities representing populations of organisms dynamically interacting with their local environment. This point of view would be strongly maintained throughout all Simpson's later writings.

Preservational Bias

Simpson noted that the more abundant and better preserved mammal remains came from the three quarries (Gidley, Silberling, and Scarritt). Other fossils could be found throughout the area, particularly in the finer-grained shales and only rarely in the coarser-grained sandstones, which lay at the surface after long weathering and wind erosion. Simpson discussed at some length the state of preservation of the fossils and their distribution, recognizing that accurate interpretation of the assemblages required understanding how they were preserved. "The ideal conditions at such localities are deep weathering *in situ* without erosion or surface drift or wash, combined with gentle [wind] deflation, which removes the small weathered shale particles but leaves the larger or heavier fossils. . . . The quarry localities are those where fossils are so concentrated in a local pocket that it is profitable to work the bed as a whole and recover fossils in place. . . . Were it not for its three principal quarries, . . . this [fossil] field would be of relatively little importance. . . . An outstanding characteristic of the field . . . is the fragmentary nature of the material. . . . In the hundreds of specimens collected, there are so far known only four or five mammal specimens complete enough to be called skulls, and only two of these really adequately reveal most of the skull structure. Only about 10 specimens include associated upper and lower teeth, and only three any surely associated limb bones. Nothing approaching a complete skeleton has ever been found" (29–30).

Simpson paid special notice to the manner in which fossils were preserved in the Gidley Quarry, which provided the great bulk of the Fort Union mammals. Taking an approach that today would be called taphonomic, Simpson described in considerable detail "the unusual occurrence of the fossils and the indications of the possible conditions surrounding death and burial of [the] animals" (63). He concluded that the fossils and the nature of the sediments suggest "that the deposit was formed in sluggish water, perhaps a swampy stream course, ox-bow lake, or bayou. . . . Regardless of whether the mammals came here to drink, swam into the water, dropped from trees, or were occasionally washed in, it seems likely that the breaking and scattering of their bones, and perhaps commonly their deaths also, were the result of activities of the carnivorous fishes and reptiles. Such a history would probably explain the small ratio of bones to teeth . . . the many clean breaks, lack of association, and also . . . maceration, without apparent weathering . . . between death and burial" (63–64).

This acknowledgment of the role that preservational bias played in the resultant formation of fossil assemblages allowed Simpson to differentiate between what he later called "minor discontinuities of record" and "major systematic discontinuities of record," the former due to adventitious vagaries of geological and human sampling, whereas the latter result from significant biologic events like the rapid evolution of small, localized populations whereby a good or continuous fossil record is precluded.[50]

* * *

In the Fort Union fossil monograph, Simpson pulled together a number of different approaches to the interpretation of fossils that he had been pursuing more or less piecemeal throughout the early and middle 1930s. Whatever impact and influence the monograph may have had on other workers, it certainly was — for him, at least — a milestone, a point of no return as it were, for his own future development with respect to evolutionary theory. Immediately upon completing the monograph in early 1936, he turned his attention to two short papers[51] that were "of no great importance in themselves but of biographical interest because they marked the opening of two doors. One was on superspecific variation in nature and in classification. . . . The other opening door had a similar background but opened into a different field . . . the theoretical aspects of evolution. . . . Now age thirty-five and more than ten years out of graduate school I dared to take modest personal steps in the direction of principles and theory. I was planning and working toward larger steps."[52]

No sooner were these two papers delivered at professional meetings in the fall of 1936 than Simpson completed the manuscript for *Quantitative Zoology* with Anne Roe. Then in the spring of 1938 he started his most important and enduring book, *Tempo and Mode in Evolution*, which he completed in the fall of 1942, just before entering military service. All these works can be seen to have their factual and theoretical roots in the Fort Union Field monograph.

CHAPTER 6

ON SPECIES

[Speciation] is the basic structure of the web of life, the ever-present detail of the fabric of evolution.
— G. G. Simpson

In *Tempo and Mode in Evolution*, Simpson provided arguments from the fossil record to support the "modern evolutionary synthesis" during the latter's consolidation from 1936 to 1947.[1] More than thirty years later, Simpson retrospectively summarized the book's argument: "[T]he thesis, in briefest form, is that the history of life, as indicated by the available fossil record, is consistent with the evolutionary processes of genetic mutation and variation, guided toward adaptation of populations by natural selection, and furthermore that this approach can substantially enhance evolutionary theory, especially in such matters as rates of evolution, modes of adaptation, and histories of taxa, particularly at superspecific levels."[2]

Simpson was, unremittingly and unequivocally thereafter, a major public advocate of the evolutionary synthesis, particularly in defending one of its key axioms — namely, that the microevolutionary processes operating within populations, as observed in ecological time, were necessary and sufficient in explaining macroevolutionary paleontologic phenomena, as displayed at supraspecific levels, especially families and higher, in geological time.[3] Simpson was well positioned to assert this claim because, as a student of fossil mammals, he worked at the three levels of taxonomy: the alpha level, where he defined many new paleontologic species; the beta level, where he was concerned with their phylogenetic classification; and the gamma level, where he interpreted the species' broader biological meaning.[4] Indeed, his first series of published papers in 1925 at age twenty-three, involved all three levels of taxonomy, for there he proposed a new classification of the Mesozoic mammalian order Triconodonta, defined new genera and species within the order, and offered his interpretation of the evolution of early mammalian molars from reptilian ancestors.[5]

In this chapter I discuss the sequential development of Simpson's perception of "species," before, during, and after the consolidation of the evolutionary synthesis for some insight on how one participant in the synthesis responded to and contributed to its formulation by focusing on one of its important concepts. My discussion is more or less chronological because there was clearly a shifting over time of how Simpson thought about species: from types to populations; as operational ecologic units; as initial, transitional steps in the subsequent formation of higher categories; and as fundamental units of evolution. However, there was no sharp demarcation as Simpson moved from one perspective to the next; on the contrary, each perspective grew out of earlier ones and so their boundaries are blurred. There is also the practical difficulty of determining the precise moment when Simpson actually entered a new phase in his thinking about species.

Paleontologists like Niles Eldredge and Stephen Gould give due credit to Simpson's role in the making of the evolutionary synthesis, but provide a different reading with respect to Simpson's concept of species and its role in evolution. For example, Eldredge in a detailed analysis of Simpson's *Tempo and Mode* states that "Simpson . . . claims that speciation leads nowhere."[6] Gould argues that Simpson "regards the process of speciation by splitting largely as a parceling out of existing variation, not as a way of incorporating evolutionary novelty."[7] Ernst Mayr has been even more strongly dismissive with specific reference to Simpson on species, for he has claimed that, "Paleontologists, when studying macroevolution, traditionally never come to grips with the problem of the origin of taxa or types that evolved . . . higher . . . or experienced radiations. Simpson (1944), for example, makes no reference to species or speciation in his *Tempo and Mode in Evolution*. . . . Simpson was not alone in his neglect of the problem of speciation. Most paleontologists, during the 100 years after the publication of the *Origin [of Species]* entirely ignored the problem of the origin of organic diversity."[8]

My purpose is to document in more detail Simpson's view of species as species and how his conclusions about them contributed to the modern evolutionary synthesis.

From Types to Statistical Inferences

During the time of his doctoral dissertation at Yale, 1923 to 1926, until his rising awareness of population genetics in the early 1930s, Simpson defined species in the then time-honored way, namely using a few diagnostic characters that qualitatively differentiated the new from the already known. In the two "Mesozoic mammal" monographs that resulted from his Yale dissertation and his year of postdoctoral work at the British Museum, Simpson restudied over three hundred individual specimens — mostly jaws and teeth, with a few skulls and limb bones that were usually worn or crushed — assignable to one hundred species of which he defined more than one-third (thirty-five) as new.[9] For most species descriptions Simpson had only one or two specimens, sometimes several, and rarely as many as a dozen. Hence his materials did not usually allow for much quantitative analysis (fig. 6.1). In determining if a specimen in question belonged to an already-defined species or justified a new one, Simpson qualitatively compared its hard-part anatomy to that of existing type specimens.[10] However, if there were enough data for comparison, Simpson did make detailed measurements and presented them in tabular form (fig. 6.2). Given the nature of his materials, Simpson's species' discriminations, even if well-informed, were necessarily qualitative. His methodology, however, was typical not only for paleontologists working with fossilized skeletal remains but also for the many biologists who used dead, preserved specimens as the basis for their determination of species.

In the mid-1930s Simpson made a major shift away from the descriptive statistics (i.e., recording the statistical means of a given anatomical character) of the sort used in his Mesozoic mammals work to inferential statistics (i.e., using the t-test for determining the probability that two samples were from the same statistical population) when he began to view his few fossil specimens as small samples from which the larger biological population might be inferred. That is, rather than strict comparison — whether quantitative or qualitative — of individual specimens to known type specimens, Simpson began using simple

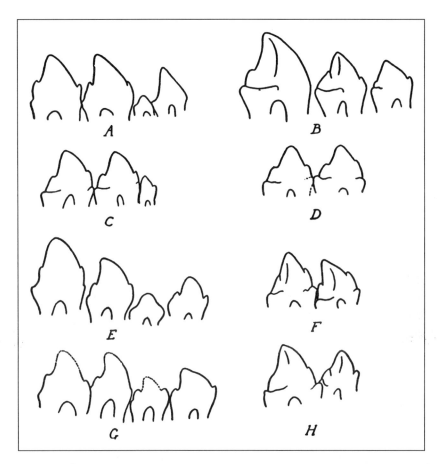

FIGURE 6.1 Simpson's drawings of the lower premolars of eight different specimens of the early mammal genus, *Docodon*, to illustrate qualitatively their individual variability. (Simpson 1929a, "American Mesozoic Mammalia," fig. 38; © Peabody Museum of Natural History, Yale University)

statistical analysis in his taxonomic methodology to infer the characters of the biological population from which his specimens were drawn. His writings of the late 1930s and early 1940s thus began implementing the populational as against the typological concept of species, which as noted above was to become a major element of the synthesis.

In 1937 Simpson published two papers and one monograph that illustrate well his conscious effort to bring statistical analysis and populational concepts to the interpretation of fossils. Several factors came together to encourage Simpson's more elaborate use of statistics. First, the

	C. serratus. Size taken as 100.	C. scindens. Size 112.	C. minor. Size 100.	C. falconeri. Size 150.
$\frac{\text{Length } P_4}{\text{Length } P_3} =$	1.54	1.82	2.00	2.20
Projections on shearing edges of P_3 and P_4, 3 and 6, respectively.		3 and 6	4 and 7	4 and 8
P_4 higher relative to length than in other species.		—	—	—
Antero-internal cusp of M_1 small.		A little larger.	Much reduced or absent.	Unknown.

FIGURE 6.2 Chief difference among four species of *Ctenacodon*, a Jurassic multituberculate mammal. Note the combination of qualitative and quantitative variables to characterize the species. (Simpson 1928a, *A Catalogue of the Mesozoic Mammalia*, 36. Courtesy The Natural History Museum, London)

fossil collections he was now studying contained more abundant material, usually providing a number of specimens for each presumed species so that simple bivariate statistical analysis could be performed on them. At least equally significant was the influence of his future wife, Anne Roe, whose own research in clinical psychology involved studies using small-sample statistics. As Roe herself has remarked, "We engaged in many sometimes critical discussions. He maintained that psychology [was] not a science, while I countered that paleontologists who dealt with samples of varying sizes, usually very small, should do more with their data statistically than just note means and ranges. Ultimately we wrote a text on statistics applicable to biological data, *Quantitative Zoology*. This was really the first attempt to apply statistical methods to field material, living or fossil, that was comprehensible to someone not mathematically trained."[11] Finally, one should note that Simpson had always had great facility with mathematical manipulation, so it required no particular effort for him to become adept with statistics.[12]

At Harvard's tercentenary celebration in December 1936, Simpson presented a paper that was published the following year as "Patterns of Phyletic Evolution." In this paper Simpson noted that "paleontological psychology has been to direct attention, to a disabling extent, to the individual specimen, and the number of publications on fossil mammals in which a suite of specimens has been correctly studied as representative of a group can be almost counted on the fingers of one hand."[13] Lest he seem overly critical, Simpson prefaced his comments by observing that while "I may appear in the unfavorable light of a Daniel come to judgment without the precaution of bringing any credentials; I can only say that my criticisms are directed as much against my own work as against that of anyone else" (305).

In this short article, Simpson wished to address the problem that "boils down to the old question as to what a species is," noting that he was taught that a 15 percent difference in any linear dimension justified recognition of a different species. "The fact is, as everyone who has studied variation should know, that there is hardly any geographic race, let alone any species, in which some variants are not more than 15 percent larger in some dimensions than are others. On the other hand, a difference in the average dimensions of two groups of individuals may be a perfectly valid and highly significant specific character, even though the difference is less than 15 percent" (307).

Simpson went on to create some data that show a morphologic character of a hypothetical fossil increasing in dimensions through a vertical sequence of rock strata, and then he indicated how the data might be qualitatively interpreted in several ways. For example, the data might record the decline over time of one smaller species and the rise of another larger species; or perhaps there was just one species in which the character in question was increasing through time; or possibly one larger species gradually diverged from the smaller ancestral species. Simpson then explained how one could choose among the three alternative interpretations by simple statistical analysis of the data, stratum-by-stratum, through the rock sequence. In particular, frequency distributions of the character in question from different, successive horizons would indicate which interpretation was correct (see fig. 6.3). Toward the end of his career, Simpson credited this paper as marking his "abandonment of the typological thinking of my college teachers and started aiming me toward statistical biometry and the deeper investigation of evolutionary theory and taxonomic stance."[14]

Whereas the Harvard tercentenary paper used hypothetical data, Simpson published that same year another paper in which he applied the same statistical reasoning to real data from various late Paleocene and early Eocene fossil mammals.[15] One set of his data shows an increase in size of the lower first molar of a genus of an archaic ungulate through successively younger geologic formations (fig. 6.4). Walter Granger, Simpson's senior American Museum colleague, had previously recognized three species, in order of size. Simpson, however, after calculating means, standard deviations, and coefficients of variation for the whole collection, concludes that, "It appears extremely probable . . . there is a single species, variable in all horizons but slowly increasing in size with passage of time. . . . This concrete example suggests that many of the supposed cases of . . . orthogenetic evolution

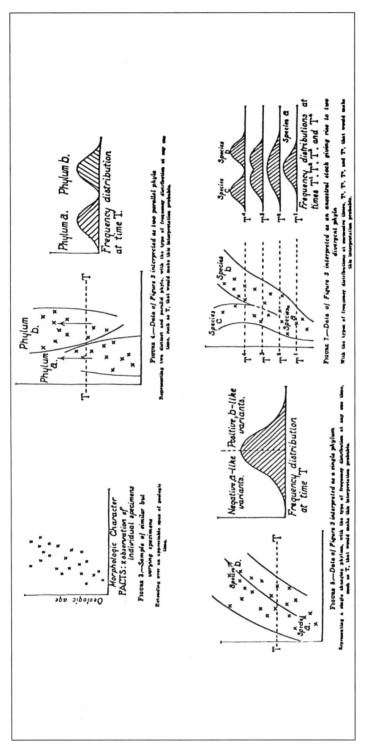

FIGURE 6.3 Hypothetical fossil data showing systematically varying morphologic character through time (*upper left*). In the three accompanying graphs, Simpson indicates how one can discriminate among the three possible interpretations of these data by using simple statistical analysis: Two species overlapping in time with one species decreasing in abundance as the other increases in abundance (*upper right*); or a single species within which the measured character changes enough over time that a second species is recognizable (*lower left*); or one species splitting into two daughter species (*lower right*). (Simpson 1937c, "Patterns of Phyletic Evolution," 307, 308, 309, 311; © Geological Society of America)

The following grouped distribution is typical of that for all the available variates of *Ectocion* not referable to *E. parvus:*

Length of M_1	Clark Fork	→ Sand Coulee	→ Gray Bull	All
5.7–6.0	1	0	0	1
6.1–6.4	6	3	1	10
6.5–6.8	4	4	7	15
6.9–7.2	1	3	2	6
7.3–7.6	0	0	2	2

Horizon	Number of Specimens	Observed Range	Mean	Standard Deviation	Coefficient of Variation
Clark Fork	12	5.9–7.1	6.5 ± 0.1	0.32 ± 0.06	4.9 ± 1.0
Sand Coulee	10	6.2–7.2	6.7 ± 0.1	0.33 ± 0.07	5.0 ± 1.1
Gray Bull	12	6.4–7.4	6.8 ± 0.1	0.39 ± 0.08	5.7 ± 1.2

FIGURE 6.4 Data for the length of the first lower premolar of thirty-four specimens of the genus *Ectocion* from three stratigraphic levels of the Upper Paleocene/Lower Eocene: Clark Fork (older), Sand Coulee, and Gray Bull (younger). (Simpson 1937b, "Notes on the Clark Fork," 20. Courtesy American Museum of Natural History)

. . . may be misinterpretations of conditions similar to these: that in fact only one variable phylum [or lineage] is present and that the several supposed phyla [or lineages] are constructed by selecting variants in the same direction from each horizon and supposing them to form separate species" (20–21).

Thus, Simpson's statistical approach allows him to "compare the groups [or populations] as such, to use the individual specimens only as representatives of a group [or population] rather than thinking of the group as secondary and the individuals as the essential units. Although the distinction may seem unduly subtle, it is in fact fundamental . . . and seems absolutely essential for the placing of paleontology on a more exact, more objective and less intuitive basis" (2).

Simpson's 1937 monograph on fossil mammals of the Paleocene Fort Union Group was another, still more extensive use of statistical analysis for species discrimination (see chapter 5).[16] One of his more elaborate treatments is that of twenty-two lower jaws of three genera of multituberculates (fig. 6.5). Simpson demonstrates that if just one character is used to determine species, the results would contradict species-

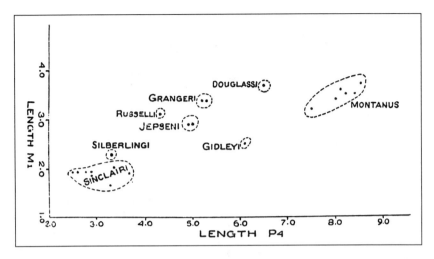

FIGURE 6.5 Scatter diagram of twenty-two specimens of multituberculate mammals, plotting length of lower first molar versus length of lower fourth premolar, that provides the basis for Simpson's recognition of eight species. (Simpson 1937a, "The Fort Union of the Crazy Mountain Field, Montana," 75. Courtesy Smithsonian Institution, National Museum of Natural History)

delineation using a different, single character. Instead, Simpson measured nineteen primary traits (including lengths and widths of molars and premolars, their numbers of cusps and serrations, and the presence and character of incisors) and calculated a variety of statistics. "After consideration of all these primary and secondary data, it was clear that of the eight groups finally achieved and checked each represents a variable morphological unit, that the variation in each is not greater than commonly occurs in natural species, but that no two [groups] can be combined without producing a unit statistically heterogeneous and morphologically much more variable than a species. The biological conclusion is thus that eight species are present."[17]

In a short theoretical paper in 1940, Simpson codified his concept of the species as a taxonomic category, whereby he emphasized that all specimens are equally valuable and necessary in inferring the living population from which they are but a sample.[18] Simpson wished to reduce the importance given to the single "type specimen" in defining a species, not only because it could not, even in principle, accurately represent the full variability of a natural species, but also because it reinforced the essentialist concept of species, which deliberately ignored such variation. Although Simpson did not explicitly invoke population genetics, this paper is informed by the view that species are inherently variable and thus no single individual can express completely the full character of a species: "this concept [of a taxonomic species] corresponds more or less with a real thing in nature, a group of individual animals that are truly related in a way that makes them a natural unit of a certain approximate scope."[19]

From Statistics to Ecology

In his Fort Union monograph Simpson had already anticipated this view that a species "corresponds with a real thing in nature" by demonstrating how two species of *Metachriacus*, an archaic ungulate, "intergrade [statistically], but . . . are separated by provenience, the two coming from different horizons and localities. . . . They are thus interpreted as closely allied but separable species."[20] From histograms of lower second molar lengths (fig. 6.6), Simpson had concluded from their slightly over-

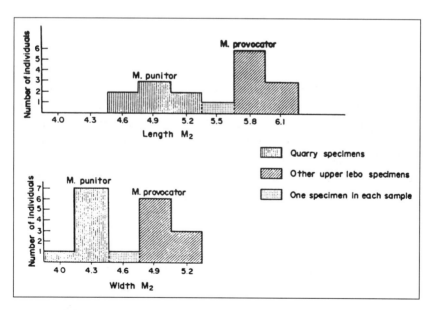

FIGURE 6.6 Lengths and widths of the lower second molar of two species of an archaic ungulate, *Metachriacus*, from different localities of the Lebo beds within the Fort Union. (Simpson 1937a "The Fort Union of the Crazy Mountain Field, Montana," 66. Courtesy Smithsonian Institution, National Museum of Natural History)

lapping bimodal distributions that the close yet distinctive morphology of the two allopatric species demonstrated the "law of ecological incompatibility" because, given that both species filled the same ecologic niche (as interpreted from their similar dentition), they would not occur in the same place (as indicated by their separate occurrences).[21] It was this sort of reasoning about species in nature — as defined by population geneticists and field biologists — that Simpson was now beginning to write about more extensively — namely, species as groups of interbreeding, genetically variable individuals interacting with particular environments.

With the publication of *Tempo and Mode in Evolution*, Simpson entered the next and enduring phase of his forceful advocacy for the synthesis by arguing that larger-scale evolutionary patterns of higher taxa ("macroevolution") depended upon various ecologic opportunities prevailing during speciation ("microevolution"). As a paleontologist, Simpson not only studied the results of microevolution, namely species, but he was more generally concerned with the origins, adaptive radiations, and extinctions of higher taxa (chiefly genera, families, and orders), that is, with macroevolutionary phenomena.[22] In *Tempo and Mode* Simpson argued cogently that the evolutionary mechanisms observed by population geneticists and field biologists were not only necessary but also quite sufficient to explain macroevolution.

Simpson chose the title *Tempo and Mode in Evolution* to emphasize his own claim that it is the differing rates and styles in microevolution that explain diverse macroevolutionary phenomena. The crux of Simpson's argument lay in his chapter on "Organism and Environment" (ch. 6), where he uses microevolutionary processes of population genetics and field biology against an environmental backdrop that he called the "adaptive grid."[23] Depending upon where particular species find themselves within the adaptive grid and the characteristic environmental opportunities thus available to them, organisms may exhibit widely varying evolutionary rates. For example, species in narrow and stable adaptive zones would be most likely to show slower evolutionary rates after initial adaptation and radiation in the zone. Species in adaptive zones that were expanding, contracting, or changing in one direction would show, respectively, diversification, extinction, or increased specialization. Because the boundaries between adaptive zones are discontinuous, the most interesting situations would be those where a species moved from one adaptive zone to another, for here evolutionary rates would have to be very high if the species were to be successful in bridging the gap between zones, what Simpson called "quantum evolution" (206ff.).

Irrespective of the validity of Simpson's explanations, the relevant point here is that his argument depended upon viewing species as populations of individuals having differing degrees of inherited variation and thereby responding adaptively (or not) to the local environment. For Simpson the outcome of these microevolutionary events occurring in ecological time display themselves as macroevolutionary phenomena when aggregated over geologic time. Or in Simpson's words, "how necessary it is to consider evolution in terms of interaction in organism and environment . . . [it is] relationships in the organism-environment system that bear most directly on the main themes of tempo and mode" (180).

Although a commonplace but hardly universal conclusion today, Simpson's claim that microevolution could fully explain macroevolution was quite radical for a paleontologist when made more than a half century ago. Moreover, the nagging problem of the systematic absence in the macroevolutionary (fossil) record of taxa bridging gaps between adaptive zones — direct fossil evidence for the origins of higher taxa, for example — was, in fact, to be expected given very high evolutionary rates occurring in small populations in geographically localized areas. Thus, with one stroke of the pen Simpson brought paleontology into the mainstream of evolutionary biology by demonstrating that the fossil record could best be interpreted in accord with contemporary population genetics and natural history. Simpson tested the claims of the geneticists and the field biologists against paleontological knowledge and found each compatible with the other.

From Species to Higher Categories

As this discussion so far suggests, Simpson considered himself not only competent as a paleontologist but equally competent as an evolutionary theorist. As discussed in preceding chapters, his first formal venture into broader evolutionary theory also occurred in that watershed year of 1937 with the publication of "Supra-specific Variation in Nature and Classification."[24] Although this article focused on supra-specific phenomena, Simpson alluded to his developing notion of how to go about recognizing fossil species in the first instance. "In some cases it has been now possible to bring together large collections all of one geologic age and from one locality . . . [and] it is possible to recognize and delimit natural minimum groups, conventionally recognized as species. . . . The general principle [is] to recognize a sample of a population in which many variations appear, as in all populations, but in which these variations were arising in a single interbreeding community."[25]

Particularly relevant for the discussion of "supra-specific variation" was Simpson's view of the role of species in evolution, for he asserted that evolution results not only by species transforming linearly through time but also by species splitting again and again — or in present-day terminology, by anagenesis and cladogenesis, respectively (fig. 6.7).[26] According to Simpson, the evolution of species "has been both radiate and linear. The effect of radiation is usually achieved by dichotomies. . . . [A] higher category consists essentially of a number of

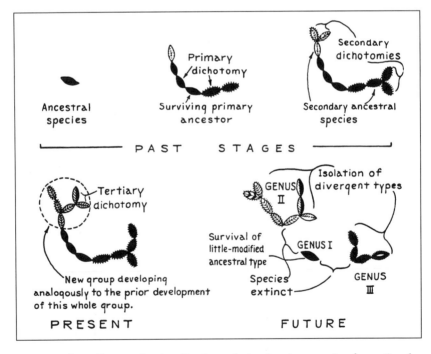

FIGURE 6.7 Simpson's visualization of step-by-step species formation by branching, from the past into the present, with future disappearance of some of the links thereby creating apparent gaps between genera. (Simpson 1937d, "Supra-specific Variation in Nature," 252; © University of Chicago Press)

such lines radiating, or better diverging, by successive dichotomy from an originally unified ancestry. . . . [T]here seem to be two flaws in the classic figure of phylogeny as a tree. . . . At any time, wherever the cross-section be taken, life consisted of species. There was no time when there existed only large branches."[27]

This is, I believe, the first published, explicit claim by Simpson that microevolution — branching and linear evolution at the species level — accounts for macroevolution as recorded in higher, supraspecific categories. Over the years, Simpson steadfastly maintained this claim, although elaborating on it and expressing it more forcefully once the synthesis had jelled: "I . . . repeat my conviction that the basic processes are the same at all levels of evolution, from local populations to phyla, although the circumstances leading to higher categories are special and the cumulative results of the basic processes are characteristically different at different levels."[28]

Species as an Evolutionary Concept

Besides naming, classifying, and interpreting the biological meaning of his fossil species (the three levels of taxonomy), Simpson was also interested in their systematics, which, as he defined it, meant inquiring further into their kinds and diversity as well as into any and all relationships they might have among other organisms.[29] Systematics thus naturally led him into the larger-scale dimensions of geological space and time, that is, paleobiogeography and macroevolution.

With respect to paleobiogeography Simpson developed a theory of historical biogeography, which claimed that ordinary migrations of species and chance dispersal of individuals could account for past geographic distributions, especially during the 65-million-year-long Cenozoic Era, when mammals increased greatly in kinds and numbers. For Simpson it was, once again, a question of small-scale interactions of individuals within species with their local environment that produced the larger-scale, paleobiogeographic result. Consequently, Simpson chose to explain past distributions of fossil biotas in terms of mobile organisms on stable continents rather than stable organisms on mobile continents. Predictably, therefore, Simpson was a fervent and authoritative opponent to Wegenerian continental drift, even well into the plate tectonics revolution of the 1960s.[30]

The ultimate, if not logical, result of Simpson's thinking about species was his "evolutionary" definition of species, first offered in 1951 in his paper entitled "The Species Concept," and later slightly emended in 1961 in his book *Principles of Animal Taxonomy*.[31] As he defined it, "An evolutionary species is a lineage (an ancestral-descendant sequence of populations) evolving separately from others and with its own unitary evolutionary role and tendencies."[32] This definition is sufficiently abstract that it seems difficult to apply in a given real situation when determining species. However, Simpson often made the point that a "definition defines" and so is a separate issue from evaluating whether a given case meets the definition. According to Simpson, ambiguity in determining if a particular something meets a definition does not necessarily make the definition itself any less valid.[33]

Simpson contrasted his evolutionary species concept with Ernst Mayr's "biological species concept," namely that "species are groups of actually or potentially interbreeding natural populations which are reproductively isolated from other such groups."[34] While in all other aspects Simpson and Mayr were in close theoretical agreement about the explanatory power of the evolutionary synthesis, each preferred his own definition of species. As might be expected from one whose initial research was in field ornithology and who later developed the concepts of allopatric speciation and the founder effect, Mayr emphasizes the fine-grained spatial relationships of populations within narrowly resolved time. Simpson, on the other hand, worked only with fossils, so his materials were intrinsically coarse-grained, both in space and especially in time. Dealing with relatively small numbers of specimens of mostly extinct species that were more widely separated geographically and stratigraphically, Simpson's interpretations were, at best, what one might call "kinematic," in that only evolutionary movement could be described, with the actual agents responsible for the movement having to be inferred. Whereas Mayr's interpretations, given the very different nature of his materials where time and space were much more finely resolved, are more nearly "dynamic," that is, causal forces can be more credibly postulated.[35] Therefore, it would have been very unlikely for Simpson-as-paleontologist to have formulated, say, the concept of allopatric speciation. Conversely, one should not expect Mayr-the-field-ornithologist to have formulated, say, the concept of quantum evolution or expounded on evolutionary rates.

Consequently, Mayr's biological species concept, located within the short range of ecologic time, gives priority to geographic space and process, and thus has a predominantly horizontal component. Mayr deals explicitly with the nature of existing species and, by inference, how new species must arise. Simpson's evolutionary species concept, by contrast, is literally embedded (stratigraphically) in the long range of geologic time and, by giving priority to phyletic lineage and pattern, has a predominantly vertical component.[36]

* * *

In summary, then, this discussion focuses on one of the ways in which Simpson participated in the "modern evolutionary synthesis" by examining his developing concept of species. In particular, we see him moving from species-as-types to species-as-populations, and next how those populations, through organism-environment interactions, might give rise to new species, some of which rapidly lead to higher taxa.

Simpson's participation in the creation of the modern synthesis is more generally evident here by his contribution to what biology historian V. B. Smocovitis termed the "quantification of evolution — the attachment of numbers to . . . nature."[37] Smocovitis rightly views such quantification as an essential ingredient for the formulation of the evolutionary synthesis; however, she neglects Simpson's role in this regard. Simpson's commitment to a positivistic philosophy often found expression in quantitative argumentation, even if the paleontological data upon which such arguments were made were limited in number and precision.[38]

CHAPTER 7

TEMPO AND MODE IN EVOLUTION

I mean to suggest that some accepted examples of actual scientific practice . . . provide models from which spring particularly coherent traditions of scientific research.
— Thomas S. Kuhn, historian of science

In addition to the numerous articles that Simpson published during the years between the award of his Yale doctorate in 1926 and the beginning of *Tempo and Mode* in 1938,[1] he also published *Attending Marvels* (1934), a popular book-length account of his first field expedition to Patagonia, and was near the completion of his first technical book, *Quantitative Zoology* (1939), coauthored with Anne Roe. As important as this published research was — it got him elected to the American Philosophical Society in 1939 and to the National Academy of Sciences two years later — there was no obvious indication of what was to come in *Tempo and Mode*.

This chapter now takes a detailed, comprehensive retrospective look at *Tempo and Mode* because it is the work that essentially defines Simpson's paleontologic career, the scientific contribution that gives the shine to whatever luster his research before and after has. I will therefore address the following issues: What specifically did *Tempo and Mode* contribute to the modern evolutionary synthesis? Was it more than a "consistency argument"? If so, what else did it do? What were the paleontological problems and misconceptions that *Tempo and Mode* resolved and clarified? Which specific writings of Simpson's before *Tempo and Mode* anticipated its content and methodology? How was *Tempo and Mode* received by contemporary biologists and paleontologists, and what influence did it have on subsequent research in paleobiology?

And, finally, to what extent is Simpson's contribution in *Tempo and Mode* within the mainstream of present-day evolutionary theory?

Two Door-Openers

In the paper "Patterns of Phyletic Evolution" (1937), Simpson applied simple univariate statistics to a hypothetical data-set of a morphologic character varying through time in order to differentiate among possible alternative interpretations: parallel evolution of related but distinct phyla (or clades); or a single phylum (or clade) of distinctly changing character; or a branching phyletic history.[2] The seven figures in the article are simple graphs of his hypothetical data; there are neither real paleontologic data nor illustrations of specific fossils. This format of presentation, later typical in *Tempo and Mode*, was highly innovative for a paleontological treatise. Underlying Simpson's approach was a shift away from typological thinking about a species to the concept of species as variable populations (as described in the previous chapter in the present volume). The argument in this paper is straightforwardly deductive: Simpson hypothesized three different cases of evolutionary history — parallel, phyletic, and branching — and then predicted how each of the resultant fossil assemblages can be recognized statistically and distinguished from the others (see fig. 6.3).

The second article that anticipates *Tempo and Mode* is "Supra-specific Variation in Nature and in Classification from the Viewpoint of Paleontology" (1937).[3] The background for this paper was supplied by Alfred C. Kinsey, who had published a monograph on the origin of higher categories in gall wasps in which he criticized what he viewed as the prevailing concepts regarding the origin and recognition of higher taxonomic categories.[4] In December 1936 several professional societies under the auspices of the American Association for the Advancement of Science sponsored a symposium on this topic, and Simpson, at the suggestion of his American Museum colleague, W. K. Gregory, rebutted Kinsey's notions from the paleontologic point of view.

In the first half of his paper Simpson summarized what was then known about the mostly South American mammalian order of extinct hoofed herbivores, the notoungulates — a group discovered by Charles Darwin, described by Richard Owen, and studied in detail by Simpson.

In particular, Simpson noted the evolutionary trends in limbs and teeth in the several suborders of notoungulates and used them to establish the following concepts: (1) that evolutionary rates vary within the order; (2) that closely analogous or parallel evolution in the dentition of a family of notoungulates (the notohippids) and also in a family of perissodactyls (the equids) is adaptive in both these unrelated lineages, thereby yielding superficial resemblances between them; (3) that it is necessary to recognize the biological reality of interbreeding populations of variable individuals; and (4) that *supra*specific differences may come about from the evolutionary segregation of such *infra*specific variations.

In his discussion of notoungulate limb and tooth evolution, Simpson drew three of the four conclusions regarding evolutionary rates that he was to repeat in *Tempo and Mode* (ch. 1), although in the latter work the example chosen was shifted from notoungulates to horses. He asserted that "differences in rate of evolution between different structures in the same phylum [or clade], between different periods in the history of one phylum, and between different phyla with a common origin . . . are among the most important factors in the differentiation of supra-specific groups."[5]

The second half of the paper turned to specific consideration of thirteen principles and concepts that Kinsey claimed had guided neontologists in their recognition of higher taxonomic categories and which he, Kinsey, then proceeded to knock down. Simpson discussed each of the thirteen points by "adding what seems to be the best or most generally held paleontological opinion on each concept."[6] In fact, this opinion is chiefly Simpson's own special insight. Hence here, as later in *Tempo and Mode*, Simpson melded the principles and methods of zoology with those used in paleontology, an approach that thoroughly permeates all his scientific work.

Although Kinsey's thirteen concepts of higher categories and Simpson's clarification or refutation of them are somewhat dated and not always relevant to the purposes of this chapter, Simpson did make several points that reveal his general approach to the interpretation of the fossil record.

Thus, Simpson concluded, contrary to Kinsey, that "Supra-specific groups do have an objective reality . . . . [T]heir status . . . is arbitrary but the groups themselves are natural" (255, 257); evolution has been both "radiate . . . usually achieved by a series of dichotomies," and "linear" with individual phyla (or clades) developing through time (259); therefore, the "classic figure of phylogeny as a tree" has two flaws. "At any time, wherever the cross-section be taken, life consisted of species. . . . There was no time when there existed only large branches" — a more apt metaphor is a "vegetative and branching figure" (259); thus, the differences between higher categories (between, say, perissodactyl and whale limb structure) "are no doubt an accretion of small steps, and the first dichotomy that was to lead eventually to whales and to perissodactyls was probably of the specific type" (263).

In his discussion Simpson enunciated a recurrent theme in *Tempo and Mode*, namely, that microevolution — branching and phyletic evolution at the species level — explains macroevolution as recorded in higher, supraspecific categories. His paper's figure 5 is reproduced here (see fig. 6.7) because it not only summarizes well the chief points Simpson made in this regard, but it also exemplified nicely his penchant for illustrating theoretical ideas in such schematic form, a characteristic found throughout all his writings. Simpson concluded the paper by observing how "characters can be lost as well as acquired" in evolution as, for example, teeth, toes, and vertebrae among the mammals. He also noted that new "species might be distinct yet not differ appreciably in any morphological character" (265). This latter point has been emphasized since to explain the apparent "stasis" of paleontological species.[7]

In summary, then, we see Simpson dealing with the "tempo" of evolution in the first half of this "door-opener" article and with the "mode" of evolution in the second half. Let us now turn to what Simpson was learning about the new genetics of the 1930s and how this influenced his work at that time.

The New Genetics

In later writings Simpson explicitly acknowledged the work of R. A. Fisher (esp. Fisher 1930), S. Wright (esp. 1931), and J. B. S. Haldane (esp. 1932) as having influenced his own views about the relationship between the new genetic interpretations of the 1930s and the paleontological record of life as he understood it from his studies of the previous decade and a half.[8] More specifically, he stated: "I was even more stimulated and owe most to [Theodosius] Dobzhansky's . . . *Genetics and the Origin of Species* [1937] . . . . That great, seminal work showed me that genetics is indeed consistent with and partially explanatory of nonrandom explanations of evolution."[9]

Simpson also noted that both the paleontologist Otto Schindewolf (1896–1971) and the geneticist Richard Goldschmidt (1878–1958) stimulated him in an opposite way by their championing a contrary saltationist interpretation of evolution, one based on large-scale, random genetic novelties.[10] Fisher, Wright, and Haldane apparently reinforced Simpson's already held opinion about "Neo-Darwinism . . . namely . . . that the directional element in evolution was natural selection . . . and that research, theoretical research, along these lines was the most fruitful."[11] What Simpson may have found particularly cogent about these three works was their detailed mathematical analysis of population genetics and the quantitative evaluation of the relative importance of genetic mutation and chromosomal linkage, migration, population size, and selection intensity. Simpson's mathematical intuition and skill allowed him not only to follow the respective arguments in all their detail but also to weigh their merits. (Recall that Simpson at this time had just about completed *Quantitative Zoology*.)

Rather than being intimidated by the mathematics of population genetics, Simpson actually relished them. These works of Fisher, Wright, and Haldane are cited more than two dozen times in *Tempo and Mode* — most of them in chapter 2 ("Determinants of Evolution"). Simpson did not merely appeal to the mathematical geneticists as authorities, but instead recast and reformulated what they said in terms understandable to the nonspecialist. Moreover, Simpson did not hesitate to add his own insights in these areas that were, strictly speaking, outside his own field of research.

Whereas Fisher and Wright made no mention of the fossil record, Haldane did allude from time to time to fossils and the history of life. Haldane remarked, for example, on some paleontological phenomena that puzzled him, including "exaggerated" development in the coiling of the Jurassic oyster *Gryphaea*, and in horns of the Oligocene archaic herbivores, the titanotheres. Despite Haldane's summary — and inadequate — treatment of the paleontological evidence, at least he saw the connection between genes and natural selection in living populations with the larger-scale, longer-term results studied by paleontologists.[12]

If *Tempo and Mode* did nothing else, it not only brought the work of the mathematical geneticists to the attention of the practicing paleontologists but, more important, made their ideas accessible in a way that was intelligible and useful to a generation of scientists more familiar with the systematics of shells and bones than with the differential equations of dominant and recessive alleles.

Genetics and the Origin of Species

As important as Fisher, Wright, and Haldane were, "there is complete agreement among the participants of the evolutionary synthesis as well as among historians that it was [Dobzhansky's *Genetics and the Origin of Species* in 1937] that heralded the beginning of the synthesis, and in fact was more responsible for it than any other."[13] This was certainly so for Simpson, who has remarked: "My own thinking along theoretical lines was nevertheless mostly along lines of historiography and organismal adaptation, in fossil and recent organisms, until the first edition of Dobzhansky's [book]. That book profoundly changed my whole outlook and started me thinking more definitely along the lines of an explanatory (causal) synthesis and less exclusively along lines more nearly traditional in paleontology."[14]

Theodosius Dobzhansky (1900–1975) began his scientific career at the University of Kiev in the Soviet Union, working on the taxonomy of lady beetles, and soon turned to experimental genetics with *Drosophila* in 1923.[15] Shortly after, in 1927, he came to the United States on a postdoctoral fellowship to work with Thomas Hunt Morgan, one of the leading American geneticists, at Columbia University. Whereas in the United States the gap in mutual understanding and respect between the laboratory geneticists and the field naturalists was great, the tradition in the Soviet Union was that the "geneticists seemed to be much more of the naturalist type than pure experimentalists."[16] It is consequently not surprising that Dobzhansky, a product of this latter tradition, "successfully integrated the naturalist's profound understanding of evolutionary problems with the knowledge he had acquired in the preceding dozen years as an experimental geneticist. Truly he was the first to build a solid bridge from the camp of the experimentalists to that of the naturalists."[17]

In his preface, Dobzhansky asserted that his book is "a discussion of the mechanism of species formation in terms of the known facts and theories of genetics. Some writers have contended that evolution involves more than species formation, that micro- and macro-evolutionary changes may be distinguished. This may or may not be true; such a duality of the evolutionary process is by no means established. In any case, a geneticist has no choice but to confine himself to the micro-evolutionary phenomena that lie within reach of his method, and see how much of evolution in general can be adequately understood on this basis."[18] After this initial framing of the question, Dobzhansky soon after averred, in the first chapter, that, "Since evolution is a change in the genetic composition of populations, the mechanisms of evolution constitute problems of population genetics. . . . Experience seems to show, however, that there is no way toward an understanding of the mechanisms of macro-evolutionary processes observable within the span of human lifetime and often controlled by man's will. For this reason we are compelled . . . to put a sign of equality between the mechanisms of macro- and micro-evolution, and, proceeding on this assumption, to push our investigations as far ahead as this working hypothesis will permit."[19] As Stephen Gould noted in his introduction to the 1982 reprint of the first edition of this seminal work, Dobzhansky was making a strong methodological claim that experimental genetics could provide the mechanisms of evolutionary processes at all levels of evolutionary phenomena, from discrete interbreeding populations within species all the way up to the major hierarchical categories of animals and plants.[20] This claim, of course, was the foundation of the modern evolutionary synthesis, and as a chief participant in that synthesis, Simpson also unequivocally subscribed to it throughout all his own writings.

According to Dobzhansky, the genetic mechanisms for evolutionary diversification operate on three levels: "mutations and chromosomal changes . . . [that] supply the raw materials; influences of selection, migration, and geographic isolation then mold the genetic structure of populations into new shapes. [It is here that the] impact of the environment produces historical changes in the living population; [then the] fixation of the diversity already attained . . . [by isolating] mechanisms encountered in nature . . . the origin and functioning of [which] constitute one of the most important problems in the genetics of populations."[21]

Subsequent chapters lucidly and cogently elaborated on each of the three levels of genetic phenomena by not only citing relevant experimental results but also, crucially, by showing how observations of many different sorts of organisms in field studies coincide with and corroborate the experimental data and conclusions. As mentioned earlier, it was this constant interplay between controlled laboratory experiment and the study of wild populations of diverse animals and plants that made Dobzhansky's treatise so convincing and so influential with his contemporaries. Moreover, in the course of his discussion of the role and importance of natural selection, Dobzhansky also wove into his overall argument the mathematical results of the theoretical studies of Fisher, Wright, and Haldane and their conclusions about the relative importance of population size and the pressures of mutation, natural selection, and migration.[22]

So with one stroke Dobzhansky integrated laboratory experiments, field observations, and mathematical models to provide a unified, coherent, and consistent argument as to how the genetics of the populations of closely related individuals can yield the tremendous diversity of life seen all around us — today as well as in the geologic past. Simpson's second chapter of *Tempo and Mode* ("Determinants of Evolution") explores these same issues and ideas of Dobzhansky, not by mere rote repetition but by Simpson's own recasting and elaboration from the point of view of the historical evolutionist treating paleontologic data.

Finally, the writings of another geneticist, Richard Goldschmidt, stimulated Simpson in a contrary way that led to his assertion that the fossil record was consistent with the new genetics. Goldschmidt had argued that only large-scale genetic changes — systemic changes or "macromutations" — due to major chromosomal rearrangements could be responsible for the appearance of new species and other higher taxa.[23] Small-scale mutations at the level of individual genes ("micromutations") were only of sufficient magnitude to account for intraspecific variation that leads to local adaptation in varying habitats. By extension, Goldschmidt could then claim that long-term evolutionary history, as recorded by fossils, was essentially the result of macromutational saltations that produced a long, temporal sequence of "hopeful monsters" of the sort whereby the first bird hatched, quite literally, from a reptile's egg.[24] For Simpson, Goldschmidt "carried to an extreme the already increasingly dubious view that evolution is mainly a matter of radical, random mutational modifications of the whole genetic system."[25] Obviously, Simpson could not subscribe to such a view, one that was quite opposite to the other genetic conclusions that he found so persuasive.

Simpson also objected to Goldschmidt's enthusiastic acceptance of orthogenesis, which argued for "evolution in a single, straight direction, . . . a fact of which the paleontologists have assembled innumerable examples. Indeed, there can be no doubt that in frequent cases evolution is of the orthogenetic type. (Data are found in all technical books on evolution . . . especially in the books written by paleontologists.)" Yet Goldschmidt did not accept either the "Lamarckian interpretation" or the "mystical one . . . an existing urge toward improvement, or a similar transcendental principle." Instead, "the explanation preferred by geneticists is the selection of mutants which deviate in a definite direction."[26] In the particular view of Goldschmidt, selection was thought to occur in the earlier ontogenetic stages of large-scale genetic novelties rather than in small-scale mutations affecting the adult condition.[27] Hence, for Simpson, Goldschmidt erred doubly — both in his genetics and in his application of them to the interpretation of the fossil record.[28]

The Contemporary Paleontology

Besides considering Simpson's own explicit acknowledgment of the factors that led up to the writing of *Tempo and Mode* — the two "door-opener" articles and his reading of the new genetics — it is necessary to summarize briefly the contemporary state of paleontological theory in the mid-1930s. This is perhaps most economically done by reviewing the writings of two leading paleontologists of the time: Henry Fairfield Osborn and Richard Swann Lull, who had published significant works of their own prior to *Tempo and Mode*, and who, as close associates of Simpson's, had undoubtedly influenced him, if only as foils for his own developing, original ideas.[29]

H. F. Osborn was founder and chairman of the department of vertebrate paleontology at the American Museum of Natural History, where he also served as museum president for some twenty-five years.[30] Osborn published a series of papers on "The Origin of Species as Revealed by Vertebrate Paleontology."[31] In these and two subsequent articles[32] that elaborated on his evolutionary theories (as inferred from his prodigious monographs on titanotheres and proboscideans), Osborn enunciated the major principles and conclusions he had derived from a lifetime study of fossils. Written in the last decade of a long, productive, and distinguished career, these articles grew out of major addresses to the scientific elite of his era: the National Academy of Sciences, the British Association for the Advancement of Science, the dedication of the new Peabody Museum of Natural History at Yale,[33] and the prestigious Sedgwick Memorial Lecture to the American Association for the Advancement of Science. In these essays, Osborn claimed that there were "three outstanding principles discovered [by himself] in paleontology. . . . First, the unbroken continuity of speciation, which becomes absolute as the gaps are filled by discovery. Second, the constitutional pre-disposition to speciate in certain predetermined directions which must be inherent in the germ-plasm of ancestral forms. Third, that these constitutional predispositions are not released except through adaptive reaction to new conditions of life; they are not, therefore, of the nature of inherent perfecting tendency, but rather, of the nature of *potentiality to appear when the need for them arises.*"[34]

Osborn subsumed these principles under the phenomenon he called "rectigradation," whereby "new specific characters . . . pass continuously from the most rudimentary and inefficient into the most efficient and highly developed stages . . . [as] confirmed in four great orders of hoofed mammals, namely the horses, the rhinoceroses, the titanotheres, and the proboscideans."[35] Subsequently, Osborn refined and renamed rectigradation as "aristogenesis . . . a creative process from the germ plasm of entirely new germinal biomechanisms . . . [that] is continuous, gradual, direct, definite in the direction of future adaptation."[36]

Osborn differentiated aristogenetic evolutionary trends from another phenomenon he named "alloiometry" — that is, "new modifications [measurable and quantifiable] in form and proportion of more or less aristogenic characters . . . [that] are subject to the interaction of the four forces of (a) physical environment, (b) living environment, (c) influences of habit, (d) predispositions and potentialities of heredity" (209).

Hence, aristogenesis is responsible for the "gradual . . . origin of new and distinct adaptations," whereas "alloiometry" is the "modification of existing adaptations as seen in changes of proportion and function" (234). Osborn explicitly denied that aristogenesis was in any way similar to vitalism or entelechy, although "what evokes an aristogene from the germ plasm is as mysterious to us as what evokes a horn rudiment in the skull of titanothere" (227). He denied that natural selection had anything to do with aristogenesis, for in the case of proboscidean evolution there is no evidence of competition between species, as well as because the rates of evolution were so rapid in these slow-breeding animals, as contrasted with fast-breeding rodents that show far less evolution during the same time interval.

Osborn cited another reason why natural selection could not be involved with aristogenesis. "What evidence is there for the initial but all essential assumption that, for example, a tiny adaptive cusp is a factor in survival, while its tiny inadaptive fellow is a factor in extinction? Not to mention the succeeding assumptions which overwhelm us *when we seek to derive definite adaptations from indefinite variations*." Osborn also denied mutation any role in evolution, for to him the randomness and discontinuity of mutation were not consistent with his observation of "continuous, gradual, direct, definite" change in the "direction of future adaptation" (230; emphasis in the original). Thus Osborn concluded that his three principles are "totally at variance with the working hypotheses of Darwin's variational natural selection or of Lamarck's inheritance of acquired characters." Moreover, the "recent authorities (Haldane, Morgan, and Huxley) offer no new hypotheses or explanations, only variations of the . . . old ones."[37]

R. S. Lull, initially trained as an entomologist, returned to graduate school at Columbia University, where he received his Ph.D. in vertebrate paleontology under H. F. Osborn. A few years later he joined the faculty of Yale University and also became associate curator at the Peabody Museum. Lull published a widely used textbook, *Organic Evolution*, which he claimed was the outcome of years of teaching Yale students.[38] Simpson remarked years later that Lull, in his course and in his textbook, discussed the various conflicting theories of evolution without declaring which one he favored.[39]

Although *Organic Evolution* predated the modern synthesis, Lull did attempt to integrate contemporary biology with current paleontological understanding. Besides devoting five chapters to genetic variation and mutation, heredity, and artificial, natural, and sexual selection (as well as one chapter to Lamarckian inheritance, to which he gave some begrudging credence, and another chapter to orthogenesis), in the part entitled "The Mechanism of Evolution," Lull also referred constantly throughout the book to principles and phenomena seen in living organisms, even in chapters ostensibly reserved for the exclusive discussion of fossil groups. In whatever other way the book may strike the modern reader as obsolete, it possessed in this one respect a remarkably fresh and up-to-date paleobiological perspective.

What, then, were Lull's views on contemporary paleontologic theory and understanding? Early in the book, Lull stated flatly that although "[s]ince Darwin's day, Evolution has been more and more generally accepted . . . [w]e are not so sure, however, as to the *modus operandi*."[40] While allowing that variation, heredity, and isolation are fundamental factors in evolution, Lull was ambivalent about how important natural selection might be in the formation of new species, and cited the contrary arguments of the mutationists, who "believe that natural selection therefore has nothing to do with species-forming, but only in . . . keeping . . . a species true to type when once it has been formed" (83–85). Later on, Lull cited other objections to natural selection, including the beginnings or rudiments of subsequent highly developed, well-adapted structures, the overspecialization of the Irish deer, and the degeneracy of certain organs (129). One suspects that Lull's own personal objection to the efficacy of natural selection in directing evolutionary change was covertly expressed when he stated, "There are those who object to natural selection because it is essentially a materialistic doctrine, depending as it does purely on the laws of chance" (130).[41]

Lull's "Part II — The Mechanisms of Evolution" concludes with a chapter entitled "Orthogenesis and Kinetogenesis," which deals with two hypotheses "opposed or supplemental to the Darwinian factor of natural selection. . . . [O]f these hypotheses that of orthogenesis stands first, although it is not an explanation of evolution but a statement of possible fact." According to Lull, "Orthogenesis . . . is the assumption that variations and hence evolutionary change occur along certain definite lines impelled by laws of which we know not the cause" (151). Thus Lull vacillates in the space of half a page of text, from considering orthogenesis as an evolutionary mechanism to a hypothesis to a possible fact to an assumption.

Orthogenesis is invoked for those phenomena "for which natural selection seemingly cannot account," such as "the beginnings of advantageous modifications, . . . [and] elimination of a character formerly useful but which, in altered circumstances, has become a . . . menace to survival" (151–52). More positive evidence for orthogenesis includes parallel evolution of South American litopterns and Northern Hemisphere equids, and "excessive structures" like enlarged horns, antlers, and tusks (153–54). These objections echo those of Lull's mentor, H. F. Osborn, and in fact Lull credited Osborn's titanothere studies as leading to such generalizations.

Lull also invoked a quasi-Lamarckian mechanism that he attributed to E. D. Cope, a mechanism called kinetogenesis, namely "that animal structures have been produced directly or indirectly, by animal movements" (155). The inferred mechanism would result from the response of the "great plasticity of bone" to the "pressure of soft parts. . . . There is, in all of this, abundant evidence of the effects of use and disuse," presumably with the hard parts — as seen by the paleontologist — changing in accord with newly evolved soft-part activity and behavior. "The stimulus is stress due to disharmony between an organism and its environment, and . . . kinetogenesis is the result of the effort on the part of the organism to overcome this lack of harmony" (164).

Lull's hesitation as to the efficacy of natural selection occurs again, although more obliquely, in later chapters of *Organic Evolution*. For example, in the chapter entitled "Recapitulation, Racial Old Age, and Extinction," he explains various paleontological phenomena such as phyletic size increase, elaborate spines and antlers, "physical degeneracy" such as tooth loss, and nearly universal extinction all as symptoms of "racial senility" (179ff.). While aware of the role of changing physical and biologic environments in bringing about extinction, Lull also believes it due to "internal causes" — that he does not identify — which generate "inadaptive structures" (191ff.).

The next section of *Organic Evolution* is built around the theme of adaptive radiation, from H. F. Osborn after W. D. Matthew, whereby a primitive stock of organisms diversifies very broadly as they evolve into a wide variety of habitats in some new, previously isolated or unoccupied region. Lull remarks that such adaptive radiation will lead in four general directions for terrestrial vertebrates: running, burrowing, climbing, and flying (246–48). Lull proceeds to describe vertebrate history in terms of these broad ways of life (Simpson's "adaptive zones" in *Tempo and Mode*). Thus, rather than describing and discussing vertebrate fossils taxon-by-taxon, chapter-by-chapter, Lull emphasized instead these individual adaptive lines and showed how the morphology of diverse groups, both living and fossil, operates in each. The treatment was functional and ecological, and shifted constantly back and forth between extant and extinct examples.

The remainder of Lull's text then applied the ideas and observations regarding adaptive radiation to selected groups of organisms, chapter by chapter, including cephalopods, insects, primitive vertebrates, dinosaurs, archaic and modern carnivorous mammals, cetaceans, proboscideans, horses and camels, South American mammals, and early humans. Lull thus developed a roughly chronological overview of animal life-history, emphasizing for each selected group their "place in nature," form and function, habitat and habit, particular specializations, and ontogenetic and phylogenetic history. His treatment was anything but dry-bones paleontology.

Lull concluded *Organic Evolution* with a chapter on "The Pulse of Life," wherein he argued that the "laws of chance may not be invoked by way of explanation. The geologic changes and pulse of life stand to each other in relation of cause and effect. This statement does not, however, imply the acceptance of the Lamarckian factor any more than that of natural selection. . . . [T]he fundamental principle remains that changing environmental conditions stimulate the sluggish evolutionary stream to quickened movement. . . . We have found the immediate influence to be one of climate. . . . Back of these climatic changes lies . . . earth shrinkage . . . which produces mountain ranges and enlarges the lands. Thus it will be seen that the most momentous changes, so far as life is concerned, may have, geologically speaking, a very simple basic cause" (693).

Although throughout this almost 700-page treatise Lull paid his respects to what we would today call microevolutionary phenomena, one cannot help but come away with the definite impression that, for Lull, evolution is the response, in some hitherto unexplained way, of organisms, more or less passive, to active, extrinsic, and global environmental change. What determines evolution for Lull — and for that matter for H. F. Osborn — is *not* "variability, rate of mutation, character of mutations, length of generations, size of population, and natural selection," as they will for Simpson in *Tempo and Mode*.

Before concluding this selective review of the state of paleontological theory prior to the publication of *Tempo and Mode*,[42] the influence of Otto Schindewolf ought to be mentioned. Simpson has credited this German invertebrate paleontologist as being "the first person — the first paleontologist — to perceive that there should somehow be an attempt at synthesis between genetics and paleontology."[43] However, Simpson did not accept Schindewolf's synthesis, which "adopted the . . . entirely unacceptable view that evolution is essentially random and involves mainly mutations effecting changes in early ontogeny."[44] Schindewolf's book, *Paläontologie, Entwicklungslehre, und Genetik*, published in 1936, in fact took quite the opposite view to that which Simpson was to adopt in *Tempo and Mode*, especially Schindewolf's "view that macroevolution had a different basis from microevolution." Instead, Schindewolf stimulated Simpson to study "the fossil record particularly with a view to determining which view is right."[45] "The fact that Schindewolf's views did not agree with my own interpretations of the fossil record or with what I had so far gathered about evolutionary genetics made me resolve to go further into these matters."[46] Thus, Schindewolf the paleontologist, like Goldschmidt the geneticist, helped focus for Simpson the degree to which majority opinion in contemporary genetics could explain his own reading of the fossil record.[47]

Tempo and Mode in Evolution

Simpson noted in the preface to *Tempo and Mode* that he began the manuscript in the spring of 1938. Looking back now, one can perhaps see why it was that Simpson undertook such an important task at that particular time. Clearly the time was right in that, as just summarized, genetic theory and observations were making significant new claims in such a way that an inevitable tension was developing between what was becoming known about the sources, transmission, and selective value of genetic novelties in living populations of animals and plants versus what paleontologists claimed to be the driving mechanisms for the evolution and diversification of organisms in the fossil record.

In addition, Simpson himself was an experienced and prolific researcher who had spent some dozen years, previously, in thinking and writing about intraspecific and interspecific variation within fossil species (mostly mammalian), as well as in tracing out their morphological transformations over geologic time. However, rather than adopting a strictly inductive process whereby his previous observations and conclusions about fossil vertebrates generated, on their own merits, the ideas and interpretations found in *Tempo and Mode*, Simpson instead thoroughly assimilated the results of the new genetics and then applied them to his own reading of the fossil record. It is crucial to emphasize this point: namely, that Simpson took a body of evidence and theory from a completely different field — microevolutionary genetics — and applied it in a totally original and skillful way to the macroevolution of fossils.[48]

Simpson wrote and completed the manuscript for *Tempo and Mode* over the next four years. In the late fall of 1942, he entered military service for a period of almost two years. During that hiatus, his wife Anne

Roe and his American Museum colleague Edwin Colbert saw the manuscript through to final publication in 1944. Simpson by no means spent all of those four years working on this manuscript. His bibliography for the period between 1939 and 1944 shows eighty-three separate entries. These include the first edition of *Quantitative Zoology*, coauthored with Anne Roe; a book-length ethnographic study of the primitive Kamarakoto Indians, whom he observed while doing fieldwork in Venezuela in the spring of 1939; a monograph on the beginnings of North American vertebrate paleontology, for which he was later given the Lewis Prize of the American Philosophical Society; several short contributions to Osborn's second volume of his study of proboscideans; and some two dozen longer journal articles — dealing with, among other things, the development of South American marsupials; the earliest primates; the role of the individual in evolution; the function of saberlike canines in carnivores; the description of a new Chilean Miocene sloth; and methods in taxonomic nomenclature.[49] Also during this period Simpson was completing his classic monograph, "The Principles of Classification and the Classification of Mammals" (1945), which was submitted for publication shortly after the completion of *Tempo and Mode* and just prior to his departure for military duty.

Prelude to the Argument

In the introduction, Simpson laid out the methodological assumptions that he was making and briefly outlined the nature of the book.[50] He began by asserting that his would be a "synthesis" between paleontology and genetics (xv). In particular, he proposed to focus on "phenogenetics[51] . . . how hereditary characters achieve their material expression" in living organisms and on "population genetics . . . because the paleontologist is learning to think in terms of populations . . . and . . . on the meaning of changes in populations" (xvi). Immediately upon bringing up the whole question of paleontology vis-à-vis genetics, Simpson — like Dobzhansky (1937) as previously noted — made a plea for the microevolutionary approach: "One cannot directly study heredity in fossils, but one can assume that some, if not all, of its mechanisms were the same as those revealed by recent organisms in the laboratory" (xvii). Given the great temporal sweep of the fossil record as well as its complex natural environments at any one interval of time, the paleontologist can "learn whether the principles determined in the laboratory are indeed valid in the larger field [of time and space], whether additional principles must be invoked and, if so, what they are" (xvii).

In two areas of inquiry the paleontologist has a special advantage in contributing to these issues, continued Simpson. First, because paleontologists observe significant morphological change over time, they can discuss rates of (morphologic) evolution — if and how they might vary — or what Simpson would thereafter refer to as the "tempo" of evolution. Second, paleontologists can also discuss "how populations . . . differentiated, . . . how they passed from one way of living to another or failed to do so, to examine the figurative outline of the stream of life and the circumstances surrounding each characteristic element in that pattern" — or, in short, the "mode" of evolution (xviii).

There is another important innovation that Simpson incorporated in the latter part of that last quotation. As we will see, throughout *Tempo and Mode* Simpson would continually relate rates and patterns of evolution to the environmental circumstances in which organisms found themselves. Thus, rather than relying on internal forces driving organisms toward "perfection" or "racial senescence," as did so many of his paleontological contemporaries, Simpson would instead attribute evolutionary changes in organisms to ecological opportunities and demands that are external to the organisms. The changes result from the adaptive interplay of populations of varying phenotypes and constantly shifting environments, an interplay mediated for the most part by natural selection (xviii).

Simpson concluded the introduction with an explicit warning to the reader who might expect the usual trotting out of descriptive morphology, labored taxonomic decisions, and enumeration of one or more empirical rules, all as usually found in most standard paleontological treatises. On the contrary, fossils as fossils were to be discussed by Simpson only as examples of the "how" and "why" of evolutionary processes and not as centerpieces for a mere description of the "what" of evolution (xviii). In the closing sentence, Simpson could not resist pointing out that his study "suggests new ways of looking at facts and new sorts of fact to look for" (xviii).

Rates of Evolution

Simpson opened his treatise with a discussion of the phenomenon most apparent to the student of fossils, namely the transformation of hard-part morphology over geologic time. He led off with the intriguing question, "How fast, as a matter of fact, do animals evolve in nature?" (3). By beginning with the most objective, least controversial contribution that paleontology can make to the subject of organic evolution — that organisms as represented by their hard parts (bones, teeth, shells, and so on) display temporal changes in morphology — Simpson executed a brilliant tactical maneuver. He immediately grabbed the attention and interest of the nonpaleontological evolutionist by suggesting that fossils, after all, *do* have something important to contribute to evolutionary theory. He was also appealing to paleontologists, who at least agree on this single point of morphological change over time, however else they might be divided into schools, factions, and research programs. And if we are to consider change over time, we are, by definition, considering rates of evolution.

However simple it was to pose the question, Simpson quickly observed that the actual measurements of evolutionary rates can only be approximated, because neither morphological change nor geologic time can be recorded in precise absolute terms. In fact, morphological change itself is an indirect reflection, presumably, of underlying genetic change — of real "evolution" — that is ultimately inaccessible. But we can monitor morphological change — and hence genetic change — by observing the transformation of morphology, whether of a single character or a combination of characters, up to and including the whole organism. These changes may be followed, temporally, within a single ancestor-descendant line ("phyletic evolutionary rate") or may be averaged across a wider set of related lines ("group evolutionary rate"). The scale used for measuring these changes can be geologic time, either absolutely determined by radiometric methods (which in the late 1930s were still very rough) or relatively determined by the physical stratigraphy of the fossil-enclosing rock strata (which, it is to be assumed, is time-related). Morphological rate changes can also be compared with each other to yield relative intergroup rates.

Turning then to data from recent and fossil horses, by means of nontraditional (for paleontologists at least) logarithmic graphs, allometric plots, and basic statistics, Simpson arrived at his first set of conclusions, or "basic theorems, concerning rates of evolution: (1) The rate of evolution of one character may be a function of another character and not genetically separable even though the rates are not equal. (2) The rate of evolution of any character or combination of characters may change markedly at any time in phyletic evolution, even though the direction of evolution remains the same. (3) The rates of evolution of two or more characters within a single phylum [or clade] may change independently. (4) Two phyla of common ancestry may become differentiated by differences in rates of evolution of different characters, without any marked qualitative differences or differences in direction of evolution" (12).[52]

These conclusions thoroughly undermined and contradicted the contemporary opinions of some other paleontologists who claimed that a trend, once started, gained its own "momentum" irrespective of its survival value, and often led to "racial senescence" and extinction. Although reserving the explicit interpretation of his conclusions for later in the book — because Simpson wisely did not wish to muddy demonstrable fact with arguable interpretation — these conclusions required some factors external to the organisms (like varying selection intensity and direction, varying environmental opportunity, and so on) instead of internal driving mechanisms (such as "entelechy," "vitalism," or a "tendency toward perfection"). Obviously, before entertaining his ideas about evolutionary modes, Simpson had first to establish what, in fact, had occurred in evolutionary history.

Simpson then went on to consider evolutionary rates from the taxonomic point of view by measuring morphological differentiation in terms of taxonomic proliferation. After outlining the difficulties and pitfalls of the methodology, Simpson argued that the number of genera per million years will give reasonable estimates of evolutionary rates. From the examples he discussed, he concluded that horses and chalicotheres (extinct, clawed ungulates related to horses) evolved at about the same rate, and that both groups of these perissodactyls (a mammalian order of odd-toed, hoofed herbivores that includes modern horses, tapirs, and rhinoceroses) evolved much faster, perhaps three to four times faster, than early ammonite cephalopods had evolved.

Simpson closed the chapter with a discussion of how the distribution of the times of first and last appearances of fossil genera, together with their implied temporal duration, could also indirectly measure the rate of evolution. From such data he introduced the notion of "survivorship," as displayed by survivorship curves from which one can determine, for example, mean survivorship, i.e., the mean duration of a genus within a larger taxon. Using data from marine bivalves and land-mammal carnivores, he calculated mean generic survivorships of 78 and 6.5 million years, respectively, and therefore stated, "It is safe to say that carnivores have evolved, on an average, some ten times as fast as [bivalves]" (25).[53]

The breakthrough Simpson made in this initial chapter was to establish quantitatively that distinct and different rates of evolution exist among a wide variety of fossil groups, a variety of rates that calls for explanation. By this chapter alone he made an important contribution to the modern evolutionary synthesis by introducing data and ideas drawn *not* from laboratory experiments or from field studies with living animals and plants, but rather from an altogether different data-set from ancient and largely extinct organisms that span eons of the earth's history. Moreover, this data-set, and Simpson's conclusions about it as summarized here, are still valid today, even though we can refine the data and the conclusions on the basis of what we have learned in the half century since he wrote the chapter. It is clear that Simpson's contribution to the modern synthesis, from the evidence of this one chapter, certainly is more than a "consistency argument" whereby it has been claimed that Simpson merely showed that the data of paleontology were consistent with the new genetics.

Determinants of Evolution

Having established that the fossil record shows varying evolutionary rates of morphological transformation, Simpson in this the longest chapter of *Tempo and Mode* (comprising almost one-third of the book) discussed what he judged to be the most important factors in effective evolutionary change. He recast and elaborated on what the new genetics had to say on the subject and, significantly, introduced paleontological evidence to buttress the conclusions. It is important to emphasize that Simpson by no means swallowed uncritically what the geneticist authorities had to say. On the contrary, he fully evaluated their interpretations, modified them where he thought necessary — often in light of fossil data — and without hesitation offered his own insight and understanding.[54] His tone, however, was neither dogmatic nor self-important. For example, when referring to his analysis of the Jurassic ammonite *Kosmoceras*, he said, "In terms of evolutionary processes the most probable (although not the only possible) interpretation of these facts appears to be as follows . . ." (33).

The factors Simpson believed to be most important for the "rate and pattern of evolution are variability, rate of mutation, character of mutations, length of generations, size of populations, and natural selection" (30). Each factor was then discussed in detail, usually on a scale more familiar to paleontologists, by referring to specific fossil examples. Simpson ended the chapter with an analysis of horse evolution framed in the "adaptive landscape" metaphor of Wright (1932), with adaptive peaks and valleys that integrate natural selection, morphological variation, and environmental opportunity.

VARIABILITY

Simpson began his discussion of variability and variation by defining them as "the potentiality and the reality, respectively, of differences between individuals within a possibly or actively interbreeding population at any one time" (32). This definition shows that Simpson was approaching species as variable populations rather than as types, and is one of the hallmarks of the synthetic approach. Like Darwin a century before him, Simpson recognized the fundamental importance of population variation as the raw material for evolution. By citing the example of the notoungulate *Henricosbornia lophiodonta*, which he had previously discussed in one of the "door-opener papers"[55] that preceded *Tempo and Mode*, Simpson noted that it "showed extraordinary structural variability [in its dentition]. Some of the differences between individuals within a single group of these primitive animals are analogous to and some homologous with differences segregated in other and in allied more advanced types and then characterizing distinct species, genera, or even families" (32).

Another instance in which some intragroup variation is segregated, isolated, and subsequently morphologically differentiated into a new species is provided by the Jurassic ammonite *Kosmoceras*. Simpson interpreted the published data on this fossil in terms similar to Mayr's "founder principle" (although he did not actually use that term) whereby only a part of the initial variation is sampled by a much smaller, isolated population that, in turn, eventually becomes a new species and then may display a range of variability comparable to that of the older, parental stock.

In both these fossil examples, Simpson saw that "segregation or selection of intragroup variability can give rise to new groups [and that] . . . [t]his process is important and typical in speciation and lower levels of differentiation" (41), although he later acknowledged that this factor alone "cannot suffice for the maintenance of high or moderate rates of evolution over considerable periods of time or for the rise of higher taxonomic categories" (93).

The degree to which variation per se drives evolution was tested by Simpson by calculating Pearson coefficients of variation for Tertiary horse genera, for various extinct mammalian orders, and for species of condylarths (primitive early Tertiary ungulates). He found neither that evolution depletes variation nor that more variable groups evolve more rapidly than less variable groups. These findings — again, derived from fossils — led Simpson to conclude that "concentration [of research] on low taxonomic levels, such as are alone available in experimentation, has perhaps led some geneticists to overemphasize the role of variability and to assume a too constant and direct correlation of this with the rate of evolution" (37). Again, this conclusion hardly sounds like paleontology showing itself as merely consistent with genetics.

Simpson did agree with the general notion that degenerate characters that have become functionless will display much wider variability than those that are not (another of Darwin's observations). He noted that in an extinct species of Paleocene multituberculate mammals the functionless upper premolars show high coefficients of variation (about three times the norm), whereas the functioning teeth in the same jaws have about the variability seen in other mammals (39). Finally, Simpson suggested that slowly evolving taxa that have changed little over geologic time (e.g., crocodiles, opossums, armadillos, and tapirs) are no less variable than more rapidly evolved, allied lines such as lizards, kangaroos, sloths, and horses (39).

MUTATION

Simpson emphasized the points made by Dobzhansky (1937) with respect to the frequency, size, phenotypic expression, and viability of mutation — both changes in single genes and in chromosomal arrangement — although rephrasing the argument in his own words. (He also cited other contemporary geneticists, e.g., Timoféeff-Ressovsky, Sturtevant, and Waddington.) Although mutations are necessary to supply heritable variations as the raw materials for sustained evolution, "it does not follow that rates of evolution are proportional to rates of mutation" (94). Referring to the fossil record of Cenozoic horses, Simpson made a series of simple and reasonable calculations using the conclusions of the new genetics that a given equine morphological character (e.g., the length of the outer crest of the cheek teeth) could be readily accomplished by some three hundred suitable mutations between early Eocene *Hyracotherium* and modern *Equus*. "If the change in any one character from [the former to the latter] is divided into 300 steps, these . . . are almost imperceptibly small and are almost incomparably less than the amount of intragroup variation at any one time" (46).

After thus arguing that rate of mutation by itself does not drive evolution and that a moderate or even a low rate can provide sufficient variation for the long-term effects seen by paleontologists, Simpson then addressed the issue of "whether the rise of new characters is continuous or discontinuous" (49). Simpson here paraphrased his mentor, W. D. Matthew, who made a "more important distinction [as to] whether differences between successive populations (successive intergroup variation) transcend the limits of differences within a single population (intra-group variation), however broad or narrow those limits may be. . . . Because of the nature of the [paleontological] record, it is quite impossible to prove . . . that saltation [or large, single, discontinuous phenotypic change] never occurs. There is, however, abundant and incontrovertible paleontological proof that saltation does not always occur, i.e., that continuity (with Matthew's meaning) or gradual intergradation commonly occurs between what are certainly good species and genera" (50, 58). Given the context of the foregoing, Simpson was claiming ancestor-descendant intergradation in particular characters even though, when considered in toto, the species within a phyletic series will be quite distinct from one another.

Simpson was well aware of the subtleties of mutation and its phenotypic expression. For example, he cited studies where separate species are nearly indistinguishable morphologically, yet are known to have distinctly different chromosomal arrangements — essentially what Mayr subsequently called "sibling species."[56] Simpson also avoided the oversimplification of postulating a necessary one-to-one relation between mutation and phenotypic expression. "A more likely hypothesis is that the mutation . . . introduced a new tendency in a complex developmental field and that the phenotypic expression fluctuated under the influence of other genes related to the same field and of extrinsic factors influencing development" (61).

LENGTH OF GENERATIONS

Simpson noted that given that "[i]n normal evolution the discrete steps occur between generations . . . [a]ll the temporal processes of evolution are more naturally measured in generations than in direct time units. The effect . . . should be an inverse relation between rate of evolution and length of generations" (62)[57] And yet, Simpson continued, this relation in fact appears not to hold when we examine the fossil record: short-generation opossums have evolved much more slowly than long-generation elephants, and the generally shorter-generation invertebrates evolved more slowly than the generally longer-generation vertebrates. (Recall above, for example, how Simpson demonstrated that bivalve molluscan genera have about ten times the "survivorship" of terrestrial carnivore genera.)

Although "no such correlation [between length of generation and evolutionary rate] is evident in the record and it is concluded that this factor [of generation length] is relatively ineffective" (95), Simpson did suggest that generation length may "have some influence on the rate of small-scale geographic and ecologic differentiation" of low taxonomic categories such as subspecies and species.

SIZE OF POPULATION

Simpson asserted that the size of a population is "one of the dominant factors in determining tempo and mode of evolution" (95). Citing the main results of Fisher (1930), Haldane (1932), and especially Wright (1931), Simpson recapitulated the interactions of migration, mutation, and selection within different-sized populations of interbreeding individuals. Simpson asserted that "these studies . . . are most pertinent . . . for it is they that have made apparent the essential role of population size as a determinant both of rates and patterns of evolution" (65–66). Using the notation of Wright (1931), Simpson showed the various possible outcomes as these parameters change in value.[58]

Simpson enumerated the chief conclusions regarding the evolutionary significance of varying population size as follows: "In summary, very large populations may differentiate rapidly, but their sustained evolution will be at moderate or slow rates and will be mainly adaptive. Populations of intermediate size provide the best conditions for sustained progressive and branching evolution, adaptive in its main lines, but accompanied by inadaptive fluctuations, especially in characters of little selective importance. Small populations will be virtually incapable of differentiation or branching and will often be dominated by random inadaptive trends and peculiarly liable to extinction, but will be capable of the most rapid evolution as long as this is not cut short by extinction" (70–71).[59]

Turning to fossils, Simpson noted that very few estimates had been made of their effective breeding population sizes. He reviewed two examples — the Permo-Triassic reptiles of South Africa, and horses and primitive primates of the early Tertiary of North America — and demonstrated how difficult and uncertain such hypothetical calculations are. He observed that most of "the principles and theories of evolution advanced by paleontologists have been based upon groups with long, continuous, and relatively abundant records, hence, on groups that probably had large breeding populations" (73). But the "recent work on relationship of population size to evolution shows that principles so based cannot reasonably be accepted as generalizations . . . [although] they are valid descriptions of evolutionary processes in particular groups at certain times, but as far as these facts can show, the processes may have been quite different or have had quite different results in the more numerous and in some respects more important groups that are poorly recorded as fossils (74).

Simpson then concluded this section with a strong call for using the results of paleontological studies in interpreting the experimental conclusions of the biologist: "Recognition of this limitation [of an incomplete fossil record of past life] does not make paleontological data less useful in the study of evolutionary theory. On the contrary, it permits better coordination of paleontologically and experimentally developed theories, which have often been radically discrepant. It gives a better basis both for the checking of neobiological observations against the fossil record and for the interpretation of that record in harmony with what is known of the genetics and other physiological processes of living animals" (74).

SELECTION

Simpson closed this chapter on the primary controls in evolution with his longest discussion of all on the role, intensity, and direction of natural selection. He first summarized the current evidence for and arguments against the three major presynthesis alternatives to selection in directing evolution: neo-Lamarckian inheritance; vitalism or inherent tendency; and preadaptation whereby newly evolved successful organisms arise randomly by mutation.[60] Referring back to earlier parts of the chapter, Simpson claimed that selection does not act directly upon random and spontaneous mutations but rather upon "structures developed under the influence of hereditary factors, but without a point correspondence with the latter" (78). Thus, Simpson asserted, owing to the existence of pleiotropic genes and polygenic characters, selection for or against one structure may result in the concomitant survival or elimination of another, simply by genetic association. Hence, not only can "inadaptive" structures survive, but structures with little or no initial adaptive value may get started in this way (78–79). Simpson viewed selection as a "truly creative factor in evolution . . . [because] it determines which among the millions of possible types of organisms will actually arise" (80).

Citing once again the work of Fisher, Haldane, and Wright, Simpson then noted "that surprisingly slight selective advantage may control the direction of evolution . . . thus answering the objection that many characters developed under the supposed influence of selection have selective advantages too slight to have any effect" (82). He readily acknowledged that the theoretical genetic models also indicate that there is a relationship between selection intensity (s) and size of interbreeding population (N), such that in an appropriately sized population — where the product sN is around 0.5 (Wright) or 1.0 (Haldane) — "random effects and mutation rates may become dominant" (82).

Selection not only has intensity but it also has direction, emphasized Simpson. And it is here that Simpson's discussion focused on his more original thesis, for he subdivided selection into centripetal (stabilizing), centrifugal (differentiating), and linear (directional). Again using the adaptive landscape metaphor of Wright (1932), Simpson reviewed horse evolution and found it to be "readily explicable in these terms" (90–91). "In the Eocene browsing and grazing represented for the Equidae two well separated adaptive peaks, but only the browsing peak was occupied" (see fig. 7.1). "As animals became larger [owing to an overall selective advantage of larger size in these hoofed herbivores] — throughout the Oligocene, especially — the browsing peak moved toward the grazing peak, because some of the secondary adaptations to large size [such as high-crowned molars] were incidentally in the direction of the grazing adaptation. . . . In the late Oligocene and early Miocene the two peaks were close enough . . . that some of the variant animals were on the saddle between the two peaks. These animals were relatively ill-adapted and subject to centrifugal selection in two directions. . . . A segment of the population broke away, structurally, under this selection pressure, climbed the grazing peak . . . and by the end of the Miocene occupied its summit. Subsequent com-

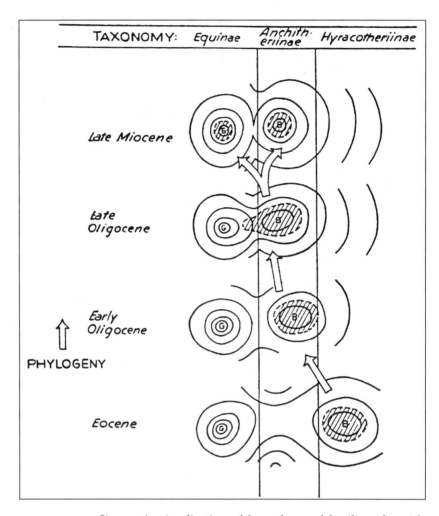

FIGURE 7.1 Simpson's visualization of how three subfamilies of equids evolved during the Tertiary Period from the hyracotherines, early browsers of the Eocene Epoch, into the grazing equines. Natural selection for overall larger size, and hence larger cheek teeth, permitted the browsing peak (B) of the conceptualized adaptive landscape to press more closely upon the grazing peak (G) in the Oligocene Epoch. Still larger variants of late Oligocene browsers found themselves in the inadaptive valleys between peaks. Further selection of these forms allowed them to occupy the adjacent grazing peak and thus give rise to the later, widespread grazing equids. (Simpson 1944b, *Tempo and Mode in Evolution*, fig. 13. Courtesy Anne Roe Simpson Trust)

petition on both sides from the two well-adapted groups caused the intermediate, inadaptive animals on the saddle to become extinct" (92–93).

With this chapter-ending example, Simpson integrated the factors he believed to be essential in determining evolution and gave particular importance to natural selection operating on phenotypic characters adaptively suited to environmental opportunities. Not only was this a useful way to conclude by bringing together all the elements of the chapter, but it also revealed for the first time Simpson's view of evolutionary processes over geologic time. The ideas and theories here were to expand and resonate throughout all his subsequent theoretical writings.

Micro-, Macro-, and Mega-Evolution

Having established that evolutionary rates determined from fossils do vary (ch. 1) and after reviewing the consensus of contemporary geneticists about evolutionary mechanisms (ch. 2), Simpson returned to a consideration of the fossil record and the morphological discontinuities seen among discrete taxa from species on up to phyla. He opened the chapter by defining microevolution, macroevolution, and mega-evolution in terms of the levels of discontinuity among populations of organisms, expressed taxonomically. Microevolution involves differentiation but no discontinuities within species, and it leads to geographic races and subspecies. Macroevolution involves differentiation and discontinuity within species and leads to new species and genera. Mega-evolution involves differentiation and still greater discontinuities at the higher taxonomic levels of families, orders, and classes.

Simpson then posed the question whether the processes responsible for these various levels of differentiation with partial or complete discontinuity are all due to the same evolutionary factors, merely operating at separate degrees of intensity, or whether there are other, entirely different factors at work at different levels of evolution. He noted that with respect to micro- versus macroevolution, "the great majority of geneticists and zoologists believe that the distinction is only in degree and combination" (97). Moreover, Simpson affirmed, geneticists and zoologists can only speak about micro- and macroevolution. For "[w]hat a paleontologist would call large-scale [or mega-]evolution . . . seems almost unnoted by the experimentalist" (98).

Paleontologists, of course, observe all levels of discontinuities among their fossils, for the rocks of the earth's surface do not contain the complete temporal and spatial, physical record of life's history and evolution. These discontinuities, said Simpson, are either minor (species and genera) or major (families, orders, and classes). Simpson called the minor gaps "casual" and attributed them to sampling problems inherent in the preservation and discovery of ancient life in the fossil state, and not to the result of fundamentally different biological factors operating in the differentiation of subspecies versus that of species and genera. According to Simpson, there are enough well-studied examples, such as the fossil record of horses, that show beyond reasonable doubt that the horizontal discontinuity between species, genera, and at least the next higher categories can arise by a process that is continuous vertically and that new types on these taxonomic levels often arise gradually at rates and in ways that are comparable to some sorts of subspecific differentiation and have greater results only because they have had long duration (99).

Simpson provided three additional examples of how sampling artifacts can yield a discontinuous fossil record of processes that differentiate populations gradually and continuously. The first referred back to the Jurassic ammonite *Kosmoceras* studied by Roland Brinkmann, who explained jumps in certain morphological characters through time as being the result of gaps in sedimentation and hence in the preservation of intermediate specimens.[61]

The second example of minor discontinuity within the fossil record comprises some fifty meters of unfossiliferous strata that separate the early Paleocene Puerco fauna from somewhat younger Paleocene Torrejon fossils. Fifteen genera from four mammalian orders in each of these formations have closely related or immediately ancestral and descendant species. "This is remarkably good evidence of the sort used to support the belief that genera and sharply distinct species regularly arise by saltation, but that hypothesis is not really necessary to explain the facts" (102), for the subsequently discovered Dragon fauna is "intermediate between the known Puerco and Torrejon faunas in age . . . [and] it includes members of eleven of the fifteen lines. . . . Nine of these are morphologically transitional between Puerco and Torrejon types" (102).

Simpson's third example again comes from the Tertiary horse fossil record. Here he shows that what was originally interpreted as an apparent saltational history of horse evolution in Europe was later explained as discontinuous sampling of essentially continuously evolving North American horses by episodic migration from the one continent to the other (103–104).

As concerns these minor gaps in the fossil record, Simpson argued "that there is no paleontological evidence that really tends to prove that there is any [discontinuity]. On these levels [of discontinuity] everything is consistent with the postulate that we are sampling what were once continuous sequences" (105).

Simpson next turned to what he called major systematic discontinuities of the record, where "essentially continuous transitional sequences are not merely rare, but virtually absent." He continued: "their absence is so nearly universal that it cannot, offhand, be imputed entirely to chance and does require some attempt at special explanation, as has been felt by most paleontologists" (105–106). To illustrate his point, Simpson referred once more to fossil examples with which he had had extensive firsthand experience, in this case, Mesozoic and early Cenozoic mammals. He noted that each of the thirty-two extinct and extant orders of mammals contains the "earliest and most primitive known members . . . [that] already have the basic ordinal characters, and in no case is an approximately continuous sequence from one order to another known. In most cases the break is so sharp and the gap so large that the origin of the order is speculative and much disputed" (106).[62]

Simpson claimed that these systematic gaps, which are seen not only among mammalian orders but also in other vertebrates and even invertebrates, share seven characteristics briefly summarized as follows (109–114):

1. The missing forms are inferred to be small animals, compared with their best-known contemporaries and with their descendants.
2. New groups invariably represent a major ecological adaptive change.
3. Whereas the ancestral group that gives rise to the new higher category is abundant and widespread in the fossil record, the new category itself is usually extremely rare.
4. The gaps are not always absolute, for they may be subdivided but not filled by isolated fossil discoveries, such as *Archaeopteryx* and *Archaeornis*, that lie between the fully perfected birds of the Tertiary and the Mesozoic reptiles.

5. There is always a considerable period of time that corresponds with the gap in morphology, taxonomy, and phylogeny.
6. The appearances of new groups with unrecorded origin-sequences frequently coincide with climaxes in the earth's history — crises of mountain-building and land emergence, although the coincidence is not as extensive as some contemporary historical geology textbooks might make it appear. In fact, thought Simpson, the coincidence is greatest between extinctions rather than origins of new groups.
7. Among higher categories the missing ecological links are not to be found at the base of some dichotomous branching point, but within a single group already ecologically distinct. For example, Simpson argued, proto-mammalian reptiles became geographically and ecologically subdivided into a large number of populations, most of which became extinct, but one or possibly more of which survived and subsequently made the transition to full mammalian physiology and morphology.

Simpson viewed the major, systematic gaps as the result of several factors, intrinsically crucial, in the origin of species that evolve rapidly from the parental stock and diverge widely in ecology and behavior, in structure and morphology, and thereby produce taxa with higher levels of discontinuity (families, orders, classes). These factors — small, geographically restricted populations testing new environments and habitats quite different from the ancestral one, with rapid selection and adaptation occurring — would necessarily further bias the fossil record, which even under the best of circumstances, as shown by the minor casual gaps, is a poor recorder of the evolutionary continuum.

Simpson thus summarized, it "seems probable that these sequences [of rapidly evolving organisms] are subject to some other limitations common to all of them and sufficient, together with the subsidiary factors already mentioned [those that produce the minor gaps], to make the chances of finding them extraordinarily small" (117). "In small populations undergoing pronounced shifts in environment and ecology, much higher rates of evolution are possible and much greater fluctuation in rate is probable. From their very nature, such groups do not leave good or continuous fossil records, and it is certainly unwarranted to conclude from this deficiency of record that great fluctuations in rate do not occur" (119). Hence, Simpson concluded, major gaps are real in the sense that they cannot be explained away by simple appeal to the inherent bias of fossil preservation and discovery. Yet we are not forced to accept the hypothesis of saltation, as Goldschmidt and Schindewolf argued, to explain these sharp biologic and taxonomic discontinuities. Rather, we can use the principles and concepts of the new genetics, as developed by the experimentalists, naturalists, and theoreticians, to interpret these major gaps of record. As shown in figure 7.2, Simpson summarized his argument graphically in such a way that the horizontal discontinuity among taxa arises by a process that is continuous vertically; and the smaller the populations and faster the evolutionary rate, the sharper the resulting discontinuity. Simpson thus made the direct connection between paleontology and genetics, and argued that each discipline can support and illuminate the other. "The paleontological picture of large ancestral populations splitting up into many small groups, which usually became extinct, but sometimes evolve rapidly into radically new types, is also the genetic picture of the situation most likely to produce such new types. These conditions would involve maximum selection, also a strong factor in rapidity of evolution after the more random preadaptive phase that probably initiated each of these major transitions" (121–22).[63] (See fig. 7.2.)

Simpson further argued that it is not unusually high rates of mutation that cause this rapid evolution, but new ecological opportunities, for "the groups affected are those whose environment is made unstable or radically altered by physical events . . . [that] set the stage for preadaptation, for intense selection, for necessary changes in adaptation or extinction, for attenuation and fragmentation of populations — in other words, for just those conditions that do lead to mega-evolution according to the theory here developed" (122).

Simpson emphasized that the "materials of evolution and the factors inducing and directing it are also believed to be the same at all levels [micro-, macro-, and mega-] and to differ in mega-evolution only in combination and intensity" (124). Thus, the transitional stages in each of these three levels of evolution start out as subspecies, but in the case of mega-evolution the process of subspeciation takes advantage of "a varied ecological terrain, . . . shifting instable environmental conditions, [and] ecological zones available to the newly arising types" (124). Simpson was to elaborate on the importance of environmental opportunity in "Organism and Environment" (ch. 6), which provided the basis for the concluding "Modes of Evolution" (ch. 7), wherein Simpson presented in comprehensive form his highly original and provocative

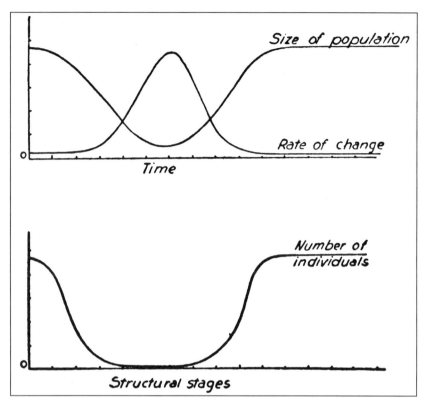

FIGURE 7.2 Simpson's graphic explanation of how varying population size and evolutionary rate (*upper*) can produce an apparent gap between ancestor-descendant linkages (*lower*), despite the fact that the evolutionary process is necessarily continuous from generation to generation. (Simpson 1944b, *Tempo and Mode in Evolution*, fig. 17. Courtesy Anne Roe Simpson Trust)

views on speciation, phyletic evolution, and quantum evolution. Before turning to these topics of evolutionary mode, however, he first had to complete his treatment of evolutionary tempo in the two preceding chapters.

Low-Rate and High-Rate Lines

In the fourth chapter Simpson returned to the subject of his first chapter, that of rates of evolution as observed among fossil taxa. Simpson's long opening chapter had carefully established that there are indeed varying evolutionary rates; in chapter 4 he focused, first, on the nature of the distribution of rates and, second, he interpreted the evolutionary factors that account for rates that are much higher or much lower than the usual ranges of rates. Although previous paleontologists had made the routine observation that some organisms have apparently evolved faster or slower than the general norm, Simpson asserted that his work was the first to examine whether there are, in fact, some rates that are distinctly slower or faster *and* what the causes and circumstances of such exceptional rates might be. Once again, we see Simpson approaching the paleontological record on its own terms — examining its elements and making generalizations about them — and then interpreting his results in the light of the evolutionary factors recognized as particularly important by modern geneticists. Surely this implies more originality and insight than is suggested by the assertion that Simpson merely showed the data of paleontology to be consistent with genetic theory, and that this was his only contribution to the modern evolutionary synthesis.

Simpson used the order of Primates as a simple illustration of the notion that he was to develop in more detail — namely, that within a larger taxon the evolutionary rates of its member taxa will vary. Thus, among extant primates the lemurs have, on average, evolved more slowly than most other primate taxa, whereas apes and humans have, on average, evolved more quickly. But this sort of variation between slow and fast evolutionary rates was of less interest to Simpson than those instances where there seemed to be exceptionally slow, or exceptionally fast, rates — rates that lie beyond the normal range of usual rate distribution. Groups that have had such exceptionally slow evolutionary rates include the "classic examples" of lingulid brachiopods, limulid arthropods, crocodiles, opossums, and coelacanth fish. Other groups, too, after a very rapid origin, have also evolved relatively little since their inception: e.g., turtles, bats, armadillos, rabbits, and whales.

Simpson recognized that the examples just given are qualitative and impressionistic in character and therefore limited in their potential to elucidate much of any significance about evolutionary mechanisms. To remedy this deficiency and to sharpen the focus of the ensuing discussion, he returned to the survivorship curves of his first chapter as a way of establishing a "standard of reference" of evolutionary rates for individual groups of related taxa. Although the absolute values of evolutionary rates will differ from group to group, Simpson asserted that the overall patterns would be similar, on the basis of his extrapolations from the marine bivalves and the land mammal carnivores: strongly peaked, negatively skewed, normal curves of distribution. (Notice once again Simpson's penchant for statistical description and analysis; although fairly simple in concept, this approach was radical for a paleontological treatise.) "The peakedness . . . means that among many allied phyla most of them tend to evolve at rather similar rates. . . . [T]he asymmetry means that within a group of allied phyla rates higher than the mode are less common than those below the mode" (128). "Both tendencies," noted Simpson in giving credit to one of his mentors, "were suggested by Matthew . . . in less explicit terms and as subjective opinions, but opinion backed by almost unparalleled knowledge" (129).

Simpson pointed out that some bivalve genera (e.g., *Nucula*, *Leda*, *Modiola*, *Pteria*, *Lima*, and *Ostrea* among extant taxa) have distinctly slower evolutionary rates than the rest of the class. These latter, non-standard, low-evolutionary-rate taxa he called "bradytelic" in contrast to the standard rate, or "horotelic," bivalve genera. (Land mammal carnivores, on the other hand, are as a group horotelic only.) Moreover, among the bradytelic lines, Simpson noted that none has become extinct: "they seem to be virtually immortal" (133). However, some bradytelic lines (e.g., the oyster *Ostrea*) have given rise to horotelic lines (e.g., the fast-evolving Jurassic *Gryphaea*). Even though Simpson could not demonstrate faster-than-usual taxa from either the carnivore or bivalve data — owing to the inherent structure of survivorship curves — he did argue that there are such "tachytelic" lines that evolve at significantly higher evolutionary rates than is common for the group to which they belong. Such evolutionary bursts or surges, he claimed, are of sufficiently short duration and occur among such small, geographically restricted populations that direct paleontological evidence for them is rare or lacking, yet still inferable from known ancestors and descendants. (Recall the discussion on mega-evolution in Simpson's previous chapter.) It should be emphasized that Simpson explicitly recognized bradytelic and tachytelic taxa as *not* simply the slow and fast end-members of the horotelic distribution of a group's evolutionary rates. Rather, he viewed the bradytelic and tachytelic taxa as distinct and separate, and their evolutionary rate as significantly slower or faster, respectively, than the slowest or fastest horotelic line.

Simpson, having presented his reasons for general factors governing tachytelic evolution in the preceding chapter, turned next to consideration of the factors responsible for bradytely. Because this sort of analysis was so novel, Simpson saw his discussion as setting up a "series of hypotheses as to factors that may be involved and to offer these as lines of attack and for future proof or rejection" (136). In general, Simpson viewed bradytelic lines as having large breeding populations that are adapted to long-lived, continuous environments that experience little variation, or if variable, as having regular and periodic variation. Bradytelic groups are not primitive when they initially appear; in fact, they are progressive and usually result from rapid evolution. Once locked in to a regularly variable or unchanging, long-lived habitat, there is no longer any impetus to change further. Bradytelic lines may provide offshoots that subsequently become horotelic or tachytelic. Consequently, among related lineages the less specialized are likely to be the oldest and either slow horotelic or bradytelic, and they tend to survive longer (148). As with the discussion of mega-evolution in his third chapter and the factors conducive to it, Simpson grounded his explanation for bradytely in ecological terms that included population size, environmental stability, breadth of adaptation, stabilizing selection, and so on, rather than invoking some internal or inherent evolutionary regulator to which earlier paleontologists had resorted.

Simpson concluded this chapter with reference to the Caenolestoidea, a group of South American marsupial mammals with which he was himself quite familiar. This superfamily displays bradytelic, horotelic, and tachytelic lineages along with the more primitive, generalized ancestral line that is surviving today. It apparently gave rise to three other lines, each more advanced than the other, especially in dentition, that then subsequently became extinct in reverse order. "All four [lineages] appear fully formed and as if of equal antiquity in the record, but each grade [of specialization] could be and in all probability was derived from the next lower grade by a sudden shift in adaptive type. The Polydolopidae, most specialized or most narrowly adapted and least tolerant in ecological relationships, became extinct in the Eocene and the Abderitinae and Palaeothentinae disappeared during the Miocene. The Caenolestinae survive today and are essentially living Paleocene or Cretaceous mammals (142–43).[64]

Inertia, Trend, and Momentum

Of all the varied phenomena displayed within the fossil record, there are two that have perhaps most gripped the attention of paleontologists and biologists: long-term, supposedly undeviating trends in morphological transformation and, related to those, the common occurrence of trends leading to extinction. By using analogies from the mechanical principles of physics, these phenomena were often explained as the result of inertia and momentum. That is, some evolutionary advances, once started, continued without deviation over long periods of geologic time, usually ending in the total disappearance of the lineage; or conversely, other lineages, once established, evolved very little or not at all thereafter.

As noted above, R. S. Lull's *Organic Evolution* cited "overspecialization," "excessive structures," and "degeneracy of organs," as examples of such evolutionary inertia and momentum that could not be adequately explained by natural selection and that, therefore, required some other special evolutionary force, such as "orthogenesis." Simpson in this his fifth chapter picked up his discussion of evolutionary rates by first determining if such phenomena truly exist — i.e., long-term, straight-line trends, some of which terminate in extinction — and then considering how such events can be reconciled with the evolutionary factors so far considered.

After citing the opinions and views of various students of the subject, together with the pros and cons regarding the terminology introduced by them, Simpson acknowledged that there is a sufficiency of examples of long-continued morphological trends as read from fossils. He suggested that they be referred to by the purely descriptive term "rectilinear evolution," in order to avoid any a priori assumptions as to its causes that might be implied by such alternative terms as "aristogenesis," "programme evolution," or "orthogenesis." In what is more or less an aside, Simpson then broadly categorized the three schools of evolutionists and their corresponding explanations for rectilinear evolution as follows: (1) the neo-Lamarckian explanation that the external environment exerts its influence directly on the internal mechanisms of change within the organisms; (2) the neo-Darwinian explanation that invokes natural selection acting upon randomly generated, spontaneous mutations; or (3) the Vitalist–Inherent Tendency explanation that attributes such linear change to the operation of some special property or principle inherent in living matter without reference to environmental influence, in either a neo-Lamarckian or neo-Darwinian sense.

While admitting that rectilinear evolution has occurred, Simpson cautioned that it is far from universal and that, in fact, evolution as observed by the paleontologist involves frequent changes in direction, as shown by many branching lines. Moreover, there are many examples of reversed trends as exhibited, for example, by secondary simplification of ammonite cephalopod suture patterns, canine tooth reduction in felids, and dwarfing in elephants. And most important for his overall argument in the book, Simpson asserted that rectilinear evolution is "almost without exception drawn from groups with large populations evolving at moderate rates (probably always horotelic) radiating on low levels within a defined ecological sphere or following one ecological zone" (153).

Although rectilinear evolution is thus not characteristic of evolution as a whole, it does occur and warrants explanation, concluded Simpson. He saw two genetic factors that could generate such linearity. In the first place, heredity is mainly a conservative phenomenon and in groups that show rectilinearity the offspring are much like their parents "in thousands of characters and differ only in a few characters that are really concerned in the long-term progression of the group" (154). A second genetic factor is the recognition that not every conceivable mutation has an exactly equal probability of occurrence, and therefore mutation must have some net direction, however slight. Thus, Simpson argued, some "definite limitation of possible trends is inherent in the mechanism of mutation. . . . The conservative factor of heredity greatly limits the possible avenues of evolution for any given type of organism. . . . [S]pecies are not constructed *de novo*" (154).[65]

Simpson next turned to a discussion of horse evolution, which in his opinion is the most widely cited example of "orthogenesis," in order to show that in fact it is not, although some horse characters do indicate overall, general long-term trends. Referring back to the conclusions that he reached in the opening chapter, Simpson reminds us that "rates of evolution were not constant in the Equidae, but varied markedly from character to character, from time to time, and from phylum to phylum [that is, from clade to clade]" (158). To the uninitiated, horse evolution may appear straight-line from Eocene *Hyracotherium* ("Eohippus") to modern *Equus*, yet, Simpson noted, "the Equidae show considerable phyletic branching, with at least twelve branches, aside from the direct *Equus* ancestry. . . . Rate of differentiation is an important element in equid branching. . . . [M]ost characters fluctuate, diverge, and even become reversed [owing] to varying and localized ecological conditions" (158).

Those characters among the equids that display an overall trend — even though certainly not orthogenetic — include trends toward larger size, changes in skull and limb proportions, increase in brain size, pad-foot to spring-foot, reduction of digits, molarization of premolars, changes in molar pattern, and low-crowned to high-crowned teeth. "It may be agreed that there are rectilinear elements involved, but they are certainly less widespread and less persistent than is usually asserted for this classical example of [presumed] orthogenesis" (163). "A dispassionate survey of many of the phenomena of orthogenesis, so called, strongly suggests that much of the rectilinearity of evolution is a product rather of the tendency of the minds of scientists to move in straight lines than of a tendency for nature to do so" (164).

So much for inertia and trends. What about momentum? "One of the most conspicuous contributions of paleontology to evolutionary theory has been the suggestion that phyla evolving in a straight line, in a direction originally adaptive, may acquire momentum that may carry them beyond the optimum, and that may even cause their extinction. . . . Opinion on this point is generally dubious and sometimes wholly subjective" (170). Simpson cited various paleontological examples for which momentum has been invoked, including those of Paleozoic graptolites, the Jurassic oyster *Gryphaea*, Permian therapsids, Tertiary titanotheres, Pleistocene saber-toothed cats, and Irish deer. The promotion of evolutionary momentum by paleontologists was so successful, noted Simpson, that even geneticists such as J. B. S. Haldane and C. H. Waddington felt that no theory of evolution, let alone of natural selection, could explain them.

Simpson then proceeded to do just that, after first observing "the impossibility of deciding categorically that a given degree of [morphological] development is really disadvantageous" (171). He argued, instead, that "the most general explanation of these so-called examples has nothing to do with momentum. . . . [Rather] a phylum [or clade] has become highly and adaptively specialized and a change in selective influences has made the specialization secondarily inadaptive" (176). Other sources of momentum effects may include "selection of juveniles or young adults favoring characters disadvantageous to old adults," or "simultaneous selection on correlated characters such that the optimum for the two together is reached after one has passed its separate optimum" (178).

Thus, in this one chapter, Simpson single-handedly laid to rest, virtually once and for all, the previous notions of straight-line evolution and evolutionary momentum that were suggested by certain real, but misunderstood, features of the fossil record. He accomplished this by showing, first, that the earlier explanations were incorrect or inadequate in themselves and, second, that these paleontological phenomena could be readily explained in the framework of the new genetics, particularly by recognition of the effects of natural selection acting upon a complex phenotype.

Organism and Environment

"The aspects of tempo and mode that have now been discussed give little support to the extreme dictum that all evolution is primarily adaptive; but they show how necessary it is to consider evolution in terms of interaction in organisms and environment" (180). Thus did Simpson begin this crucial, penultimate chapter. It is crucial because before "proceeding to the final aim of this work, an analysis of the principal elements in the pattern of evolution (Chapter VII), . . . the relationships in the organism-environment system that bear most directly on the main themes of tempo and mode" had to be examined (180). Besides being methodologically crucial, Simpson's chapter 6 was also historically crucial, in that Simpson introduced both a rationale and a metaphor to interpret evolution as observed on the paleontological scales of macroevolution and mega-evolution with the understanding gained from the genetics of microevolution. Simpson's melding of the new genetics with the fossil record occupied center stage in the evolutionary synthesis for the next half century.

Simpson began his discussion by defining adaptation in terms of differential survival and consequent differential increase in numbers of individuals, essentially in terms of what present-day evolutionists refer to as "Darwinian fitness." Simpson noted that although this definition did not precisely define adaptation, "it is the essential pragmatic test of the evolutionary significance of adaptation" (180), because survival and increase are correlated with adaptation under most, though not necessarily all, circumstances. Simpson then went on to examine adaptation in its various guises by differentiating possible (that is, genotypic) versus existing (that is, phenotypic) organism-environment interactions, or adaptation on the individual level, and how these adaptations can change dynamically through mutation or acclimatization. He made a similar distinction for populations of individuals in their potential (genotypic) versus realized (phenotypic) environmental interaction and how these adaptations, too, can change dynamically by natural selection and evolution, respectively. Such distinctions thus permit us to recognize that, both for organisms and for environments, there are "prospective" and "realized" functions. "The environment determines what prospective [or potential] functions of the organism will be realized, and the heredity of the organism determines what prospective [or potential] functions of the environment will be realized" (184).

This formulation then allows preadaptation to be defined as the "existence of a prospective function prior to its realization" in a new environment, and the "direct development of adaptations in one environment may be preadaptive in another. This is repeatedly attested by change of function in animal history" (186). Subsequent selection results in improved efficiency and effectiveness of the new function. Simpson used the example of the crossopterygian lobed-fin as a preadaptation for the tetrapod limb, for "in addition to being an excellent adaptive structure for locomotion of a body suspended in water, it was also capable of supporting and pushing the body on a hard substratum. Then it became preadaptive to a new environment . . . with respect to the crucial transition from swimming adaptation to walking adaptation. . . . Millions of years of postadaptation were necessary before such a nearly perfect means of terrestrial locomotion as the horse limb was evolved" (186–87).

By such reasoning Simpson demonstrated that "preadaptation and postadaptation are phases of a single process in which selection is a conditioning factor throughout" (188). This conclusion supported and reinforced the earlier arguments that Simpson had made for evolutionary processes being continuous in nature, even though the paleontological record will usually yield discontinuous evidence of them.

Simpson next introduced his concept of the "adaptive zone," which includes not only the physical conditions of existence but also all the biological conditions such as competitors, food, and enemies as well as other members of the species and even the individual organism itself. Simpson made clear that he viewed the adaptive zone as multidimensional even if for purposes of discussion only a few, especially relevant, dimensions were to be considered at a particular time for a given set of organisms.[66] Not only do the organisms within adaptive zones evolve, but the physical conditions change as well, owing to the varying configurations of the earth's surface. Consequently, if the constituent organisms and physical environments evolve, so also must the adaptive zones. Therefore, "the course of adaptive history may be pictorialized as a mobile series of ecological zones with time as one dimension," and this Simpson calls the "adaptive grid" (191–92).[67]

Organisms moving into an adaptive zone occupy the lower, ecologically broader, portions of the adaptive grid because their adaptation is less specific, according to Simpson. With time, increasing specialization carries the organisms into higher, ecologically narrower, bands of the grid. "The whole pattern of nature is so varied that this and other features of the adaptive grid can be considered only tendencies, and all have exceptions," cautioned Simpson (192). To occupy an adaptive zone, three factors must concur. The zone, of course, must exist — that is, "it must represent a possible mode of life. . . . This might be called the preadaptation of the environment" (193). The zone must be either unoccupied or be occupied by less well-adapted forms that cannot keep out the invaders. Finally, the organisms about to invade it must have a parental stock that impinges closely enough upon it to permit some variants to survive within it. This is preadaptation of the organism.

All of the foregoing definitions and discussion come directly to the crux, which was to be further explicated and elaborated in the final, concluding chapter — namely, the concept that differing rates of evolution — bradytelic, horotelic, and tachytelic — reflect differing patterns of evolution within the adaptive grid (fig. 7.3).

Simpson, simply and skillfully, thus inextricably bound the tempo of evolution with the mode, or pattern, of evolution within the matrix of organism-environment interactions. And all can be conceptualized by the adaptive grid, with its adaptive zones plotted over time. Evolution within a broad and stable adaptive zone is minimal and yields bradytelic groups. Evolution within an adaptive zone that is itself shifting in time (because either its constituent biota or the physical conditions that define it are changing) produces horotelic groups. Evolution from one adaptive zone to another results in tachytelic groups making the unstable transition.

Given this interpretation of evolutionary patterns and rates, we should expect the fossil record that we in fact do find, asserted Simpson: few, if any, fossils recording the steps between adaptive zones; abundant fossils tracking shifts within an adaptive zone; and occasional "living fossils" that, once adapted to a broad, unchanging zone, remain virtually unchanged. Moreover, fossils that suggest rectilinear trends, or

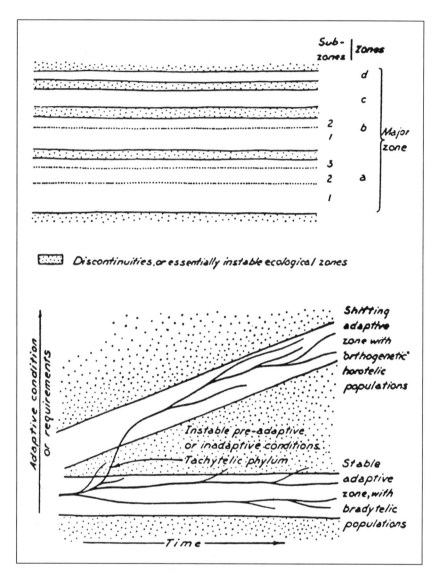

FIGURE 7.3 Simpson's "adaptive grid" (upper diagram) that provides the framework (lower diagram) of continuous, increasingly narrower, ecologically restricted adaptive zones, from bottom to top, responsible for varying rates of evolution-very slow ("bradytelic"), slow to fast ("horotelic"), and very rapid ("tachytelic")-in stable and shifting zones as well as across zones, respectively. (Simpson 1944b, *Tempo and Mode in Evolution*, figs. 26, 28. Courtesy Anne Roe Simpson Trust)

even so-called orthogenesis, are in fact samples of "successive structural steps, rather than direct phyletic transitions," as groups invade a new adaptive zone and subsequently move into narrower and more specialized zones marginal to it (194). "The steps preserved in the record represent the relatively abundant, relatively static populations of successively occupied adaptive zones, while the rare, rapidly changing transitional populations are, as a rule, absent from the record" (195). These latter produce what Simpson referred to in his third chapter as the "major systematic discontinuities of record."

Having now explained organism-environment interactions in the framework of adaptive zones and adaptive grids, Simpson was at last ready to consider the modes of organic evolution that he characterized as "one valid classification of basic descriptive evolutionary phenomena" (197). Interestingly, these modes are distinguished chiefly by evolutionary tempo: characteristic rates and patterns are mutually interrelated, each producing its own typical fossil record.

Modes of Evolution

As his preceding discussion intended to demonstrate, the evolutionary forces or factors will produce "quite different sorts of evolutionary patterns," claimed Simpson. "These patterns are protean." Despite their "seemingly infinite variety . . . there run three major styles, the basic modes of evolution." "Analysis destroys the compound," warned Simpson, so that to attribute one evolutionary mode or another to a specific pattern "may not be an adequate description of the whole . . . [yet] it will be agreed that the analysis, if correct, has revealed something of fundamental importance" (197). So while Simpson would proceed with his analysis, he wanted it "to be remembered that any [evolutionary] sequence at any [taxonomic] level is almost sure to manifest a combination of modes and that a single mode perhaps does not occur in nature in wholly pure form" (197).

This introductory caution in the final chapter of *Tempo and Mode* is summed up in the double sense possessed by the word "mode," a word Simpson no doubt chose for that very reason, because it not only means a style, fashion, or form but also, in the statistical sense, "a center of clustering or piling up of observations."[68] Hence, upon examining the full range of evolutionary phenomena available to the paleontologist and the geneticist, Simpson distinguished three modes where observations cluster or pile up, each of which reflects a particular style of evolution: speciation, phyletic evolution, and quantum evolution.

Each of the three main modes of evolution has its own set of characteristics that include the taxonomic levels involved, the types of adaptation and their relation to the adaptive grid, the direction and evolutionary equilibrium achieved, the size and proportion of morphologic changes, the population size, and the associated rate of evolution, whether bradytelic, horotelic, or tachytelic (fig. 7.4).

SPECIATION

"The process typical of this mode of evolution is the local differentiation of two or more groups within a widespread population," thereby forming geographic races or subspecies (199). Subsequent genetic isolation may split the groups into species, genera, or "somewhat higher units." Such evolution, Simpson noted, is "one more of degree than kind . . . [that] occurs within one [adaptive] zone . . . with differentiation of a population over a zone into subzonal units, . . . [or] the development or fanning out of a population across adjacent subzones of one zone. . . . In both cases the adaptive factor is adjustment to relatively minor differences in local conditions. . . . The direction of evolution in speciation . . . is shifting, erratic, and not typically linear" (199–200) (fig. 7.5). "This sort of differentiation draws mainly on the store of pre-existing variability in the [moderate-sized] population. . . . Speciation, in this sense, is more likely a matter of changing proportions of alleles than of absolute genetic distinctions, although the latter also occur. . . . As a sustained phenomenon, this mode implies moderate average evolutionary rates over long periods, probably comparable to the horotelic distribution of phyletic rates. . . . If these processes are long continued, the mode changes and grades into that of phyletic evolution" (201–202).

Simpson pointed out that the speciation mode of evolution — what is today also referred to as cladogenesis — is "almost the only mode accessible for study by experimental biology, neozoology, and genetics. It embraces almost all the dynamic evolutionary phenomena subject to direct experimental attack" (202). He remarked, further, that although also subject to paleontological study, such research was only just beginning, so that we could not yet know the extent of its possible contribution.

CHARACTERISTICS OF THE MAIN MODES OF EVOLUTION

Mode	Typical Taxonomic Level	Relation to Adaptive Grid	Adaptive Type	Direction	Typical Pattern	Stability	Variability	Typical Morphological Changes	Typical Population Involved	Usual Rate Distribution
Speciation	Low: sub-species, species, genera, etc.	Subzonal	Local adaptation and random segregation	Shifting, often essentially reversible	Multiple branching and anastomosis	Series of temporary equilibria, with great flexibility in minor adjustments	May be temporarily depleted and periodically restored	Minor intensity; color, size, proportions, etc.	Usually moderate with imperfectly isolated subdivisions	Erratic or comparable to horotelic rates
Phyletic evolution	Middle: genera, subfamilies, families, etc.	Zonal	Postadaptation and secular adaptation; (little inadaptive or random change)	Commonly linear as a broad average, or following a long shifting path	Trend with long-range modal shifts among bundles of multiple isolated strands, often forked	Whole system shifting in essentially continuous equilibrium	Nearly constant in level; most new variants eliminated	Similar to speciation, but cumulatively greater in intensity; also polyisomerisms, anisomerisms, etc.	Typically large isolated units, with speciation proceeding simultaneously within units	Bradytelic and horotelic
Quantum evolution	High: families, sub-orders, orders, etc.	Interzonal	Preadaptation (often preceded by inadaptive change)	More rigidly linear, but relatively short in time	Sudden sharp shift from one position to another	Radical or relative instability with the system shifting toward an equilibrium not yet reached	May fluctuate greatly; new variants often rapidly fixed	Pronounced or radical changes in mechanical and physiological systems	Commonly small wholly isolated units	Tachytelic

FIGURE 7.4 Simpson's summary table describing the relation between evolutionary pattern or mode (*left*) with the respective rate or tempo (*right*) as well as the other major characteristics associated with each mode and tempo (*middle*). (Simpson 1944b, *Tempo and Mode in Evolution*, 216–17. Courtesy Anne Roe Simpson Trust)

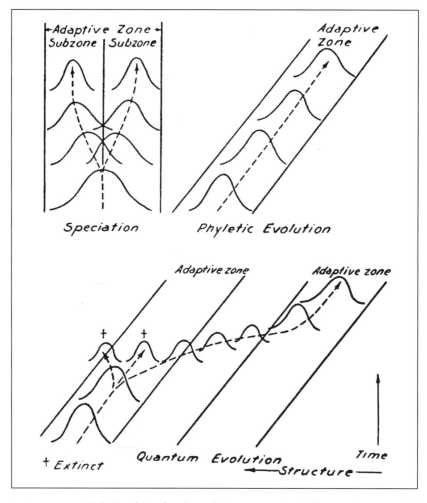

FIGURE 7.5 Relationship of each evolutionary mode to Simpson's concept of the adaptive grid with its zones and subzones. The Gaussian curves represent populations of variable individuals about some norm with respect to a key biological structure that is evolving through time. (Simpson 1944b, *Tempo and Mode in Evolution*, fig. 31. Courtesy Anne Roe Simpson Trust)

PHYLETIC EVOLUTION

This evolutionary mode "involves the sustained, directional (but not necessarily rectilinear) shift of the average characters of populations. It is not primarily the splitting up of a population [cladogenesis], but the change of the population as a whole [anagenesis].The phyletic lines that fall into patterns of this mode are composed of successive species, but . . . quite different . . . from the contemporaneous species . . . involved in speciation. . . . This mode is typically related to middle taxonomic levels, usually genera, subfamilies, and families" (202–203).

"In relation to the adaptive grid, phyletic evolution is . . . within the confines of one rather broad zone. . . . Subzonal speciation, of course, proceeds continually within the zone simultaneously with the zonal phyletic evolution." Simpson asserted that "nine tenths of the pertinent data of paleontology fall into patterns of the phyletic mode" (203). What drives this evolution is on the one hand either "perfecting adjustment to an occupied ecological position" — what Simpson earlier had called "postadaptation" — or a "slow adjustment to a shifting environment"; or, on the other hand, a "change from one environment to another, generally similar, by stages in which the adaptive equilibrium is constantly maintained" (203). Such perfecting or slow adjustment is inherently directional in appearance, and because most paleontological data are of this mode, there consequently follows the overgeneralized "dictum that all evolution has definable trend." Simpson allowed this to hold true as long as we do not interpret it to mean a rectilinear trend because, as previously explained, evolutionary trends seem to be "very rarely truly rectilinear in detail and for long periods of time" (203). Trends do, however, appear to be approximately linear when seen more broadly as the average direction of change in numerous different characters of groups that are ecologically or phylogenetically related. At all stages of phyletic evolution, the organism-environment interaction is at equilibrium, variability among populations is more or less constant, and the populations themselves are moderate to large.

Within this mode there can be several patterns. If there is early postadaptation to the adaptive zone, which then remains relatively the same for a long period of time, there will be little subsequent evolution and a bradytelic rate. If the successive populations become increasingly specialized within a stable adaptive zone, they will display horotelic rates and perhaps eventually even become extinct. If the adaptive zone itself — defined by the physical and biological conditions therein — changes, then the populations will also change, tracking the environmental shifts through time and showing horotelic evolutionary rates. Simpson cited horse evolution as the prime example of phyletic evolution.

QUANTUM EVOLUTION

"Perhaps the most important outcome of this investigation, but also the most controversial and hypothetical, is the establishment of the existence and characteristics of quantum evolution" (206). With this sentence, Simpson acknowledged that his concept of "the relatively rapid shift of a biotic population in disequilibrium to an equilibrium distinctly unlike an ancestral condition" is the most radical of all the ideas and interpretations he has presented in this most radical and innovative work. He further asserted that quantum evolution is "the dominant and most essential process in the origin of taxonomic units of relatively high rank, such as families, orders, and classes . . . [and will consequently] explain the mystery that hovers over the origins of such major groups (206). Although Simpson concentrated his discussion on the origins of higher taxa by quantum evolution, he noted that quantum evolution may also be involved in the other two modes of speciation and phyletic evolution; moreover, the latter two modes may intergrade with quantum evolution.[69]

When quantum evolution occurs so that a distinctly new adaptive type occurs — as visualized on the adaptive grid, it involves a shift from one adaptive zone to another — it gives rise to such major groups as families, suborders, and orders. Of course, it is the species, or more accurately a rapid succession of species, that make the transition, and the transition occurs in three phases or steps. First, the group involved loses the equilibrium of its ancestors or collaterals and enters an inadaptive phase. But owing to some preadaptive complex of characters, the group experiences great selection pressure and moves toward a new equilibrium. There then ensues an adaptive phase during which a new equilibrium is achieved. Simpson emphasized the instability of all three phases; in fact, he noted that the usual outcome, by far, is extinction: "In all typical cases in which the transitional stages are unstable, the vast majority of populations . . . simply become extinct" (211–12). In those rare instances where an unoccupied adaptive zone exists marginal to the one occupied by the group experiencing loss of adaptation, and if the group has preadaptive characters suitable for the new zone, then the transition may occur. Because the transitional steps are not in evolutionary equilibrium, the shift must be made rapidly, and to evolve rapidly requires small isolated populations, said Simpson. Therefore, paleontological evidence of such a shift is hardly to be expected; hence arise the major systematic gaps in the fossil record between higher taxa (fig. 7.6).

Unlike Darwin, who was forced to rationalize away the lack of fossil evidence to support his "principle of gradation" between species, Simpson viewed what he called major systematic discontinuities as positive, not negative, evidence of quantum evolution: "Major exam-

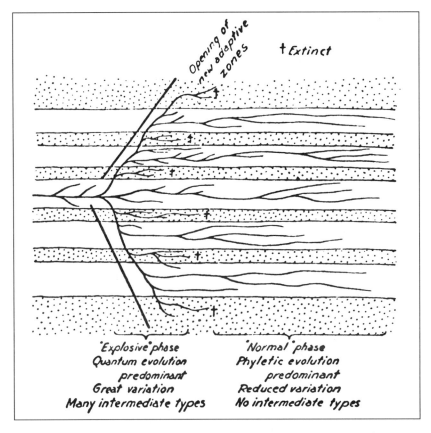

FIGURE 7.6 Simpson's illustration indicating how quantum evolution results in an "explosive" phase as viewed from the concept of the adaptive grid. As in other similar diagrams, the vertical axis represents discontinuous adaptive zones and the horizontal axis is relative time. (Simpson 1944b, *Tempo and Mode in Evolution*, fig. 35. Courtesy Anne Roe Simpson Trust)

ples of quantum evolution are never well documented — indeed this deficiency of documentation is itself good evidence for the reality of the inferred phenomenon" (209). The invasion of new adaptive zones, often made available by the extinction of its prior occupants, also explains so-called "explosive evolution," as seen in the case of, say, the Cenozoic radiation of mammals following the late Mesozoic mass extinctions of reptiles, or the South American radiation of ungulates during the Tertiary Period when isolated from the rest of the world's continents.

Simpson concluded the chapter, and the book, with a table that summarizes the various attributes and conditions typical of, but not entirely restricted to, his three modes of evolution (refer back to fig. 7.4). For each inferred mode, he showed, among other things, the usual evolutionary rate associated with it. Thus, Simpson completed his presentation as he had started it, by answering the question raised in the first chapter: "How fast, as a matter of fact, do animals evolve in nature?" Not only had he shown how fast evolution proceeds, but he had also answered the obvious follow-up question: "How come?" Figure 7.7 summarizes in schematic form the chapter-by-chapter development of the book's argument.

Reception and Impact

Initial Reviews

As might be expected, the reviews of *Tempo and Mode* by geneticists heralded it as an important bridging of the gap between genetics and

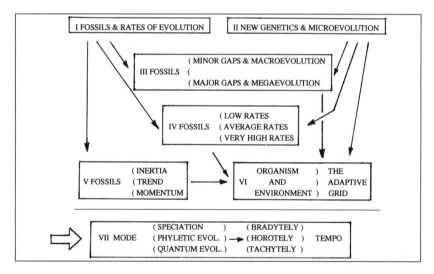

FIGURE 7.7 Summary of the chapter-by-chapter development of the argument in Simpson's *Tempo and Mode in Evolution*. Although by no means an inductive process, Simpson used the fossil record to infer significantly distinct evolutionary rates, or "tempos." He bolstered his argument, wherever possible, with data and concepts from the genetics of his day. He then introduced the "adaptive grid" to illustrate how differing ecologic opportunities yield different evolutionary patterns, or "modes." He completed his argument by indicating how the major modes of evolution produce distinctive tempos.

paleontology. As Bentley Glass remarked, "Between no branches of evolutionary biology has the chasm yawned wider."[70] Sewall Wright expressed much the same view, although he tempered his enthusiasm by noting that he found "remarkably little difference between Dr. Simpson's interpretation and that which had seemed to him [Wright] . . . the natural deduction from genetic principles." Wright moreover noted, "It would certainly be too much to say that paleontologists and geneticists as groups have reached agreement but at least contact has been established that did not exist a few years ago."[71] Theodosius Dobzhansky was the most ebullient in his praise, for he judged *Tempo and Mode* as one of those "good books which mark the end of one and beginning of another period in the development of . . . science," so that "from now on no competent discussion of the mechanisms of macroevolution can be made outside the ambit of Simpson's analysis."[72] Glass echoed the same sentiment: "It is not too much to say that hereafter no treatise on the causes and modes of evolution can ignore [*Tempo and Mode*]."[73]

Wright saw Simpson's most important contribution "from the genetic side to the reconciliation with paleontology" to be Simpson's recognition that "in general, evolution is a population problem . . . [because the] elementary evolutionary process is . . . change of gene (or chromosome) frequency."[74] Wright further noted, however, that Simpson's quantum evolutionary mode — small preadaptive populations that evolve continuously without saltation at very rapid rates into new ecologic zones — was "essentially the same idea . . . I have expressed." Wright noted too that "Simpson rejects all mystical notions" in the latter's discussion of inertia, trend, and momentum.[75] Dobzhansky viewed this as a "good debunking job" by Simpson, one wherein he "separates the grain from the chaff, and shows that several different phenomena, none of them particularly mysterious, are being confused."[76]

While these geneticists were highly laudatory of *Tempo and Mode*, they also looked forward to a more fully developed treatment of its subject. "[O]ne is tempted to suggest a much enlarged edition with a more extensive documentation would eventually be in order" (Dobzhansky, 114); "the treatment is therefore by design a limited one. . . . [T]he supporting evidence for the volume of theory is rather thin. Simpson would probably agree" (Glass, 261, 263); "this is a relatively short book. So much is condensed into its compass" (Wright, 419). Simpson clearly agreed with these opinions, for in his *Major Features of Evolution*, published nine years later, he stated in the preface that *Tempo and Mode* "was a series of technical theses speculative in part, developed in a rather personal way, and necessarily thin in background and support. This book is more rounded and complete within its scope."[77]

The reviews of the naturalists concurred with those of the geneticists: "first rank treatise . . . brilliant, but not flashy . . . fresh ideas . . . comprehensive . . . quantifies data."[78] "Simpson's great achievement . . . is the fusion of classical paleontological research with the results of modern mathematical investigations on demogenetics. . . . [The] most important contribution yet to come from paleontology on the methods by which evolution takes place."[79] "[A] bold and highly successful attempt to apply population genetics to the problems presented by the evolution, not of species, but of the larger categories of taxonomy."[80] "Mammalian paleontology forsakes its former role of merely proving that evolution has occurred and makes a definite contribution to our knowledge of the forces involved."[81]

Huxley, whose *Evolution: The Modern Synthesis* (1942) was itself both a contributor to, and a name-coiner for, this revolution in evolutionary biology, was also highly favorable to *Tempo and Mode*. "Dr. Simpson has performed the double task of reminding neozoologists of many facts of paleontology which they have tended to overlook as unfamiliar or inconvenient, and at the same time showing the possibility of accounting for them in genetic terms."[82] Like other reviewers, Huxley was taken with the notion of quantum evolution, especially with respect to its nonadaptive/preadaptive aspects. However, he questioned whether such "non-adaptation could persist for the millions of years postulated by Dr. Simpson for his periods of rapid major change. "He also was intrigued by exceptionally slow, bradytelic evolutionary rates and suggested that 'paleontology should further clarify this question . . . and the reasons for them.'"[83]

Whereas *Tempo and Mode* was well reviewed by the biological community, paleontologists virtually ignored it, neither praising nor rebutting it. For example, the *Journal of Paleontology* failed to review it at all. The only review by a paleontologist was that of Glenn Jepsen of Princeton University, a vertebrate paleontologist and colleague of Simpson's. Jepsen acknowledged explicitly the good reception the book had already received from biologists, but he believed that they missed "the most original contributions to interpretation of fact and epistemology." In this regard, Jepsen cited among other things Simpson's treatment of bradytely and tachytely; his rejection of saltation, orthogenesis, and momentum; his recognition that major systematic gaps in the fossil record cannot be explained solely as sampling accidents; and his emphasis that adaptation is a continual theme throughout the book. Jepsen also praised Simpson for his careful "differentiation between frequently confused members of concept pairs, such as symbol and object, being and becoming, explanation and description, fact and report, primitive and generalized, record and 'artifact.'" Overall, Jepsen concluded that *Tempo and Mode* "is a major contribution to the expansion and correlation of scientific reasoning in paleontology."[84]

Subsequent Citations

The "modern evolutionary synthesis" was considered "consolidated" at the January 1947 Princeton conference (see note 15), which brought together a wide variety of evolutionary biologists. The history of the conference is interesting because it reflects the growing impetus for the synthesis as well as its diverse sources. It is also one of those infrequent instances where a committee, formed to address a problem head-on, actually produced a useful and lasting result.

At the 1941 annual meeting of the Geological Society of America, Walter Bucher, professor of geology at Columbia University, gave a talk reviewing the role of the National Research Council, under the auspices of the National Academy of Sciences. He underlined the importance of interdisciplinary communication across the "borderland fields" of geology and its allied sciences of chemistry, physics, and biology. With respect to the latter, Bucher noted that the "lack of systematic joint counseling of paleontologists with biologists is conspicuous." He continued, "'Evolution' is still full of riddles. The disciples of the young and vigorous experimental science of genetics have little time for, and less interest in, the body of solid fact which the older, undaunted historical disciplines of morphology, taxonomy, and paleontology are accumulating. Insofar as the disciples of the latter exhibit interest in genetics, it is mingled with distrust in the long-range significance of the former's findings. Here is a field that requires joint counseling."[85]

A year later, a small informal study group that included geneticists, paleontologists, and systematists organized itself at Columbia University, and Simpson became its overall chairman. (This was in October 1942, a few months after Simpson had completed the manuscript of *Tempo and Mode* and two months before he entered the army.) The study group eventually expanded into larger Eastern and Western U.S. divisions, and throughout the rest of World War II it met, discussed, corresponded, and exchanged preprints and even reading lists among its members, some of whom were soon called up for military duty — including Simpson.

Two important results came from this Committee of Common Problems of Genetics, Paleontology, and Systematics, which was formally established in February 1943 within the National Research Council. The first was the formation of the Society for the Study of Evolution in 1946, with a quarterly journal, *Evolution*, that still flourishes today. (In fact, this society also collaterally descended from another, prewar group called the Society for the Study of Speciation, whose informal leader, Ernst Mayr, and some of its other members overlapped with the later committee on genetics, paleontology, and systematics.)[86]

THE PRINCETON VOLUME

Two years later the collected papers of the 1947 Princeton conference were published under the editorship of Jepsen, Simpson, and Mayr (*Genetics, Paleontology, and Evolution*, 1949). As Mayr has recollected, "the degree of consensus among evolutionists was again tested. . . . [I]t was almost impossible to get a controversy going, so far-reaching was the basic agreement among the participants. . . . [E]volutionary biology was no longer split into two noncommunicating camps [of experimental geneticists and naturalists]. . . . [T]he conference constitutes the most convincing documentation that a synthesis had occurred during the preceding decade. All participants endorsed the gradualness of evolution, the preeminent importance of natural selection, and the populational aspect of the origin of diversity."[87] Of the resulting publication, Simpson has noted that "the book was another major contribution to the development of the synthetic theory of evolution."[88]

This somewhat lengthy background for "the Princeton volume" — as it came to be called — is necessary to validate the claim that even before *Tempo and Mode* Simpson was obviously judged by his peers and colleagues as one who could contribute much to their efforts to resolve "common problems of genetics, paleontology, and systematics." Moreover, Simpson fulfilled that promise, for when the conference volume was published, there were more than two dozen references by eight different contributors to his *Tempo and Mode* (which had been published during the interim between the committee's formation and the Princeton conference). Simpson's book, in fact, is referred to more than any other contributions made to the synthesis, including those of Fisher, Haldane, and Wright, or even Dobzhansky, Huxley, and Mayr. (Darwin himself rated but six citations in five articles.) The bulk of the references to *Tempo and Mode* relate to Simpson's measurement and interpretation of varying evolutionary rates, to his concept of quantum evolution, to his emphasis on organism-environment interaction (especially the notion of the adaptive zone), and to his quantitative approach to paleontological data that make them useful to experimentalists.

For example, Dwight Davis, an anatomist, claimed that Simpson had "emphasized that evolution is gradual, that it is often sustained and directional, that progressive adaptation is typical, and that major adaptations often appear suddenly. These are major features of the mode of evolution that could hardly be demonstrated in any other way. Only recently have attempts to reconcile such empirical facts with the data of genetics met with approval." Davis did take issue with Simpson over several points — for example, Simpson's too vigorous dismissal of Schindewolf's saltationism or what he meant by inadaptive evolution — but there was no serious disagreement. Davis observed that "the fact that the observed data of paleontology conform with the neobiological theory enhances the probable validity of the theory, and certainly no theory that does not conform can be accepted." Davis continued his comments with the insight that "the facts of paleontology conform equally well with the other interpretations that have been discredited by neobiological work, e.g., divine creation, innate developmental processes, Lamarckism, etc., and paleontology by itself can neither prove nor refute such ideas."[89]

G. Ledyard Stebbins's *Variation and Evolution in Plants* (1950), based on the Jesup Lectures he gave at Columbia University in 1946, made another classic contribution to the synthesis. He asserted that the "complete agreement with Simpson's conclusions in all points yet mentioned indicates that, like him, we botanists should look for the principal factors governing rates of evolution in the organism-environment relationship."[90]

Sewall Wright used Simpson's bradytely, horotely, and tachytely in his discussion of what happens when the achievement of a major adaptation passes a threshold so that an adaptive radiation ensues either because the adaptation is so broad and generally successful, thereby conferring widespread ecological success, or because of an absence of competition from a similarly endowed organism. "This rapid evolution in the origin and adaptive radiation of a higher category constitutes the tachytely of Simpson. [It is] followed by secondary and tertiary radiations of progressively lower scope. . . . There is slow and more or less orthogenetic advance, the horotely of Simpson. . . . Ultimately forms may become bound each to a single [adaptive] peak, with no other peak attainable and evolution virtually ceases, the bradytely of Simpson."[91] The geneticist and the paleontologist were at long last communicating with each other.

THE GREAT AEPPS

The consolidation of the modern synthesis by the late 1940s can be marked in other ways — for instance, by the publication of treatises that incorporated its new ideas and interpretations. In 1949, for example, the *Principles of Animal Ecology* was published, authored by the Chicago school of ecologists — W. C. Allee, A. E. Emerson, O. Park, T. Park, and K. P. Schmidt (whose last initials made up "the Great AEPPS"). The volume became a classic in its own right and only yielded the field in the 1970s, when the ecological renascence produced a new generation of textbooks. Such a resurgence in ecology was to be expected, given the importance that the synthesis placed on organism-environment interaction, rather than on earlier, discredited nonecological factors of "inherent tendency" or "mutationism." Given the ascendance of natural selection and adaptation, renewed interest in ecology would necessarily occur.

Principles of Animal Ecology referred heavily and about equally to the major contributors to the synthesis, namely, Dobzhansky, Huxley, Mayr, and Simpson, as well as, of course, to the genetical foundations provided by Fisher, Haldane, and Wright. (Stebbins's *Variation and Evolution in Plants* was still a year in the offing.) In particular, Simpson's *Tempo and Mode* is referred to two dozen times in part 5 ("Ecology and Evolution"), which was written by Alfred Emerson, an authority on social insects, especially termites, and who was professor of zoology at the University of Chicago. Besides the by now familiar citations of Simpson's views on evolutionary rates, gaps in the fossil record, and his debunking of orthogenesis, most of the references concerned Simpson's formulation and discussion of "evolutionary determinants," particularly population size and variability, the nature and rate of mutation, and natural selection. What is most striking is that Simpson is cited as an authority not just as regards the paleontological evidence but also on evolutionary theory. His views were thus given equal billing and weight with those of the experimental geneticists and field naturalists. This is certainly further evidence that *Tempo and Mode* was judged by Simpson's colleagues as being far more than just a demonstration that paleontology is consistent with the new genetics. In fact, one gets the distinct impression that Emerson found Simpson's handling of the genetical arguments more accessible to the nonquantitative, field-oriented naturalist, for whom the book was presumably written, than any arguments by the geneticists themselves. It might be assumed that such secondary sources as this one cast a still wider net for the developing synthesis. If one did not understand the mathematical arguments of a Wright, one could nevertheless get the essential reasoning and results from a Simpson directly, or as recast, indirectly, by the Great AEPPS.

STEBBINS AND DOBZHANSKY

The botanical contribution to the synthesis came primarily from G. L. Stebbins's *Variation and Evolution in Plants*, published in 1950 in the same Columbia Biological Series that had published Dobzhansky (1937), Mayr (1942), and Simpson (1944). Stebbins cited *Tempo and Mode* some dozen times, mostly with respect to evolutionary rates and trends, the role of natural selection, and so-called orthogenesis. For example, Stebbins noted that Simpson "has presented a convincing analysis of evolutionary rates . . . and in general the paleobotanical evidence can do little more than support the principles which he has deduced." And, with respect to plants, "the gaps in the fossil record are as important as the fossils themselves."[92]

In the section of his book entitled "The Possible Basis for Differential Evolutionary Rates," Stebbins reported the same sort of phenomena in plants that Simpson had observed in animals: very slow evolution in some plant taxa, such as *Gingko, Sequoia,* and *Thyrsopteris* (the latter a Japanese fern that has a close Jurassic look-alike), as well as a very rapid evolution in other taxa, most typically the early angiosperms. Stebbins agreed with Simpson that the "causes [of bradytely and tachytely] are to be sought in the nature of the relationship between the organism and the environment. This conclusion is amply borne out by the evidence from plants." Stebbins demurred, however — as did Huxley in his review of *Tempo and Mode* — with respect to what "rapid" evolution as shown by fossils might actually mean in genetic terms. Rather than requiring a few million years, Stebbins ventured that, under the right conditions, a new plant species "could evolve in fifty to a hundred generations."[93]

Dobzhansky's third edition of *Genetics and the Origin of Species*, published in 1951, was the first that could take into account *Tempo and Mode*. Dobzhansky did then refer to it, but only a couple of times (in respect to orthoselection and evolutionary rates), as he preferred instead to cite Simpson's later *Meaning of Evolution* (1949), which was a more popular, less technical treatment of much of the content of *Tempo and Mode* (as well as including a third part that deals with "Evolution, Humanity, and Ethics"). Among other things, Dobzhansky cited the latter work for Simpson's views on orthoselection versus orthogenesis, "that all evolutionary changes are compounded of microevolutionary ones," and the adaptive importance of cooperation in organisms. As had Stebbins, Dobzhansky quoted Simpson as one of a number of authorities and students of evolutionary theory — not of just fossils per se — to support his own conclusions on these issues.[94]

In summary, then, *Tempo and Mode* was not only written as a treatise on evolution, but it was also *read* that way. Of course, given his interests and expertise, Simpson quite naturally couched his discussion in terms of fossils, but he by no means restricted himself to them. Thus the central theme of Simpson's treatise — tempo and mode in evolution — is indeed aptly summed up in the title. Although Simpson stated his case by using fossil evidence, he went far beyond mere inductive conclusions based on that evidence. Instead, he made broad generalizations, formulated testable hypotheses, and corrected previous paleontological misinterpretations. Clearly, he fulfilled his avowed "purpose to discuss the 'how' and — as nearly as the mystery can be approached — the 'why' of evolution, not the 'what.'"[95]

By even the most casual and crude measures, Simpson's *Tempo and Mode* had a continuing impact on current discussions of evolutionary theory, either by direct reference to his ideas and hypotheses as presented therein or indirectly by reference to the subsequent expansion and revision made in *The Major Features of Evolution*. For example, Mayr's *The Growth of Biological Thought* cites Simpson twenty times, in comparison with Dobzhansky (nine), Fisher (four), Haldane (five), J. Huxley (seven), Stebbins (three), and Wright (five). Only Mayr citing Mayr (twenty-eight times) surpasses Simpson. Admittedly these numbers greatly oversimplify the issue, if only because Simpson has written so much on such diverse topics within biology that any history of the field will give him fair prominence. Thus, Mayr has included not only Simpson's views on evolutionary theory but also his views on the philosophy of biology, on taxonomy and systematics, and on biogeography, among other subjects.

It should be emphasized that when Simpson is cited with respect to evolutionary theory, the authors consider his views on variable rates of evolution, quantum evolution, the adaptive zone, and the nature of evolutionary trends, as well as his debunking of momentum, racial senescence, saltationism, orthogenesis, and so on. His paleontological studies per se — with the frequent exception of the horses — are not usually cited. Students of evolutionary theory explicitly acknowledge what Simpson has to say about the "how" of evolutionary mechanisms, much less his views about the "what." (One has to delve into the writings of paleontologists for that.)

Final Perspective

In an interesting essay, Rachel Laudan has argued that within geology there is a tension between natural history and natural philosophy: that is, between a reconstruction of the earth's history and an understanding of the processes by which that history unfolds.[96] She has asserted that the history of geology shows periodic swings from one point of view to another, swings triggered by the discovery of new techniques ("adventitious disciplinary change") or by the introduction of new methodological positions ("stipulative disciplinary change"). An example of the first was the development of sensitive magnetometers that could measure very low-level magnetic forces, which catalyzed the revolution in plate tectonics theory. An example of the second was the introduction of the uniformitarian methodology found in Charles Lyell's *Principles of Geology*.

Laudan's analysis can certainly be extended to paleontology, which exhibits the continual tension between the description and classification of fossils, and the tracing out of their history vis-à-vis the discovery and understanding of the processes underlying that history. Despite the periodic natural philosophy approaches to paleontology of Lamarck, Cuvier, Darwin, Cope, and Osborn, the natural history emphasis has been dominant. Using Laudan's phrasing, we can view Simpson's *Tempo and Mode* as a prime example of stipulatory disciplinary change, for it demonstrated how paleontologists might study the fossil record in terms of the processes that operate upon populations of individuals within species interacting with their physical and biologic environment. Although the processes identified were obviously based upon the laboratory and field data of experimentalists and naturalists, Simpson indicated how such short-term processes could operate over much longer geologic time.

So it *is* true that Simpson showed the fossil record to be "consistent" with population genetics; and yet his demonstration was certainly more than a mere extrapolation or clarification. His contribution, on the contrary, resulted in a new understanding (or at the very least, a new interpretation) of just how microevolutionary phenomena express themselves — in varying tempos and varying modes of evolution — throughout real, historical time. It would be fatuous to assume that, given the new genetics, anyone at all could otherwise have made such sense of the fossil record.

It is no coincidence that the shift in emphasis in much paleontological research from "natural history" to "natural philosophy," however slow it has been in coming about, dates from the publication of *Tempo and Mode*. Simpson himself was the center of theorizing in paleontology, but he was by no means alone. One can trace an expanding process-oriented paleontological literature, following World War II, to be seen initially in the journal *Evolution*, and later with the founding of *Paleobiology*. (Even the term *paleobiology*, which now so frequently substitutes for the essentially synonymous term *paleontology*, is a kind of code word that signals that shift in focus and emphasis in fossil research.)

We have then another reason to emphasize the importance of *Tempo and Mode*, for it indicated and exemplified new, useful, and interesting ways to study fossil remains. Even if the conclusions reached in that volume had been badly flawed, one would still find it a crucial contribution in terms of its role in advancing a novel methodology for paleontological inquiry. This is the point made, I believe, in historian and philosopher of science Dudley Shapere's remarks when he wrote: "And in the case of the synthetic theory of evolution, perhaps what we ought to be asking is not whether paleontology (or any other discipline) was brought into some tight deductive unification with Darwinian-Mendelian theory but rather to what extent the removal of barriers to seeing paleontological data in Darwinian-Mendelian terms, and the consequent application of the latter terms to those data, reoriented the kinds of questions asked, the kinds of research engaged in, the expectations about the sorts of things one should expect to find, and so forth, in paleontology and other fields."[97]

Simpson was by no means unaware of the originality of his approach, for as noted earlier in this chapter, he concluded his introduction to *Tempo and Mode in Evolution* with his own appraisal of the work, an appraisal as valid today as when it was written some six decades ago: "The one merit that is claimed for this study is that it suggests new ways of looking at facts and new sorts of facts to look for."[98]

CHAPTER 8

MENTOR FOR PALEOANTHROPOLOGY

Our species is a twig among others, but it is for us the most important twig.

— G. G. Simpson

The influence of Simpson upon paleoanthropology provides a well-documented historical example of how one scientific discipline can impact upon another, bringing the latter quickly "up to speed" without having to retrace ground covered by the former. Although further immediate progress in the field may be cast in the terms of the mentor-discipline, its subsequent maturation is independent of it. Paleoanthropologists were bystanders during the formulation of the evolutionary synthesis (1936–1947). After World War II, the younger paleoanthropologists looked to Simpson as one of several mentors regarding the implications of the synthesis for their own discipline. But why Simpson? Having earlier defined the superfamily Hominoidea (1931) to hold the Pongidae and Hominidae families and monographed lower primate fossils (e.g., "Studies on the Earliest Primates," 1940), Simpson's "Principles of Classification and a Classification of Mammals" (1945) further solidified his reputation as a mammalian systematist. Simpson's *Meaning of Evolution* (1949) was widely read as an introduction to the synthesis, while his *Tempo and Mode in Evolution* (1944) made accessible the synthesis's more complex aspects.[1]

Consequently, in the 1950s and 1960s paleoanthropologists invited Simpson to participate in their symposia (e.g., "Some Principles of Historical Biology Bearing on Human Origins," 1951; "The Meaning of Taxonomic Statements," 1963), used his books as classroom texts, and cited his publications to support claims for their own work. Later in the 1960s, Simpson moved from mentor to apologist, as the paleoanthropologists were by then well familiar with the synthesis and incorporated its theoretical concepts in their interpretations of the many newly discovered hominoid fossils. Simpson now took special care to celebrate these results in his more general, less technical writings, acting as a forceful apologist for the materialistic view of human origins (e.g., "The Biological Nature of Man," 1966; "The Evolutionary Concept of Man," 1972). During the 1970s, Simpson's influence waned, and he became just another practitioner at the margin of the discipline. However, anthropologists acknowledged Simpson's earlier impact as, for example, when he was invited, yet again, to address them at the fiftieth anniversary celebration of the American Association of Physical Anthropologists in 1981.

The Evolutionary Synthesis

Although there may have been individuals within the discipline so inclined, paleoanthropology did not actively participate in the evolutionary synthesis. Perhaps the chief reason for this lack of disciplinary involvement was that physical anthropologists of that era had been trained as comparative anatomists or medical doctors, and hence they were unable to recognize either the relevance or the cogency of the arguments of the population geneticists during the early 1930s, which precipitated the synthesis. Even those formally trained in the discipline, like Earnest A. Hooton (1887–1954) of Harvard, the "dominant figure intellectually in physical anthropology in America for more than 40 years, . . . had little use for and no real knowledge of the emerging field of genetics."[2]

One measure of the thorough lack of active participation by physical anthropologists in the evolutionary synthesis was that the British biologist J. B. S. Haldane presented the only paper in section VII ("Human Evolution") at the Princeton conference. Entitled "Human Evolution: Past and Future," Haldane's discussion was more social biology than physical anthropology.[3] Moreover, no physical anthropologist contributed to the retrospective volume on the history of the development of the evolutionary synthesis, although one anthropologist, Irven De Vore, did attend the second of the two conferences.[4]

When the younger anthropologists returned from World War II, including several of Hooton's former students, they began to catch up on what they had missed while away in military service.[5] Among other things, they were soon reading the literature of the synthesis, and before long were citing the writings of Dobzhansky, Mayr, Wright, and Simpson. In this chapter I focus on the role Simpson played as mentor to paleoanthropology in the 1950s and 1960s, and indicate why and how Simpson had the special influence he did. By concentrating here on Simpson, I do not mean to imply that others — like Dobzhansky, Mayr, and Wright — did not contribute at least as much as Simpson, but rather merely wish to indicate the nature of Simpson's impact on paleoanthropology.[6]

Simpson's Credentials

In the area of human evolution Simpson had solid credentials because he had not only worked extensively with mammalian fossils, but he also had a sound grasp of the new evolutionary theory — indeed he had formulated some of that theory himself.[7] Simpson had the additional advantage that he had no hominid fossil bones of his own so he was not promoting any particular anthropological point of view, although he had described lower primates as part of his work on early Cenozoic mammalian assemblages. Thus anthropologists could invoke Simpson as a neutral third party in order to bolster their own theories. Finally, Simpson's underlying philosophy of materialistic positivism appealed to the rising generation of physical anthropologists, who wished to dissociate themselves from the teleological and essentialist bias that infected much contemporary paleoanthropology.[8]

Simpson, of course, had several other points in his favor with the paleoanthropologists. In 1931 he had published an abbreviated classification of mammals where for the primates he claims to have "followed the recent views of W. K. Gregory," with the exception of coining a new taxon, the superfamily Hominoidea, wherein he placed the families Pongidae and Hominidae.[9] In 1940 Simpson published a monograph that tentatively discussed the affinities and distribution of the then five known families of Paleocene and Eocene primates to each other, and to the Primates in general.[10]

Before leaving for military service himself in late 1942, Simpson completed the manuscripts for both *Tempo and Mode in Evolution* and "The Principles of Classification and a Classification of Mammals." *Tempo and Mode* made accessible to nongeneticists the more complex concepts and arguments of the synthesis that were grounded in the new observations and concepts of population genetics and field biology. Simpson argued that the short-term microevolutionary processes of population genetics, operating within a given local environment, were sufficient to explain the long-term macroevolutionary phenomena of the paleontologist. For Simpson it was the evolutionary results recorded by fossils that provided an empirical testing ground for population genetics. The classification monograph not only provided a rationale and a context for primate taxonomy and systematics, including a fuller discussion of his notion of the Hominoidea, but presented as well a précis of Simpson's philosophy of classification in light of the evolutionary synthesis. With respect to the Primates, Simpson claimed: "As to the relationships of the two suborders, the Anthropoidea were almost surely derived from early members of the Prosimii. Given the variety of each and the survival of both, this does not argue against considering their separation as the primary or most important (not necessarily the first) dichotomy of the Primates."[11]

Simpson followed the lead of W. K. Gregory "that the gibbons, apes, and man are a unit," and thus he erected the superfamily Hominoidea to contain them, first in 1931 and then reaffirmed in 1945. But he disagreed with Gregory that the hominids themselves should not be in a separate family: "[On] the basis of using the usual diagnostic characters, such as teeth, viewed with complete objectivity, this union seems warranted. I nevertheless reject it, for two reasons: (a) mentality is also a zoological character to be weighed in classification and evidently entitling man to some distinction, without leaning over backward to minimize our own importance, and (b) there is not the slightest chance that zoologists and teachers generally, however convinced of man's consanguinity with the apes, will agree on the didactic or practical use of one family embracing both."[12]

In 1949, Simpson's *Meaning of Evolution* served as yet another introduction to the evolutionary synthesis, especially for those who were neither paleontologists nor biologists. Consequently, the book was widely adopted in beginning physical anthropology courses, and thus many of the new generation of anthropologists cut their first scientific teeth on Simpson.[13] Of special interest to anthropologists was chapter 7 ("The

History of Primates"), wherein Simpson discusses the order within the overall framework of neo-Darwinian evolution, but first criticizes some previous efforts: "Primate classification has been the diversion of so many students unfamiliar with the classification of other animals that it is, frankly, a mess." For the primates as a whole, Simpson notes that there is "no apparent new grade or type of organization such as is usually seen in the origin of a group that later rises to importance. Among the primates such distinctions came later, and only to some of the branches in their radiation." Simpson recognizes "four general levels and types of structure . . . of unequal fundamental variety or taxonomic value." These are the prosimians, the ceboids, the cercopithecoids, and the hominoids, and for each he cites several key characters. Thus, "among the primates as a whole, displacement and expansion into new ecological positions seem to account for contractions and expansions with relatively little replacement of one group by another." Each level and type reflect four major radiations and therefore the "four main primate groups do not represent four successive steps, each leading to the other."[14]

The influences between Simpson and paleoanthropology were not all in one direction, however. In fact, at times they were symbiotic. For example, Matt Cartmill argues that the "most influential convert to Le Gros Clark's vision of the primates was George Gaylord Simpson, whose classic works of the next two decades reiterated the ideas of Wilfrid Le Gros Clark and lent them the stamp of Simpson's great authority as a paleontologist and evolutionary theorist."[15] At other times, Simpson was obviously derivative. In the revised edition of *The Meaning of Evolution*, Simpson follows a "strong [paleoanthropological] consensus" in recognizing that the australopithecines "belong to the human family, not only hominoids (a term including apes) but hominids (a term excluding apes and all but our closest relatives)," and thus he was modifying his earlier opinion — also following the paleoanthropological consensus of the time — that the australopithecines were "merely another branch from the original Miocene radiation or may, with far less probability, prove to be actual human ancestors."[16]

Simpson as Mentor

In 1950, Sherwood Washburn (1911–2000), one of the younger anthropologists most influenced by Simpson, and the geneticist Theodosius

Dobzhansky organized a symposium at Cold Spring Harbor on the "Origin and Evolution of Man." Washburn is generally credited as being a major force in creating the "new physical anthropology" (his phrase) and introducing physical anthropologists to the literature and individuals that formulated the evolutionary synthesis.[17]

At the symposium Simpson gave an invited keynote paper entitled "Some Principles of Historical Biology Bearing on Human Origins," where he emphasized the following points: that paleoanthropologists could not solely rely upon comparative morphological studies of living primates to determine their evolutionary history, but rather that the additional dimensions of time, geography, and environment had to complement morphology; the nature of, and distinction between, the phenomena of parallel and convergent evolution; what paleontologists mean by the irreversibility of evolution; and that instead of orthogenesis, evolutionary history recorded patterns of evolution that varied both in direction and rate.[18] Of course, these were the same themes addressed in both *Tempo and Mode* and *The Meaning of Evolution*, but more sharply focused for his anthropological audience. William Howells, who chaired the session in which Simpson participated, gave the benediction to Simpson's paper by noting that "Simpson calls for a belated testing of our ideas by the principles of paleontology."[19] This paper of Simpson's is perhaps the one most often cited by paleoanthropologists in the ensuing decade.

At a subsequent Cold Spring Harbor symposium in 1959, Simpson emphasized one of the dominant themes of the evolutionary synthesis — namely, that microevolutionary processes of the population geneticists could explain macroevolutionary phenomena of the paleontologists. Therefore, no appeal to special processes unique to paleoanthropologic phenomena need be invoked. This, of course, was a claim that Simpson made more generally for all fossils: that all the fossil record was consistent with the recently articulated principles and concepts of the new genetics. "I . . . repeat my conviction that the basic processes are the same at all levels of evolution, from local populations to phyla, although the circumstances leading to higher levels are special and the cumulative results of the basic processes are characteristically different at different levels."[20]

In the same paper, Simpson focused on several higher mammalian taxa (including rodents, lagomorphs, carnivores, tubulidentates, the classes of vertebrates, and early mammals as well as primates) to demonstrate that differing circumstances of environmental opportunity and biological antecedents could yield higher taxa having differing degrees of specialization and diversity. Simpson concluded his discussion by examining the order of Primates and argued that primate history had its own peculiar characteristics, with the successive radiations of prosimians, simians, pongids, and hominids representing, respectively, increasingly specialized, less diversified groups of organisms (figs. 8.1 and 8.2): "The primates provide another example of the grade concept. . . . [I]t arose by general adaptive improvement and not by any more or less clear-cut single basic adaptation. . . . [P]rimate history can be drawn both in terms of radiations and of grade. The most primitive Primates are distinguished only arbitrarily from primitive Insectivora. . . . The distinction of grades in this representation of the history is admittedly biased by human interest in *Homo*. . . . [A]s the grades designated here are followed upward [in time] they become much less diversified and also less widely distinct from adjacent grades. Those facts are reflected in most classifications . . . by assigning the successive grades lower categorical rank."[21]

Simpson repeated these same ideas in *Principles of Animal Taxonomy* (1961), but with more theoretical background. In his chapter on "Higher Categories," Simpson discusses the primates as but just one example of a "higher category" and within the larger context of a theory of systematics. Here as elsewhere, by deliberately subsuming Primates in general, and hominoids in particular, under the broader principles of the evolutionary synthesis, Simpson was making a statement about human origins merely being a small, and fairly typical, part of the larger whole. "It is a frequent phenomenon in evolution that whole groups of animals, with numerous separate lineages, tend to progress with considerable parallelism through a sequence of adaptive zones or of increasingly effective organizational levels. . . . In some instances the progression is fairly steady, as in that of bony fishes toward the teleost level in the Mesozoic. In others there is comparatively rapid change from one level to another with subsequent stabilization and lesser diversification at each level after it is attained. Then there are more or less clearly definable steps, as between browsing and grazing horses or between prosimians, monkeys, apes, and men. It is to such steps that the concept of [evolutionary] grade most clearly applies. . . . The Hominidae . . . deserve separate grade recognition for their eventual development of fully upright posture, tool-making, greatest intelligence,

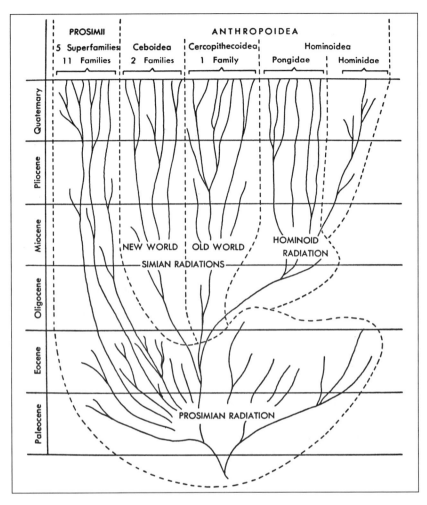

FIGURE 8.1 Simpson's depiction of primate evolution as a series of adaptive radiations, each less diverse and more specialized than the one that preceded it. (Simpson 1961c, *The Principles of Animal Taxonomy*, 213; © 1961 Columbia University Press)

and speech — perhaps in that order."[22] Washburn has remarked that Simpson "had a real role at these meetings of not only participating as one contributor among many, but also more generally educating people with respect to the evolutionary synthesis. But he was very careful not to appear to be doing so. He avoided being condescending or teachery. He could be very severe when someone went off the deep end. For ex-

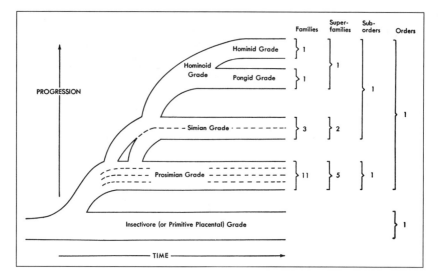

FIGURE 8.2 Simpson's interpretation of each primate adaptive radiation represented as an increase in evolutionary grade. (Simpson 1961b, *The Principles of Animal Taxonomy*, 215; © 1961 Columbia University Press)

ample, I garbled something Sewall Wright had written and Simpson corrected me and then added, 'Get it right!' I noticed that Simpson acted like an elder statesman, judge, helper with respect to the modern synthesis — a judicious catalyst who was very effective. I'm not sure if he consciously assumed this role or not, but he did give the impression of being ten years older than he was."[23]

Simpson as Critic

By the 1960s, paleoanthropology was fully incorporating the evolutionary synthesis as the theoretical grounding for its research. Simpson's role as mentor now shifted from that of doting parent to Dutch uncle. At a New York Academy of Sciences symposium in 1962 where papers discussed how comparative anatomy, immunology, cytogenetics, behavior, and sociobiology could contribute to the understanding of the relationships between nonhuman primates to human evolution, Simpson gave the concluding lecture. He expressed some dismay that, contrary to his hope, he was unable to present an updated version of his 1945 classification of the primates because "study of much of the new information has, however, convinced me that the very accumulation of data that makes systematic revision of the order necessary has also made it impossible for the present. . . . [P]rimate biochemistry and . . . primate behavior . . . obviously bear on classification . . . but they also . . . are still inadequate to be used uniformly throughout the order as major sets of taxonomic criteria."[24]

However, after acknowledging that new discoveries precluded what he had originally hoped to do, Simpson sounded a warning that, although there were many recent interesting studies of primate biochemistry and behavior, they were as unlikely to be taxonomic touchstones as were other putative "key" characters — despite "long and futile search" — in unraveling human evolution. "The addition of these new kinds of [biochemical] data will enormously increase the biological validity of classification, but it will not replace the older. . . . [And because] behavior is particularly labile . . . , reaction ranges are generally large . . . , farthest removed from the genetic substrate, . . . [and] also highly subject to convergence . . . , studies of behavior have not yet made a single concrete, acceptable contribution to primate classification . . . [although] it has been demonstrated in other groups, especially . . . birds" (503, 504).

Simpson went on to repeat what for him was a lifelong verity: that, while classification and phylogenetic interpretation were closely related, they were not identical. Moreover, he pointed out that all such interpretations must be based on the populational, not typological, concept of species because "the biological significance of taxa is evolutionary, the most biologically significant arrangement of taxa, that is, classification, would have an evolutionary basis. . . . [W]hat is actually being classified is not individual specimens, types, or an abstraction of characters in common, but reproducing and varying populations . . . which are inferred . . . from more or less adequate samples. . . . The new biochemical, karyological, behavioral, and other nonanatomical data will be particularly significant for . . . important details at levels below the family" (506, 507).

Simpson foreshadowed this populational viewpoint earlier in his review section of the 1945 classification of mammals when he asserted that all specimens of hominids, fossil and recent, could well be placed in the genus *Homo*, and that *Sinanthropus* and *Pithecanthropus* were "clearly synonymous by zoological criteria." In that same classification, Simpson also expressed some doubt about whether the apelike jaw of

"Eoanthropus" belonged to the skull, presumably reflecting the contemporary doubts about this taxon. Franz Weidenreich (1873–1948) made the same points himself in 1940 (summarized in his Table II, 382). While Simpson does not cite Weidenreich or others as authorities on these points, he must have been aware of the debates taking place in the late 1930s about them.[25]

Careful not to give undue importance to fossil evidence, Simpson did remind his audience that fossils would continue to provide crucial insight to the timing and sequence of human evolution. "[A]lthough myself a paleontologist, I disagree with a few taxonomists who think that classification adequately consistent with phylogeny can be achieved only if ancestral relationships are fully documented by fossils. [However,] the fossil record is essential for the establishment of some of the interpretive principles, . . . the time dimension [and] the temporal sequence of specializations. For instance, . . . elongation of the arms in gibbons was their last specialization and not a basic one, or . . . that in hominids upright posture preceded brain specialization. . . . The fossil record also reveals . . . extinct groups. . . . This is notably true of the primates, [which] poses serious problems in accommodating the extinct lineages in a classification also useful and significant for the living forms. The problem is particularly significant for the now extremely varied known Paleocene and Eocene forms, and questions are raised as to how the whole order Primates is to be delimited and defined."[26]

In 1963 Simpson was again invited by Sherwood Washburn to a conference on "Classification and Human Evolution," sponsored by the Wenner-Gren Anthropological Foundation, where he continued admonishing paleoanthropologists about their taxonomic methods. Some years later Simpson remarked that he had wanted to show "exactly what it means when you name a group of organisms, just what is implied for instance in calling man *Homo sapiens*."[27] "As a linguistic approach to the terminology of taxonomy and the nomenclature of classification [the paper] was a novelty. It was well received at the conference and has influenced subsequent work by anthropologists."[28] Such criticism was no doubt well received because Simpson was an authoritative outsider, whereas an insider's critique — however justified — might have been judged as simply self-serving.

Owing to a number of hominoid and hominid discoveries during the previous several years, considerable confusion had accumulated at the level of genera and species, particularly because of the many solecisms introduced by Louis Leakey (1903–1972). In the lead paper of the conference volume, Simpson indicated how one should go about naming taxa by proceeding very deliberately from individual specimens, to the population that they presumably sample, to the taxon the population is inferred to belong to, and then to the hierarchic assignment one gives to that taxon (fig. 8.3). "Discussion at the conference repeatedly illustrated the need for employing and distinguishing the naming sets. The ambiguity and clumsiness of usual references to particular specimens and populations were especially evident. For example no clear and simple way was found for designating the various specimens from Olduvai Bed I that are believed not to belong to the taxon called *Zinjanthropus boisei* by Leakey."[29]

As Simpson warmed to his subject, he made no bones about criticizing some of the current practices within paleoanthropology: "Much of the complexity and lack of agreement in nomenclature in this field does not, however, stem from ignorance or flouting of formal procedures but from differences of opinion that cannot be settled by rule or fiat. . . . This must be almost the only field of science in which those who do not know and follow the established norms have so frequently had the temerity and opportunity to publish research that is, in this respect, incompetent" (5).

Simpson also argued against the persistent, covert typological thinking still found among some anthropologists. "The 'definition' [of a taxon] has often been only a description of an individual 'type' with no regard for or even apparent consciousness of the fact that taxa are populations. . . . It is of course also true that the significance of differences between any two specimens has almost invariably come to be enormously exaggerated by one authority or another in this field. Here the lack is not so much lack of taxonomic grammar as lack of taxonomic common sense or experience. Many fossil hominids have been described and named by workers with no other experience in taxonomy" (6).

Simpson further emphasized that more than one classification may legitimately ensue, depending upon prior phylogenetic interpretation. Using the Hominidae as an example, he showed how three classifications might result, depending on how the fossil data were interpreted (fig. 8.4). Aggregating the fossil discoveries by time and skeletal character, one could arrive at one arrangement (A, upper left). By using morphology strictly alone, without consideration of temporal sequence, one might recognize "an arbitrary number of fixed, distinctive

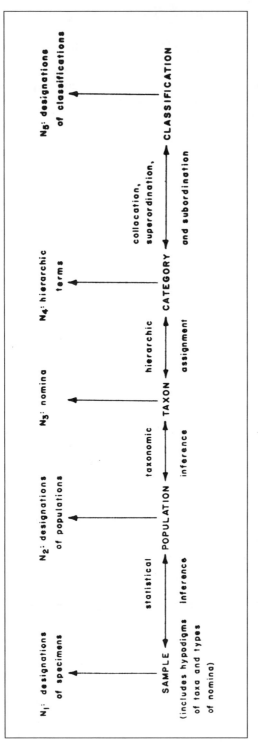

FIGURE 8.3 Steps in taxonomic procedure showing process (arrows), name sets (N), and the referents to which the names apply (capitals). Simpson notes that the procedure logically moves from left to right, but "in practice no one operation can be carried out without reference to the others. These arrows are therefore pointing both ways." (Simpson 1963a:2; reprinted from Sherwood L. Washburn, ed., *Classification and Human Evolution*, Viking Fund Publications in Anthropology, No. 37, 1963, by permission of the Wenner-Gren Foundation for Anthropological Research, Inc., New York, New York)

types in a morphogenetic field" (B, lower left). Or one might interpret that hominids have belonged to a single interbreeding population since the early Pleistocene, and thus represent only one genetical species at any one time (C, upper right). Or perhaps one might assume recurrent speciation through time, producing two or more distinct, contemporaneous species, of which only one of the last two has survived (D, lower right). Simpson gave his view how each of the interpretations might be favored, and why. He also made clear that it was unlikely that there ever would be universal agreement on classification. "Even if the objective data were complete — indeed especially if they were complete — classification would have many arbitrary elements. Complete agreement as to what is to be said is thus chimerical and indeed may not even be desirable. We can, however, agree how to say what we mean" (18).

Simpson concluded his discussion with a "classification I now favor" of the Hominoidea, a superfamily that included the three families of oreopithecids, pongids, and hominids. The only major changes from his 1945 classification were placing *Australopithecus* in the hominids instead of the pongids and erecting a separate family for *Oreopithecus* that he had earlier placed within an uncertain subfamily of the cercopithecids. (Of course, following the contemporary consensus, Simpson excluded "Eoanthropus" altogether, which he had included in the 1945 classification.) Simpson stuck to this interpretation for the rest of his life, for in his last paleontology text his hominoid phylogeny is topologically almost identical, the chief difference being that he shows the hylobatids coming out of a dryopithecine complex instead of a more ancestral "unidentified source."[30]

In his discussion of the classification he now favored, Simpson placed the genus *Gorilla* in synonymy with *Pan*: "Merely listing characters that demonstrate the self-evident fact of their distinctness does not necessarily suffice to maintain their time-honored generic separation, and at present I prefer to consider both chimpanzees and gorillas as species of *Pan*. . . . Placing all the African apes in *Pan* permits classification to express the clear fact that they are much more closely related to each other than to any species of other genera." In this latter portion of his paper, Simpson also used the karyologic and serologic evidence presented at the conference for separating apes (*Pan* and *Gorilla*) from orangs (*Pongo*) as well as from gibbons (*Hylobates*), and having the apes issuing from the hominid line rather than the gibbon line. He summa-

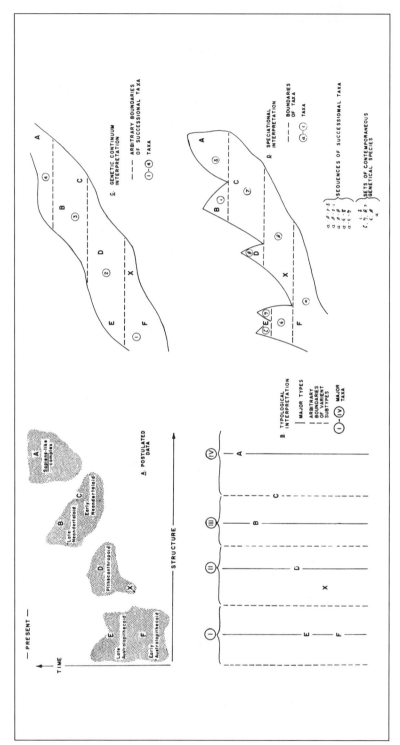

FIGURE 8.4 Simpson showing how a set of postulated hominid data (A) may yield three possible interpretations-typological (B), genetic continuum (C), and speciational (D). (Simpson 1963a:13; reprinted from Sherwood L. Washburn, ed., *Classification and Human Evolution*, Viking Fund Publications in Anthropology, No. 37, 1963, by permission of the Wenner-Gren Foundation for Anthropological Research, Inc., New York, New York)

rized these relationships in a figure that shows "by combination of dendrogram and adaptive grid, my present views as to the affinities and the adaptive or structural-functional relationships of the living hominoids."[31] Thus, following his own earlier advice about the complementary value of such evidence, Simpson used molecular and genetic data together with the more traditional morphologic information in interpreting these relationships.

Simpson as Apologist

By the mid-1960s Simpson's role among anthropologists began to wane because what he and others had been promoting about the evolutionary synthesis had been incorporated into the anthropological mainstream. Moreover, the new generation of anthropologists, fully educated within the new theory, had taken over. Whatever authority and expertise Simpson had brought to paleoanthropology now shifted to bringing the knowledge of modern paleoanthropology to a nonprofessional audience of educated lay people.

The culmination of his role as mentor was the reprinting of Simpson's chapter ("The Study of Evolution: Methods and Present Status of Theory") from his and Anne Roe's 1958 volume *Behavior and Evolution* as the opening chapter in William Howell's 1962 collection of essays, *Ideas on Human Evolution*. With respect to the latter book, Howells remarked in his preface that "ideas have changed largely because of two things: new fossils and new theoretical understandings," with the chapter by Simpson providing, according to Howells, "a brief and simple review of certain main points of the modern synthetic theory of evolution, written by one of its principal architects."[32]

Soon after, Simpson's role became more that of an apologist for, and explicator of, the materialistic view of human origins. In his 1966 *Science* article, "The Biological Nature of Man," Simpson promoted anthropology's modern views about human origins to a nonanthropological audience. The oral versions of this paper were given at two public lectures in universities at opposite ends of the country as well as at the University of Paris when Simpson received an honorary doctorate there.

By now Simpson was making the obvious points a paleontologist would make about human evolution — namely, that humans evolved from apes that were also ancestral to chimpanzees and gorillas; that divergence from the latter was marked early by bipedalism and upright posture as well as by dental evolution; and that only after all that did the human brain begin to enlarge. But Simpson went further and obliquely addressed the racial issues then being debated in the United States. He stated as strongly and authoritatively as he could that the human species does include quite diverse races, yet which always grade into each other without definite boundaries. He noted that neither now nor in the past has there been a pure race, but rather that races are evanescent, may change, disappear by fusion with others, or die out altogether, and that they have the same biological significance as the subspecies of other species of mammals, being originally geographic and, for the most part, adaptive. But "at the present time race has virtually no strictly biological significance because of two crucial changes. First, human adaptation . . . is now mostly cultural . . . so that the prehistoric biological adaptations have lost much of their importance. Second, tremendous increases in population size, in mobility, and in environmental changes brought about by man himself have the result that few men are now living under the conditions to which their ancestors were racially adapted. Regardless of the diversity of races . . . all men resemble one another much more than any of them differ from each other."[33]

Simpson continued his discussion by indicating that many of the nonmaterial attributes of humans — like behavior and psychology, speech and language, and an ethical and religious sense — are ultimately grounded in human physical anatomy. "We are no longer concerned with whether man evolved, because we know he did. We are still concerned with how he evolved, with what is most characteristically human about him and how those characteristics arose."[34] That this paper made an impression on at least one reader is shown by the remarks of author-mathematician Jacob Bronowski who, in reviewing in *Science* the collection of essays in which this paper was reprinted, claimed that "I can think of no other account as concise and well reasoned as this that transmits . . . the sense that man is unique as a biological unit, and not merely as an example of particular traits."[35]

Both in this and in Simpson's subsequent 1972 paper, "The Evolutionary Concept of Man" — a somewhat more technical version of the 1966 effort — Simpson also argued that contemporary anthropological interpretation of human evolution is in direct continuity with Darwin, and although acknowledging the great interest primate evolution will have for a primate, Simpson refused to give special primacy to Primates. At the end of the paper Simpson remarked: "The greatest impact of the Darwinian revolution . . . was that it completed the liberation from superstition and fear that began in the physical sciences a few centuries before. Man, too, is a natural phenomenon."[36]

One aspect of his role as explicator and defender of the prevailing paleoanthropological views regarding human origins was his ongoing critique — in essays, conferences, and book reviews — of the mystical approach to human evolution, past, present, and future taken by Pierre Teilhard de Chardin (1888–1955). Teilhard was a French Jesuit priest and physical anthropologist who had collaborated in the discovery of "Peking Man" in the late 1920s. He also wrote several quasi-philosophical books, wherein he argued for spiritual direction in cosmic evolution toward an "Omega Point." The 1959 English translation in particular of *The Phenomenon of Man* gained him, posthumously, a popular following in the United States. While fond of Teilhard personally, Simpson was an outspoken critic of both his science (sometimes) and the philosophical conclusions therefrom (always). In a 1973 article that Simpson called his "principal publication on that particular subject," Simpson dismantled Teilhard's thesis. Among other things, Simpson asserted that "as regards the actual processes of evolution, . . . Teilhard simply ignored the consensus on this subject, did not understand theories current in his own lifetime, overlooked or brushed aside practically all the evidence for them, and ostensibly based his own views on evidence that does not, in fact, logically support them. . . . I am not competent to judge the Teilhardian mystic theology, and I have nothing to say about its validity in itself. I do say that it does not follow from any of Teilhard's scientific work or any other premises acceptable as science. When so presented it is indeed a divine non sequitur."[37]

Simpson was also strongly opposed to the "naked ape" view of the origins of human behavior as promulgated in a best-selling popularization by the zoologist Desmond Morris. In a *New York Times* Sunday book review, Simpson, calling Morris's views "bad zoology and inept biology," claimed that, "Most evolutionists, geneticists, and anthropologists will feel that [Morris] overemphasizes genetic controls and underemphasizes the fact that a loosening of such controls, a broadening of genetic reaction ranges, is a fundamental element in the origin of this non-ape."[38]

Simpson also became marginally embroiled in the furor that arose with the publication of *The Origin of Races* by anthropologist Carleton

Coon (1904–1981) in 1963, which Simpson reviewed quite favorably. Simpson was in the minority that was willing to consider, at face value, Coon's claim, which he called "highly original (and strikingly unorthodox)," that the modern races (or subspecies) of humans became differentiated very early in their history within the ancestral species of *Homo erectus*, thereafter evolving in the direction of *H. sapiens*, but at different rates.[39] It is not surprising that Simpson would be open to such a hypothesis when we consider that one of the major themes of *Tempo and Mode* was precisely the issue of how related lineages of organisms might differentiate themselves over time through varying rates of evolution.[40] Moreover, Simpson had also championed the notion that the mammalian class had polyphyletic origins. That is, rather than tracing their ancestry to a single, unified taxon, the mammals had evolved mammalian characters in several distinct yet related lineages of therapsid mammal-like reptiles.[41]

Simpson was well aware of the dangers of intentional misrepresentation of Coon's views by racists, especially during the divisive national debate then taking place about the proposed passage of a civil rights law. But he found Coon's book "an honest and substantial contribution to the scientific study of races and as nearly free of aprioristic bias as any on its subject" (1963b:269). However judicious and rational Simpson thought he was, he only compounded his problems when he next asserted that the "initial and indubitable datum is that races are not equal. They are obviously different, and things that are different are not equal to each other. The basic differences are biological and evolutionary in origin."[42] Simpson did cavil with Coon over the way Coon equated evolutionary grade, defined by him morphologically, with the concept of species, so that one subspecies of *H. erectus* is postulated as living at the same time as another, different subspecies of *H. sapiens*, and yet if different species could not, by definition, interbreed.

Simpson had one final, formal interaction with paleoanthropology at a 1974 conference on primate evolution and human origins, where Simpson reviewed current schools of taxonomy (including phenetic, numerical, cladistic, and evolutionary) as well as the relationship between taxonomy and phylogeny. But little of what he had to say was either new or, more critically, had direct bearing on contemporary paleoanthropological issues, although he did acknowledge that biochemical information provided "[a]nother kind of resemblance and difference to be considered along with all the others" in determining phylogeny.[43] Despite the fact that Simpson did not revise the preprinted version circulated at the conference as he originally intended, his paper opened the resulting conference volume, presumably as much for what Simpson had contributed previously to the discipline over the years as for "his illuminating presence and participation" at the conference itself.[44] Examination of the rest of the book indicates that Simpson's presence in the volume was indeed honorific, because the level of discussion and areas of expertise had moved well beyond him. Although he had become an outsider, he must have felt some satisfaction for having contributed to the current status of the discipline.

* * *

Although speaking of Simpson's 1945 classification, the following quote by Sir Wilfrid Le Gros Clark suggests some of the reasons why, more generally, Simpson had the influence he did with paleoanthropologists: "We use [Simpson's classification] because (1) it is based on recognized authority, (2) it has the merit of simplicity, (3) it appears to reflect reasonably closely such phylogenetic relationships as can be inferred from the evidence at hand . . . , and (4) it has been provisionally accepted and recognized by other authoritative workers in the same field."[45] In short, authority, simplicity, inferences based on evidence, and acceptance by other workers seem to have carried the day for Simpson with the paleoanthropologists on most occasions.

In summary, then, this example of Simpson and paleoanthropology indicates one way in which a specific discipline is strongly influenced by the advances in another, allied discipline. Although one might suppose the transfer of principles and concepts from one to the other could occur in many different ways, in this particular instance a considerable part of the transfer was the result of Simpson's mentorship, and thus the evolutionary synthesis was more quickly and easily incorporated into modern paleoanthropology than by, say, general intellectual diffusion. Indeed, the physical anthropologists evidently felt this way, for in 1981 they invited Simpson to address the fiftieth anniversary celebration of the founding of their association, even though Simpson had not contributed to the field in any significant way during the previous decade. Because Simpson was unable to attend the meeting himself, Sherwood Washburn read Simpson's remarks in which the latter recounts his "contacts with anthropology" over the previous half century.

In the published version, Simpson added a postscript wherein he once again urged anthropologists to make the distinction between classification and phylogeny, to be suspicious of proposed panaceas in solving problems related to either, and to be especially cautious about relying too much on behavioral and molecular data in solving evolutionary problems.[46]

CHAPTER 9

WRONG FOR THE RIGHT REASONS

Wide rejection of a theory now believed correct in essence is a phenomenon of great interest for the history and philosophy of science.
— George G. Simpson

Actually most scientific problems are far better understood by studying their history than their logic.
— Ernst Mayr

In the two decades preceding the establishment of plate tectonics theory, Simpson wrote a series of papers that refuted the paleobiogeographic evidence purportedly requiring direct continental connections. So cogent was his rebuttal that drift proponents thereafter downplayed the past distribution of fossils. Simpson presented several different arguments. He restricted his discussion to Cenozoic mammals with whose evolutionary history he was most familiar; he challenged the accuracy of the proponents' data; he criticized their methodology, both with respect to its undue appeal to authority and its lack of parsimony; he used his own "coefficient of faunal similarity" to show that present-day mammal distributions on dispersed continents are comparable to those for Tertiary Period continents; and perhaps most effective was his statistical argument that transoceanic dispersal of organisms, while highly improbable for a single event, is practically certain given the many opportunities for such events over geologic time-spans. Simpson's objections arose from his own independently developed theory of historical biogeography which, by relying on mobile organisms dispersing across stable continents, was fully adequate to explain the paleobiogeographic data. Simpson thus regarded both drift theory and transoceanic land-bridges neither sufficient nor necessary to account for those data.

Background

It is well known that the theory of continental drift was firmly rejected by the great majority of American geologists for a half century after its initial formulation by Alfred Wegener (1880–1930) in 1912. Following the plate tectonics revolution of the 1960s, various authors suggested the sources and nature of the objections to drift. For example, some emphasized the lack of a suitable mechanism that would permit the less rigid continental crust ("sial") to move horizontally through more rigid oceanic crust ("sima"), or as colloquially expressed, "you can't push whipped cream through Jello." Others insisted that Wegener and his followers presented evidence for drift that was ambiguous, tenuous, or just plain wrong. Still others argued that it wasn't a question so much of unconvincing evidence or inadequate mechanism as it was that Wegener was an outsider, an uninformed astronomer/meteorologist whose lack of credentials as a geological specialist undermined his position at the outset.[1]

Simpson was preeminent among American geologists who opposed drift and, in a series of papers in the 1940s, dismantled the paleontological arguments for continental drift as well as for transoceanic land-bridges. From 1912 when Wegener first announced his theory, until the 1940s when Simpson definitively rebutted the paleontological arguments, fossil data were central to the theory of continental drift. After World War II, fossils were either de-emphasized or not mentioned at all in support of drift. On the contrary, when fossils were considered they were used to support a stabilist position.[2] Various reasons have been offered to explain Simpson's strong opposition to continental drift. Both Joel Cracraft and Malcolm McKenna of the American Museum of Natural History suggest that Simpson's views resulted from a priori stabilist assumptions.[3] Or, in view of the strong influence of W. D. Matthew, we might suppose that Simpson was following the AMNH's current party line,[4] whereby the fixed northern continents were viewed as the centers of origin for most Cenozoic mammalian groups that then migrated into the dispersed and fixed continents of the Southern Hemisphere.[5] Simpson, of course, knew of Matthew's important work on historical biogeography, *Climate and Evolution* (1939), and he was persuaded by it; but he did not accept Matthew's ideas uncritically.[6]

The purpose of this chapter is several-fold. First, I will show that Simpson's opposition to transoceanic land-bridges and continental drift resulted *not* from stabilist assumptions but from an earlier-developed theory of historical biogeography. Second, I will demonstrate that Simpson's biogeographic interpretations were independently arrived at, irrespective of his genuine admiration for Matthew. Third, I will dissect the various lines of Simpson's argument to illuminate his style of approach to a scientific problem. Fourth, I will indicate how Simpson's consideration of this issue developed in parallel with other major lines of research he was pursuing at that stage of his professional career. Finally, I will discuss Simpson's belated conversion to plate tectonics and offer my own conclusions why Simpson was misled. Necessarily, I will go over much of the same ground already well-covered by geologist Ursula Marvin in her book on the evolution of the concept of continental drift (see, especially, her discussion of the "verdict of paleontologists") and by historian of science Henry Frankel in his essay on the debate over the explanation of disjunctively distributed organisms. However, whereas Marvin and Frankel consider the broader issue of stabilist versus mobilist theories and Simpson's role therein, I invert the discussion by keeping the focus on Simpson, the paleontologist, to see how and why he came to oppose drift theory.[7]

Wegener and Fossils

One year before his death on the inland icecap of Greenland, Alfred Wegener published the last edition that he personally revised of his epoch-making book, *Die Entstehung der Kontinente und Ozeane*, the first edition of which he had published in 1915. In the initial chapter of that work Wegener informed his readers how he came to his theory of drifting continents. He recalled that the concept first occurred to him in 1910 when he was struck "by the congruence of the coastlines on either side of the Atlantic." However, he regarded the idea as "improbable" until a year later when he chanced upon a report "in which I learned for the first time of palaeontological evidence for a former land bridge between Brazil and Africa. . . . The fundamental soundness took root in my mind . . . [and o]n the 6th of January 1912 I put forward the idea [of continental drift] for the first time in an address to the Geological Association in Frankfurt am Main."[8]

Thus, we see in Wegener's own words his assertion that a key to his theory of mobile continents was similarity of fossils on continents that are, today, widely dispersed. In that same 1929 edition, Wegener devoted one whole chapter to "Palaeontological and Biological Arguments," where "we are justified in counting as favourable to drift theory all biological facts which imply that at one time unobstructed land connections lay across today's ocean basins" (98).

Among the many facts cited by Wegener was a table of data, attributed to paleontologist Theodore Arldt, that gave the "percentage of identical reptiles and mammals on each side . . . of the former land connection between Europe and North America" (100). The percentages fluctuated widely and unsystematically. Yet Wegener claimed that land connections between North America and Europe were corroborated when the values were high (29–64 percent for reptiles; 31–35 percent for mammals) during the Carboniferous, Triassic, Lower Jurassic, and Upper Cretaceous through the Lower Tertiary periods. No connections existed during the Permian (12 percent identical similarity of reptiles) and the Neogene (19–30 percent mammalian similarity). I will return to these data below when considering Simpson's response to Wegenerian arguments.

Elsewhere in his book, Wegener cited the by now familiar arguments for the Southern Hemisphere supercontinent of Gondwana based on the *Glossopteris* flora and the presumed freshwater reptile, *Mesosaurus*, of the late Paleozoic, as well as much evidence from disjunct distributions of living plants and animals, ranging from conifers to earthworms. The point I wish to make is that for Wegener fossil data were by no means a trivial part of his argument for drift. Nor were they trivial for his followers. For example, geologist Alexander Du Toit in his book also offered paleontological support for his views.[9] Hence, in light of subsequent apparent rebuttal of the fossil evidence for drift by people like Simpson, it is not surprising that the concept of continental drift was discredited.

Simpson's Response to Drift

Simpson's response to those who hypothesized continental drift, like Wegener and Du Toit, or transoceanic land-bridges, like geologists John W. Gregory and Hermann von Ihering, is contained in a series of key articles written during the 1940s and summarized in a short book, *Evolution and Geography*, published in 1953.[10] Although Simpson regularly discussed biogeography in most of his other work — whether systematic or theoretical — it is in these writings that we find his theory of historical biogeography best developed.

There are two points to be considered. First, Simpson's arguments for stable continents came out of an earlier well-formulated theory of historical biogeography. Having established his own principles of historical biogeography, he then applied them to the hypothesis of drift and land-bridges. Simpson did not argue, ad hoc, against mobilist views; rather, he tested those views in light of much broader, more inclusive concepts of the factors controlling the past distributions and abundances of terrestrial organisms.

Second, because his theory of historical biogeography was based essentially on Cenozoic mammals, his arguments against drift and land-bridges were couched in the data provided by them. Retrospectively, it may seem that Simpson was begging the question by using Cenozoic mammals because we now know that significant breakup and drift of the northern supercontinent of Laurasia and southern supercontinent of Gondwana took place in the Mesozoic Era. By Eocene time the continents were already dispersed and relatively close to their present-day positions.

In this regard it is important to recall what else Simpson was doing during this stage of his professional career. During the 1930s, Simpson published extensively on South American faunas; in 1930–31 and 1933–34, he spent two years in Patagonia on the Scarritt expeditions of the American Museum collecting Cenozoic mammals. His South American studies gave him a deep insight into the role of faunal isolation and interchange, on a continental scale, in determining the geographic character of a fauna.

During the late 1930s and early 1940s, Simpson was also working on his monograph dealing with the principles of mammal classification, by which he became thoroughly familiar with the worldwide biogeography of Cenozoic mammals.[11] Also at this time, Simpson was writing *Tempo and Mode in Evolution*, in which he made rather original arguments (at least for paleontology) for the crucial role that environment played in evolutionary history. Finally, Simpson coauthored with Anne Roe *Quantitative Zoology*, wherein he demonstrated his insistence on biostatistical rigor and robustness — qualities which, as we will see, he found sorely lacking in the Wegenerian arguments. All these projects were either completed or well on their way by the time Simpson attended the Sixth Pacific Congress in San Francisco, in the summer of 1939, where he gave a paper on "Antarctica as a Faunal Migration Route." In this paper he outlined what would become his theory of historical biogeography.[12]

Structure of Simpson's Rebuttal

Simpson's counterarguments to the drifters and land-bridgers can be grouped into three categories: those arguments that pointed out flaws of critical reasoning; those that emphasized what, today, we would call "evolutionary ecology"; and those arguments that were quantitatively formulated. Simpson used these different types of argument in various combinations in his series of articles dealing with problems of biogeography. Looking back at this body of work, we can see these basic threads throughout. Thus we learn something of Simpson's approach to a scientific question and "solving" it.

Flaws in the Drifters' Reasoning and Data

Throughout his rebuttal of the mobilist position, Simpson attacked the citation of evidence that he considered wrong or misleading. He also pointed out the lack of parsimony, especially with respect to moving continents about with free abandon in order to accommodate the presence or absence of some particular group of fossils being discussed, even though some other fossil taxa not under consideration indicated quite the opposite. Simpson also chided drifters for invoking the word of authorities in one field of geology in this area of paleontology in which they were patently not competent. The following example will illustrate well the foregoing. In the early part of this century, two distinguished American paleontologists — J. W. Gidley of the U.S. National Museum and H. F. Osborn of the American Museum — believed that some scrappy equid fossils from the southeastern United States were probably referable to the Old World, late Miocene genus, *Hipparion*. In 1919 a French paleontologist, Léonce Joleaud, claimed — incorrectly, as it turned out — that these New World *Hipparion* were nothing at all like the rest of western North American Miocene horses and, therefore, in order for the eastern North American taxa to have reached the Old

World there must have been a transoceanic land-bridge from Florida to Spain by way of the Antilles and Africa, rather than by the Bering Strait land-bridge.

Later that same year, Joleaud extended the temporal duration of his bridge from late Miocene to include early Miocene to late Pliocene. By 1924, Joleaud not only broadened his land-bridge so that it ran from Maryland to Brazil throughout most of the Tertiary, but he had also embraced continental drift to provide the intercontinental connection rather than a transoceanic land-bridge spanning two fixed continents. Moreover, to accommodate variations in faunal similarity between the New and Old Worlds, Joleaud opened and closed the Atlantic Ocean in accordion-like movements, thereby requiring multiple episodes of splitting apart of the continents whereas Wegener only required a single parting.[13]

In 1929, John W. Gregory (1864–1932), a distinguished British geologist (after whom the Gregory Rift in East Africa is named), but no paleontologist, cited Joleaud's *Hipparion*-bridge in his Presidential Address to the Geological Society of London on the "Geological History of the Atlantic Ocean." Gregory himself opposed drift, so he kept only the land-bridge notion of Joleaud without accepting the accordion-like movement that Joleaud assigned it.

Alexander Du Toit (1878–1948), in his book *Our Wandering Continents*, accepted a Brazil/Venezuela-to-Africa land-bridge, citing Gregory as the authority for this concept, even though as a drifter Du Toit would have preferred direct continental connection. Simpson traced in much more detail than I have given here the complicated path of reasoning from scrappy equid fossil evidence in Florida to the entrenched notion of the *Hipparion* land-bridge throughout much of the Tertiary, with or without lateral movement of the continents. Simpson was particularly chagrined that neither Gregory nor Du Toit formed or sought an independent opinion (his?) regarding the paleontological basis for the *Hipparion* bridge in the first place.

Another briefer example of faulty logic cited by Simpson in the drift argument goes back to Wegener himself. As noted above, Wegener provided a table of "percentage of *identical* reptiles and mammals . . . between Europe and North America."[14] Simpson went back to the original reference and discovered that the percentages referred to families and subfamilies, and thus could hardly be called "identical." Simpson expostulated: "Such looseness of thought or method amounts to egregious misrepresentation and it abounds in the literature of this perplexing topic."[15]

Evolutionary Ecology

However cogent Simpson's criticism of the flaws in the reasoning of the mobilists, the debate certainly would not have gone beyond the "is/is not" stage if Simpson had not also elaborated a broader theory of those factors that limit the distribution and abundance of terrestrial organisms on continental and intercontinental scales. If Simpson could account for past biogeographies, using established principles of evolution and ecology against a background of stable, dispersed continents, then of course there would be no need to invoke additional hypotheses requiring drifting continents or sinking land-bridges.

As just mentioned, during the time Simpson was rebutting drift he was also completing *Tempo and Mode in Evolution*, which broke with the then-contemporary paleontological opinion about the mechanisms of evolution, coming down on the side of the new genetics. Not only did Simpson not accept, but he also thoroughly debunked, various notions of contemporary paleontologists to explain evolutionary patterns seen in the fossil record. In a sense, Simpson was an iconoclast who broke with existing paleontological tradition and offered original and testable interpretations of the fossil record. One can note, then, in passing, that *Tempo and Mode* alone refutes any claim that Simpson was a crowd-follower with respect to his position on continental drift. It is unreasonable to assert that he opposed the evolutionary theories of R. S. Lull and H. F. Osborn, yet uncritically followed W. D. Matthew with respect to biogeography. Simpson always started from first principles and came to his own, independently derived, conclusions (even if wrong, as he was in this instance).

In *Tempo and Mode* Simpson specified the many ways in which organisms can be expected to interact with their environment, biologic as well as physical. Organisms not only evolved within a given "adaptive zone," but the physical parameters defining the zone also changed during geologic time. Thus, "the course of adaptive history (as read from the fossil record) may be pictorialized as a mobile series of ecological zones with time as one dimension."[16] This quote indicates Simpson's strongly held view of the role environment played in the fine details of evolution, as well as in its much broader aspects. In this regard, Simpson staked out an early claim as an important contributor to what is, today, called "evolutionary ecology."

With respect to biogeography, Simpson addressed what he identified as the "broadest problems of all in this field, the general way in which land animals tend to become distributed and in which their distribution tends to change in time."[17] This latter problem included, as well, the more specific one of "the different types of migration routes between major land areas, the way in which one type or another can be inferred from the faunal evidence, and the effect that a given type has on the faunas that use it." Simpson then hypothesized that once a new group of organisms arose in some small geographic area, it spread outward over a much broader area. How far and how fast it spread depended upon the environmental tolerance and the inherent dispersal abilities of the organism in question. Simpson conjectured the expansion and contraction of the organisms over time as environmental opportunities waxed and waned. At some later time, different parts of the populations might well be widely geographically isolated, yielding the classic "disjunct distributions" that were so readily cited by drifters and land-bridgers to support their positions, yet which could be explained in this alternative way (fig. 9.1).

Simpson acknowledged that terrestrial organisms would, necessarily, encounter barriers of one sort or another eventually, no matter how permissive the environment or inherently dispersible the organisms. Mountain ranges, deserts, seas, and ice sheets would inhibit further migration. Simpson emphasized, however, that barriers to dispersal should not be judged as all-or-nothing propositions. Rather, he recognized that land vertebrates would respond differently to each kind of barrier, according to their environmental adaptability, numbers of individuals, size and mobility of the animals, and the degree of environmental variability across the barrier and its geographic extent. Simpson stated, "Freedom of movement and its restriction are relative. . . . It is all a matter of degree, of diverse probabilities."[18] Although Simpson viewed barriers to faunal migration as a spectrum of decreasing opportunity, he thought it convenient to subdivide that spectrum into barriers of three broad sorts. *Corridors* were weak barriers, so they permitted a rather full and free interchange of animals across them. One might see different proportions of taxa at either end of a long corridor, like between central Asia and western Europe, owing to some attenuation be-

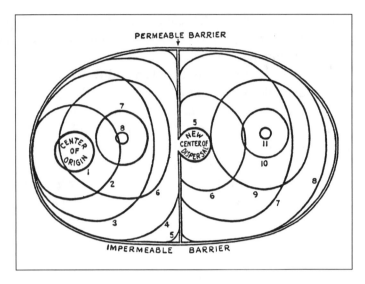

FIGURE 9.1 Simpson's schematic diagram illustrating expanding terrestrial vertebrate populations (1–4), their breaching of an ecologic barrier (5), and further expansion (5–8). During the latter, secondary expansion, the population in the primary area contracts markedly (6–8); later, the secondarily expanded population also contracts (9–11). The contracting phases result in "disjunct distribution" of the populations. (Simpson 1940b, "Mammals and Land Bridges," 145)

cause of the long distances involved (some 8,000 kms in the case of China and France). Evidence of such a corridor in the geologic past would be the widespread distribution of very similar, but not necessarily identical, terrestrial organisms.

At the other end of the spectrum were *sweepstakes routes*, where the barrier was virtually complete, and yet despite low probability, chance events such as rafting across ocean deeps and island-hopping along a broad archipelago resulted in the penetration of the barrier by only a few species. Fossil evidence for such a route would be the presence of a few similar taxa in two regions whose fossils were otherwise completely distinct from each other. Not only did the faunas differ greatly in degree of similarity when comparing corridors versus sweepstake routes, but they also differed in their ecologic balance. That is, faunas crossing a corridor would include organisms spanning a wide range of habits and habitats, whereas sweepstakes-route organisms would produce an ecologically unbalanced fauna and, of course, the migration would be in one direction only. The terrestrial faunas of Madagascar,

Australia, and the Hawaiian Islands are examples of such sweepstakes-route migrations.

Filter bridges, according to Simpson, spanned the range between corridors and sweepstakes routes; there were varying degrees of selectivity among barriers that allowed — some more, some less — different kinds of organisms to pass. Between continents, a filter bridge was typically a relatively narrow isthmus of limited environmental range and therefore a stronger impediment to faunal interchange than a corridor, but much less an impediment than a sweepstakes route. Simpson cited the present-day Isthmus of Panama as one such filter bridge. Evidence for past filters would be intermediate between the kind found for corridors and sweepstakes routes: moderately ecologically balanced faunas with some migrants from each side of the barrier.

For Simpson, then, present and past distribution of terrestrial organisms was an ecological issue: what was the nature of the barrier to be crossed, and what were the corresponding environmental tolerances of the organisms attempting to breach the barrier? Simpson viewed paleobiogeography, therefore, as a study of the potential mobility of organisms across stable continents (fig. 9.2).[19]

In addition, Simpson approached the study as a dynamic one whose elements were continuously changing, whether they were the nature of the barriers as well as their geographic breadth and temporal persistence, or of the organisms and their range of environmental response. Consequently, Simpson argued for a probabilistic analysis of barriers rather than one that insisted on all-or-nothing claims.

Quantitative Insights

In all of Simpson's scientific work there is a strong quantitative element, no doubt reflecting both his ease with mathematics and his high standards of rigor and reproducibility in scientific argument. It is not surprising, therefore, that in developing his theory of historical biogeography, Simpson would introduce a quantitative point of view.

Because much of the argument for or against direct, physical connection of continents was based on qualitative statements of faunal similarity that were imprecise and poorly reproducible, Simpson invented what he called the "coefficient of faunal similarity," a measure still widely used today. The coefficient was a simple ratio of the number of taxa (species, genera, families, or orders) in common between two fau-

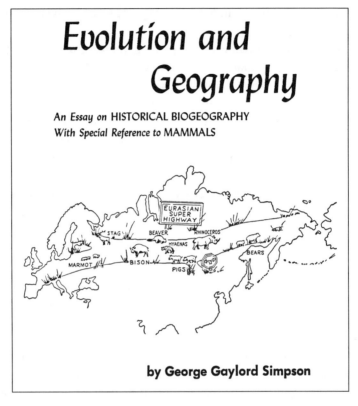

FIGURE 9.2 Cover of Simpson's essay on historical biogeography that illustrates his emphasis on the crucial role of overland dispersal in explaining the past distributions of terrestrial mammals. (Simpson 1953a, *Evolution and Geography*. Courtesy Thomas Condon State Museum of Fossils, University of Oregon)

nas (C) divided by the number of taxa in the smaller of the two (N); multiplication of the ratio by 100 converted it to a percentage, i.e., 100 (C/N_1). Although Simpson did not formally propose this measure of faunal similarity until 1947 in a short note to the journal *Evolution*, he had used it earlier (Simpson 1936a) and had discussed its derivation in a footnote (Simpson 1943).[20]

Simpson's most cogent use of his coefficient of faunal similarity was in "Mammals and the Nature of Continents," where he provided data (table 9.1) showing his calculations of the coefficient for genera of recent and fossil mammals on several continents. Simpson introduced the data by remarking "how gratuitous are the vague statements and sweeping claims of past faunal resemblances so great as to be inconsistent with the present positions of the continents."[21]

TABLE 9.1 Coefficients of Faunal Similarity, 100 C/N_1, Mammal Genera

82	Ohio/Nebraska (800 km)	⎫
67	Florida/New Mexico (1600 km)	⎬ Recent
64	France/N. China (8000 km)	
24	New Mexico/Venezuela (5300 km)	⎭

45	Eur/NA (8000 km, Greenland) — Early Eocene
15	Eur/NA (16,000 km, Asia) — "Pontian"
0	NA/SA (9600 km, Panama) — Early Pliocene

8	SA/Africa (7600 km) — Triassic Reptiles

With respect to the first value of 82 percent in the table, Simpson asserts that it recorded such a resemblance "as should frequently appear between exactly opposite points on different continents that were, according to drift theory, in former contact. . . . But *no fossil land faunas resembling each other to a degree at all comparable with this have ever been found on continents now separated.* Only evidence of this kind would be more consistent with drift theory than any other, and no such evidence is known [emphasis in original]."

Simpson noted that the next two values came from continents with distinct climatic differences and minor geographic barriers (67 percent) or marked geographic barriers (64 percent). He claimed that the values were "representative of mammalian faunas of distant parts of the same continent. . . . Resemblances of this order of magnitude may be about the least expected between continents united according to the drift theory and about the greatest to be expected between distinct continents . . . according to non-drift theories. *Resemblances of this degree are altogether exceptional among fossil vertebrate faunas of continents now distinct* [emphasis in original]." The last value for recent land mammals (24 percent) came from different faunal realms connected by an isthmian link with marked climatic and geographic barriers. According to Simpson, values "far less than this have repeatedly been given as conclusive evidence for drift or transoceanic continents. The example shows that the evidence leads to no such conclusion."

Turning to the coefficients for fossil land mammals, Simpson calculated a value for Eocene genera (45 percent): "Of two continents now separated . . . [t]his is the best fossil mammal evidence . . . of early Eocene (Sparnacian) European mammals also known from beds of the same age in North America . . . for the drift or transoceanic continent theories that has ever been found, but even this really tends more to oppose than to favor those theories. These figures appear to be consistent with full continent union only if the areas in question were very distant, more distant than drift theory postulates."

Simpson was arguing, of course, that one might expect a higher value for these Eocene genera if from the same supercontinent, considering that the generic faunal similarity between China and France today is 64 percent. Although we know now that in fact Laurasia was well split apart by the early Eocene, Wegener showed northern Laurasia reasonably intact. Du Toit also argued for a separated southern Laurasia by early Tertiary time, but in the Arctic region it was still more or less connected.[22] The similarity coefficient for the early Pliocene ("Pontian") was clearly lower still (15 percent) and, according to Simpson, argued for a long continental connection via Asia. The final mammalian generic coefficient was zero similarity for early Pliocene faunas in North and South America, just before the isthmian connection was reestablished.

Against this background of recent and fossil mammals, Simpson provided a single coefficient for generic similarity (8 percent) of "Triassic reptiles . . . for comparison and to support the incidental statement that the supposed pre-mammalian resemblances of southern faunas have been grossly exaggerated . . . [and] decidedly inconsistent with any direct union . . . of South America and Africa."

After his return from military service, Simpson returned to the subject of Eurasian–North American similarities of Cenozoic mammals, *not* to contend with the drifters and land-bridgers, but instead to develop more fully his concepts of historical biogeography. In fact, Simpson remarked that this effort was stimulated by Ernst Mayr, his American Museum colleague, who asked Simpson, "During what times in the Tertiary was a migration route between Eurasia and North America open? How long did it exist in each case? What faunal elements moved from east to west and what others from west to east? What was the climate of the bridge during each of the various times when it was open?"[23]

Simpson calculated the coefficients of similarity for scores of families and a like number of genera for fourteen intervals of Cenozoic time. His values depended upon the extensive biogeographic records he had accumulated for his 1945 monograph on mammal classification. The coefficients fluctuated almost an order of magnitude for genera (5 percent to 42 percent) and by more than half for families (52 percent to 89 percent). For both families and genera, similarities were highest in early Eocene; lowest immediately thereafter in the mid-Eocene for genera, early Oligocene for families; and generally increased throughout the rest of the Cenozoic. Simpson attributed the strong fluctuations in Holarctic faunal resemblances to a combination of factors, including direct faunal interchange, local extinction of indigenous forms, local evolution and concomitant differentiation of indigenous forms, extinction of migrants in one region, and migration of species into one region from a third (i.e., neither from North America nor Eurasia). Simpson concluded that, "All the interchanges were selective, and they apparently tended to become more selective as [Cenozoic] time went on. Not the only, but probably the most important, selective factor seems to have been climatic; the migrants generally are those groups tolerant of relatively cold climates. . . . All the faunal evidence is consistent with a single land route, the Bering bridge between Alaska and Siberia, as the sole means of mammalian interchange between Eurasia and America."[24]

Although the 1947 paper did not specifically address continental drift, it nevertheless played an important part in the debate because in it Simpson provided a rigorous set of quantitative data on faunal resemblances between continents, and adequately explained those data in terms of evolutionary and ecological principles, with no recourse at all to drift or transoceanic land-bridges. So convincing and authoritative was Simpson's explanation that his principles of historical biogeography transcended the data upon which they were based, data which, admittedly, postdated significant rifting of Laurasia. What chance then did the drifters and land-bridgers have with their much scrappier, qualitative data whose meaning was interpreted more or less solely on whether different geographic areas were in direct physical connection or not, with little or no attention paid to evolutionary or ecological factors?

Viewed in this way, the fact that Simpson's data were from Cenozoic mammals and not, say, from Carboniferous or Triassic reptiles and amphibians is *not* why the paleontological argument for drift converted so few. Instead, it was that the stabilists now had a *theory* of historical biogeography that could explain *all* important fluctuations in faunal resemblance without having to resort to moving the continents themselves about.

In 1949, at a symposium on continental drift across the South Atlantic Ocean sponsored by the American Museum of Natural History,

Simpson gave a paper on still another aspect of quantifying arguments dealing with the issue of drift.[25] He opened his paper by noting that, "Most of the large and wordy literature of historical biogeography . . . has involved postulates that are commonly unformulated and only rarely made explicit as working hypotheses. These hidden premises frequently are 'all-or-none' propositions, in the form of . . . 'either-or' dichotomies." Specifically, Simpson was referring to whether the dispersal of a group of terrestrial organisms required a land connection or not, and the inappropriate conclusion that, if dispersal does require land, then "disjunctive areas occupied by the group have been connected by continuous land." On the contrary, argued Simpson, "There is no group of organisms that cannot be dispersed across water. . . . The predictive or inferential situation is not one of absolute alternatives but one of degree. It is a matter of probability."

Simpson noted that geographic distribution, which is known, must be a function of either land connection or dispersal potential across water, essentially one equation with two unknowns. He argued that too often it had been claimed that because very low probabilities of dispersal across water (or other major geographic barriers) had been assumed to equate to zero, there then must have been some land connection (i.e., transoceanic land bridge or predrift contact of two continents). Simpson emphasized that very low probability was not equal to zero probability. In fact, even very low probabilities, if given enough opportunities or trials, could yield high probabilities of occurring at least once.

Simpson then showed from simple probability statistics how this counterintuitive result comes about, as outlined in his Table 2 (table 9.2). "If, for example, the probability, p, of a single breeding pair of rodents in any one year being dispersed to some oceanic island is one in a million ('a very low probability'), the chances of such dispersal occurring after one million years (t, trials) is, in fact, better than 60 percent ($tp = 1$, then $b = 0.63$). After 2 million years, 0.87, and after 5 million years virtually certain (.993)."

Simpson thus demonstrated to his satisfaction that, owing to the previously discounted factor of many multiple trials or opportunities for dispersal because of the long intervals of geologic time involved, very low probabilities of dispersal for a single trial were obviously not equal to zero. On the contrary, they approached one, given an appropriate length of time. That being the case, intercontinental faunal resemblances need not be the result of direct land connection, but rather

TABLE 9.2 Dispersal Probability Statistics (Simpson 1952)

tp=	1.0	2.0	5.0	10.0
b=	0.63	0.87	0.993	0.99995

p= Dispersal probability for single trial
t= Number of trials
b= Total probability of occurence at least once

the almost inevitable ability of organisms to disperse themselves over relatively short intervals of geologic time, irrespective of their inherently very low probability of dispersal in any one trial.

Simpson was aware that at first glance his argument might explain too much, in that if dispersal was eventually inevitable, how come terrestrial organisms were not more or less uniformly spread around the world's far-flung continents? Simpson returned to his previously developed ideas about the role of environment in limiting the distribution of organisms, as well as the importance of already occupied habitats resisting invasion by newly introduced migrants. Thus, Simpson was again able to emphasize the role of evolutionary ecology in clarifying paleobiogeography. The simplicity and quantitative rigor of this argument may well have been the last straw in weighing down and collapsing the use of faunal resemblances of terrestrial organisms by the drifters and land-bridgers to support their theories.

Simpson consolidated his views on historical biogeography in a small book entitled *Evolution and Geography: An Essay on Historical Biogeography with Special Reference to Mammals*.[26] Five of the six chapters developed what Simpson saw as the critical factors in determining past and present distributions of organisms. The arguments were essentially identical to the ones worked out in his earlier papers, emphasizing an ecological and quantitative point of view. In the last chapter, he specifically addressed the issue of continental drift and, not surprisingly, Simpson asserted that, "All the biogeographic features in the known history of mammals are best accounted for on the theory that the continents have had their present identities and positions and that there have been no land bridges additional to those that now exist (North America–South America and Eurasia–Africa) except for a northern Asia–North America bridge."

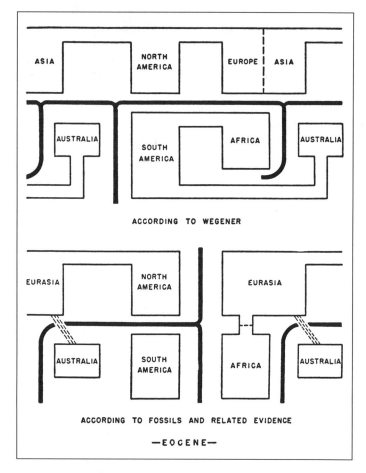

FIGURE 9.3 Simpson's "zoological test of a paleogeographic theory. The upper figure shows the connections and barriers between continental blocks in the Eocene according to Wegener's version of the theory of drifting continents. The lower figure shows the connections and barriers demanded by known faunal resemblances and differences. Wegener's ideas of Eocene geography were evidently wrong." (Simpson 1953a, *Evolution and Geography*, fig. 29. Courtesy Thomas Condon State Museum of Fossils, University of Oregon)

Simpson's penultimate figure in this volume (fig. 9.3) contrasted the Eocene world according to Wegener with one, "according to fossils and related evidence" à la Simpson. Simpson's Eocene world was virtually identical to today's world, except that Simpson showed a sweepstakes route between Southeast Asia and Australia, and a filter bridge/corridor between the New World Nearctic and the Old World Palearctic regions. Simpson's caption for this figure concluded that, "Wegener's ideas of Eocene geography were evidently wrong." In his last figure of this volume he showed the geography of the Cenozoic World, "the world of the Age of Mammals. Major features of the geographic history of mammals are best accounted for by considering the continental blocks and the main sea barriers as constants and three main filter bridges [Bering Strait, Isthmus of Panama, N. Africa–S. Europe] and one main sweepstakes route [island archipelagoes of Southeast Asia] as variables. . . . The conclusion seems to apply not only to the biogeography of mammals but also to that of all contemporaneous forms of life. It remains possible . . . that continents drifted in the Triassic or earlier, but there is little good evidence that such was the fact."[27]

As in this last sentence just quoted, and elsewhere, Simpson was certain that what he inferred about the historical biogeography for the Cenozoic Era also applied to the Mesozoic Era, even though the data for his inferences were solidly based only on Cenozoic mammals. But we should not forget that Simpson was a knowledgeable student of Mesozoic mammals — in fact the world's authority. We must assume that what he concluded for Cenozoic paleobiogeography was consonant with what he knew of Mesozoic mammalian biogeography. Therefore, we cannot claim that Simpson would have arrived at Wegenerian conclusions if only he had examined a more appropriate set of data. Simpson's opposition to drift and transoceanic land-bridges can be reduced to his preference for mobile organisms and stable continents as against mobile continents and stable organisms. After all, the mechanisms of terrestrial land-animal dispersal were known, whereas the North American geophysicists were virtually unanimous in denying any suitable mechanism for moving the continents. Although Simpson never cited geophysical authority to support his case — he couldn't very well have allowed himself advantages he denied his opponents by appealing to authority — he certainly knew of the geophysical arguments offered against horizontal continental movement.

The Plate Tectonics Revolution

So effective was Simpson's case against Wegener–Du Toit paleobiogeography, that fossil evidence for drift was thereafter de-emphasized. Du Toit himself was ambivalent with respect to their value when responding to Simpson's 1943 paper. Du Toit acknowledged that "Simpson has exposed several weaknesses in the writer's [i.e., Du Toit's] statements and arguments on this subject [of faunal relationships]." Du Toit gave an item-by-item rebuttal to Simpson, but the main weight of his argument was geological, *not* paleontological. The real impact on Du Toit can be read from the concluding remarks in his rejoinder: "The acid test of that Hypothesis [of Drift] will, it is felt, depend among other things on its ability simply and logically to account for the harvest of fossil forms yet to be unearthed."[28] Obviously, Du Toit had been pushed by Simpson to the position that, after all, fossil forms *in hand* did not provide the acid test for the hypothesis, so we must await for new fossils to be discovered to prove the case.

Du Toit was not alone in his ambivalence about the value of fossils in demonstrating drift, because throughout the 1950s and early 1960s, fossils played no significant part of the debate. On the contrary, other North American paleontologists came to Simpson's conclusions using data other than of that Cenozoic mammals. For example, U.S. Geological Survey paleontologist Preston Cloud, reviewing the principles of marine biogeography and Phanerozoic marine invertebrate fossils, concluded that, "Certainly no paleobiogeographic evidence known to me requires drift of either poles or crust." A distinguished paleobotanist, Daniel Axelrod, came to a similar conclusion after reviewing fossil plant data: "At the present time there does not appear to be any paleontologic evidence that *demands* drifting continents or wandering geographic poles; in fact, the evidence seems to militate against major movement of either sort."[29]

Another indication of the lack of enthusiasm of paleontologists for drift might be seen in some of the conferences and symposia debating drift just before the hypothesis of sea floor spreading was proposed. For example, in a collection of articles in the 1962 book entitled *Continental Drift*, edited by the British geophysicist Keith Runcorn, the only article that touched on paleontological data in any way at all is one written by Neil Opdyke, a Runcorn Ph.D. student and paleomagnetician.[30] Opdyke reviewed the sedimentary rock record for paleoclimatic indicators and included less than one page on recent and fossil reefs. Given that the book was attempting to build the strongest possible case for continental drift, it is significant that not only were fossil data virtually ignored, but that what little there was was contributed by a geophysicist, not by a paleontologist.

The key papers in the plate tectonics revolution were geological and geophysical in nature. Thus, fossils had no part whatsoever in establishing first-motion at transform faults, developing a magnetic-reversal time scale, determining the mirror-image of magnetic reversals across midoceanic ridges, or computer puzzle-fitting of continents. The concept of sea floor spreading, of course, came out of these and related observations. The data from the oceans *compelled* sea floor spreading, which in turn *compelled* continental drift. However much data from fossils may be compatible with, corroborate, or help choose among conflicting ideas about just where and when drift occurred, they are at best only suggestive of drift. Thus, Simpson was right in asserting, in effect, that fossil data do *not* compel drifting continents.

Simpson's Conversion

When did Simpson acknowledge the mobilist view of the world? During the plate tectonics revolution itself, Simpson was aware of the debate that was ensuing between stabilists and mobilists, but he still came down on the side of fixed continents. "During the latter part of geological history at least, the last hundred million years or more [i.e., back in the mid-Cretaceous], major seas and lands, the oceans and the continents, have had substantially their present identity. . . . Some will disagree with this statement, but it is the consensus."[31] And later, "Even the new paleomagnetic data, which raise serious doubts as regards earlier times, confirm that the southern continents have been at least near their present positions throughout the Cenozoic."[32] But by 1970 Simpson had recognized that "continental drift" was indeed an established fact. He still hesitated, however, in using fossil data alone to determine past continental configurations, as the following quote from a letter to *Science* indicates. Simpson was commenting on an article by geologist David Elliott and others reporting the recent discovery of the therapsid reptile *Lystrosaurus* in Antarctica: "the occurrence of *Lystrosaurus* and some other more broadly similar faunal elements in Antarctica, South Africa, and India is evidence that those crustal segments were then continuous, and they are interpreted as parts of pre-drift Gondwanaland. [The authors] also mentioned that *Lystrosaurus* and other generally similar faunal elements likewise occur in Sinkiang [China], that is, in what is now central Asia, but they did not discuss the bearing of that fact on paleogeography. Application of their reasoning would indicate that central Asia also was then in continuous (not necessarily direct) connection with Antarctica." Simpson went on to point out that drifters did not include central Asia within the Southern Hemisphere supercontinent and, therefore, "we may still have to reconsider the concept of Gondwanaland and some other paleogeographic and biogeographic points." Of course, Simpson was *not* really suggesting that China was once part of Gondwana. Rather, by indirection, he was showing how paleobiogeographic data were not sufficient, in themselves, to compel a unique solution to the problem of ancient continental position.[33] Some years later Simpson returned to this point: "There are, however, hints that the situation may not have been as simple as it seems at first sight. *Lystrosaurus* is not known from either South America or Australia, both generally believed to have been parts of a unified Gondwanaland when that reptile lived. Moreover, both the plant *Glossopteris* and the reptile *Lystrosaurus* are known from parts of Asia well north of India in regions generally believed to have been parts of Laurasia, not of Gondwanaland, when those organisms lived."[34] Edwin H. Colbert (b. 1905), one of the party that discovered Antarctic *Lystrosaurus*, was aware that the fossil did raise biogeographic questions, so he suggested a "long and circuitous route" of travel for the beast from the Southern Hemisphere to East China. And, if the reptile did range that widely, it would be comparable to the present-day geographic range of species of American and Chinese alligators, which "would seem to represent the end effects of movement between North America and Asia across the trans-Bering filter bridge in middle Tertiary time."[35] Colbert's explanation thus was reminiscent of Simpson's predrift arguments.

From the early 1970s, Simpson accepted plate tectonics and made whatever adjustments were necessary in his subsequent writings that involved paleobiogeography, although he certainly did not see plate tectonics as invalidating in any way his principles and concepts of historical biogeography.[36] In taped remarks Simpson made in August 1975, when reviewing his major publications, he alluded several times to his previously incorrect views about the stability of the continents. In particular, "I'm not really very happy about [my book, *The Geography of Evolution* (1965)]. . . . The book is sort of a mish-mash collection of essays. . . . Some . . . are all right, but some of them are badly outmoded, in fact were already becoming outmoded at the time when the book was published . . . just before it was clear that . . . plate tectonics was going to change the whole background for ideas on ancient geography and ideas on the geography of life."[37] Simpson's miscalculations in this field continued to be on his mind when writing his book-length autobiography. He remarked, "[T]he views that I supported and expounded years ago are still basically valid . . . [but] since reaching my seventies I have myself changed my mind on a basic principle of [a] . . . field of major importance to me: paleogeography." Simpson took pains to clarify the nature of his opposition to Wegener and continental drift: "The theory as Wegener advanced it rested on minimal evidence, much or all of which could also be explained by other theories, including that of stable continents. . . . I soon found (and this is still correct) that most of Wegener's supposed paleontological and biological evidence was equivocal and that some of it was simply wrong. . . . Thus in the 1930s and 1940s after lengthy investigation I concluded that the then-available real evidence of known land mammals not only did not support but opposed any effects of continental drift during the Cenozoic. . . . I did not deny the possibility of earlier effects of drift, but at that time I considered evidence for the drift theory so scanty and equivocal as to make it an unconfirmed hypothesis." Simpson indicated that the crucial data for continental drift came from magnetic reversals across ocean ridges. Like other opponents of mobile continents, he was converted by the totally different, nonpaleontologic data coming from the deep sea.[38]

Summing Up

Simpson's opposition to the fossil evidence adduced for continental drift or transoceanic land-bridges came directly from principles and concepts of historical biogeography that he was developing during the course of his own research on Cenozoic mammalian faunas. That Simpson was so effective in rebutting the drifters and land-bridgers can be directly attributed to the depth and all-inclusiveness of Simpson's notions of what factors were critical in limiting the geographic distribution and abundance of terrestrial vertebrates. Rather than arguing ad hoc, point-by-point, against the drifters, Simpson formulated a theory that was both ecologic and evolutionary, and whose application had robust quantitative predictions. The soundness of Simpson's approach is attested to, in that his theory of historical biogeography still stands, today, even in a mobilist world.[39]

Simpson's opposition to drift and land-bridges, therefore, was not the result of inherited tradition, or prior assumptions, or fear of going against the tide of established opinion. Simpson had demonstrated freshness of viewpoint and originality of thought in other areas of his research at this time that preclude the judgment that he was merely "following the crowd." On the contrary, even when the plate tectonics bandwagon was in high gear, Simpson hesitated in climbing aboard until well after the new paradigm was firmly established.

Simpson, after all, was indeed wrong about continental drift. What led him astray? Simpson's specialty was Mesozoic and Cenozoic mammals, so naturally the data upon which he based his theory of historical biogeography were those provided by mammals, especially Cenozoic terrestrial mammals whose fossil record was so rich as compared to that of Mesozoic mammals. Obviously, these fossil data would be much less likely to argue for mobile continents than, say, Permo-Triassic therapsid reptiles. One might wonder, then, if Simpson had worked with those organisms would he have come to different conclusions about drift? Most likely not. Colbert, a student of Mesozoic reptiles, was aware of the faunal resemblances of Mesozoic terrestrial tetrapods, yet he explained their distribution pretty much according to Simpson's biogeographic principles.[40] Alfred S. Romer, another distinguished vertebrate paleontologist, perhaps leaned more toward drift, but his conclusions were always equivocal on this point, as the following statement by Romer shows: "With my own interests centered on older periods, however, I have found myself weakening in my earlier beliefs as to continental fixity. . . . Although I do not believe that the evidence is — as yet — strong enough to make a very positive statement. . . . For the northern continents, the evidence from the Mesozoic is not at all decisive. . . . In the south, however, there is evidence strongly suggesting (although not proving) that in the Triassic there was free 'trans-Atlantic' faunal interchange between Africa and South America."[41]

The British paleontologist T. Stanley Westoll concluded from his study of Carboniferous freshwater fishes that their distribution was "most convincing evidence for the nonexistence of the North Atlantic basin during late Carboniferous times." Yet he concluded that these fish "can offer no decisive evidence" in favor of either "transoceanic land-bridges or continental drift," although he thinks the latter more likely than the former. Westoll did not, however, claim much more for his data than that they "provide another small factor in favor of continental drift."[42] We can never know if Simpson would have been a drifter if had he worked with older terrestrial faunas. But if other very capable paleontologists working with these faunas were equivocal or merely lukewarm about continental drift, on what basis could we conclude that Simpson would have thought differently?

There is one aspect to Simpson's research experience that may have been most decisive in his stabilist views. During the 1930s, much of his research, although by no means all, was on South American mammals, which in the Cenozoic had evolved more or less in isolation from the rest of the world's continental faunas. Simpson's research demonstrated an early Cenozoic connection with North America, presumably by the proto-Panamanian isthmus; sustained island-continent separation permitting a unique terrestrial mammalian fauna to evolve; and a late Cenozoic inter-American connection ("filter bridge") across which Neotropical and Nearctic faunas traveled and intermingled.[43] Both then and now, the history of the South American fauna during the Cenozoic Era is best explained by stable continental land masses intermittently connected by a relatively narrow isthmus, with occasional sweepstakes dispersal during times of separation.[44] Naturally, a theory of historical biogeography based upon these New World terrestrial mammals would inevitably reflect underlying stabilist phenomena. Despite his being strongly influenced by his South American research, it is still moot whether Simpson would have derived a different view of historical biogeography had he worked on, say, intercontinental biotas from predrift Gondwana.

There are two final statements to make about why Simpson held to his stabilist views. First, and paradoxically, Simpson perhaps put too much confidence in the fossil record in determining past continental configurations. Because fossils did not conclusively demonstrate drift, Simpson was therefore convinced that continental positions must have always been fixed. Ironically, even Simpson was seduced by his own incisive and inexorable logic. Second, it was not part of Simpson's research style to reconsider a problem once he had seriously thought about it and come up with a satisfying answer (to him, at least). Simpson would attack a problem by careful evaluation of the data, by deliberate and logical formulation of his argument, and by construction of whatever theory was necessary to explain his conclusions. Once having done that — and the process might be extended over a number of years, as in the case of historical biogeography — Simpson moved on to new problems and to new areas of research interest. He did not tinker, fuss over, or agonize about issues upon which he had made a well-considered scientific judgment. Therefore, after the 1950s, Simpson did not again address the issue of drift, at least not in the broad terms of his earlier work. Of course, he continued to make biogeographic statements as appropriate when writing about evolution, or describing a fauna, and so on. But he did not reopen the whole question of his views of historical biogeography in the light of the newly established mobilist theory. Quite clearly, Simpson viewed his theory of historical biogeography as essentially unaffected by continental drift.

CHAPTER 10

THE MIND'S EYE

There [is] much evidence that truly productive thinking in whatever area of cognition takes place in the realm of imagery.
— Rudolf Arnheim, psychologist

Visual modes of discourse will require a rather fundamental change of intellectual values within the history of science.
— Martin Rudwick, historian of science

A diagram is no proof! A diagram is no proof!
— Francis Toner, geometry teacher

Late in life, Simpson remarked that, "I compose my writing visually — I think visually, then translate that into words. . . . I visualize at least as much as I verbalize, perhaps more. Even in abstract theory I often visualize first & then describe in words what I saw mentally."[1] In much of his most significant theoretical work, Simpson did indeed use just such visual language to translate his more original concepts and interpretations regarding, for example, statistical inferences about evolving lineages, relationships of speciation to higher taxonomic categories, ratio diagrams of morphological dimensions, and species-density contouring. Simpson's most interesting and innovative visualizations had to do with organism-environment relationships, including adaptive landscapes, prospective and realized functions of organisms and environments, and especially the adaptive grid upon which he summarized his argument for variable rates and patterns of evolution — "tempo and mode" — in response to differing ecological opportunities available to animal and plant species.

Simpson was aware that "with certain types of material, analogical diagrams . . . proved to be enlightening for those students . . . who habitually handle abstraction by relational symbols or visual images." He claimed this was the case for himself, yet he acknowledged that for other students "who deal with theoretical or abstract concepts in other forms, such diagrams seem to have little explanatory content."[2] Simpson's interest in visual presentation, however, did not begin with his theoretical work. In fact, as a young adult he illustrated profusely his letters to his family, and in his early research he pioneered techniques — UV fluorescence and stereophotography — to enhance visually both the observation and depiction of fossil specimens.

For Simpson the visual was both an "effect," that is, something that supports a general design or intention, and an "affect," that is, an emotion as distinguished from cognition, thought, or action. Examining Simpson's use of visual language, therefore, provides some insight into both his rational and emotional states of mind. In this chapter, rather than provide all instances of Simpson's use of visual language, I discuss several examples of each of the three particular ways in which Simpson used visual language, in order of increasing abstraction (fig. 10.1). Starting with Simpson's innovative techniques for enhancing the display of what one is observing in the first place, I proceed to his creation of images that summarize many detailed observations to facilitate their interpretation, and I then conclude with his depiction of hypothetical relationships rather far removed from the primary data.

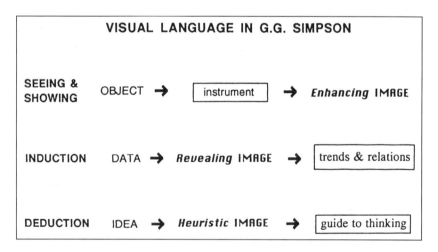

FIGURE 10.1 The kinds of images in Simpson's writings, depending upon whether he enhances an image of an object by using an instrument, forms an image that reveals trends and relations from a set of data, or develops a heuristic image that permits further conceptualization and deduction.

Showing What Is Observed

When looking back over his research, Simpson noted his penchant for visual expression in his technical work and that some of his research methods "were unusual in such applications at the time of [their publication]."[3] In fact, in Simpson's first professional publication in 1925, he asserted the value of improving the methods of seeing what one was observing in order to enhance the resultant interpretation: "Professor Marsh['s] . . . figures [of Jurassic mammals] are in general somewhat inadequate for the illustration of these almost microscopic forms. . . . [I]t is hoped that the careful redescription of Marsh's types, after thorough preparation and with the use of better optical aids . . . may help to clarify our knowledge of these tiny creatures."[4] And indeed, shortly afterward Simpson noted in another of his early papers that he was able to make better observations of the tiny jaws of two supposed mammals from the late Triassic using a high-power binocular microscope and ultraviolet fluorescence to demonstrate that both jaws, in fact, were of therapsid reptiles.[5]

Simpson also made use of stereophotography in his Yale dissertation and in his postdoctoral British Museum monograph on Mesozoic mammals, many of whose remains were very small cheek-teeth having diagnostic characters that included the size, shape, and configuration of their cusps.[6] His innovative use of stereophotography was particularly helpful because it yielded less ambiguous, three-dimensional views of these and other key characters, thereby facilitating the discrimination of species. Simpson has noted that these methods of fossil illustration "were unusual when I began to use them, and it is possible that I was the inventor of some of them, although it is always foolish to try to establish priority or to argue over it."[7]

Visual Induction

Simpson was imaginative as well in the simplifying methods he used to represent a complex set of data. For example, he invented "ratio diagrams" to display the morphology of fossils for easy comparison. After measuring the desired morphologic characters and expressing them as logarithms, Simpson then calculated the ratios of the measurements (i.e., the differences between the log measurements) of various characters and plotted them on arithmetic coordinates (fig. 10.2). The characters of a "standard morphotype," defined as "zero log difference," plots as a vertical line, and the corresponding characters of other specimens will fall to the left (if smaller) or the right (if larger); proportional differences between specimens will vary as the lines so defined differ in shape from each other.[8] Thus, a single ratio diagram can express many measured differences among characters and specimens. Moreover, the calculations required for the ratio diagrams are greatly reduced and simplified as contrasted with other biometric methods. As Simpson noted, "Related organisms commonly differ not only in size but also in proportions, and this must be kept in mind when comparing them. . . . When I undertook a revision of all the large Ice Age (Pleistocene) fossil cats that had been found in North America . . . it was necessary to use ratios in evaluating differences, and that suggested a great many calculations and serious difficulties in visualizing them overall and judging their significance. . . . As far as I know the idea had never occurred to anyone before, and certainly it was useful because it has been quite widely used ever since."[9]

Another Simpson innovation for showing in a simple way the interpretive results of a large body of data was his contouring of species density for North American mammals (fig. 10.3).[10] Using a standard publication on the distributions of species, Simpson created several hundred equal-area quadrates for mainland North America and plotted the number of recent mammal species reported within each; he then contoured the result. The contoured North American map for the 670 species allows easy recognition and interpretation of diversity gradients, such that they appear to correlate well with changes in latitude and topographic relief, as well as with proximity to and distance from continental margins.

Simpson wanted not only to display abundant and complex data in a way easy to grasp but, more importantly, to facilitate interpretation of the data. Thus he viewed such techniques of illustration "more often as concepts and methods of thought and problem-solving. In most cases the methods are brought out and simultaneously exemplified by their application to some particular research."[11] One way to think of this is to regard Simpson's "method of thought" here as a kind of visual induction where he reasons from the particular data, expressed graphically, to a general conclusion, inferred from the resulting image.

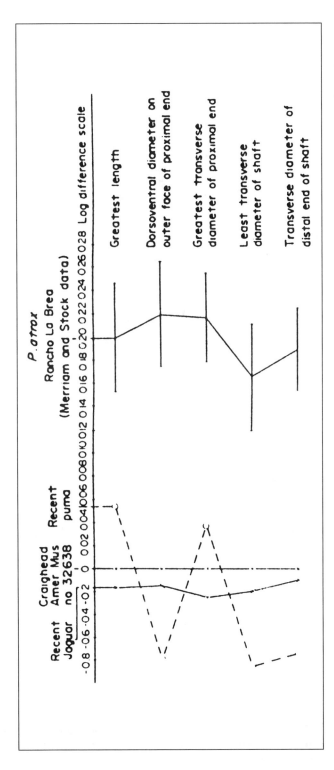

FIGURE 10.2 Ratio diagram for comparing various skeletal morphologies. (Simpson 1941c, "Large Pleistocene Felines of North America." Courtesy American Museum of Natural History)

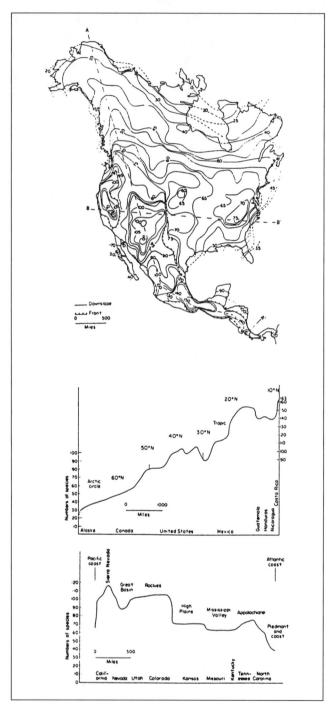

FIGURE 10.3 Species density contours of North American mammals (a), with particular densities along lines A–A' (b) and B–B' (c). (Simpson 1964a, "Species Density of North American Recent Mammals," 60, 62, 63; © Taylor and Francis, Philadelphia)

Visual Deduction

Phylogenetic Patterns

Simpson was not only aware that the image can provide the source of subsequent interpretation, as with ratio diagrams and species density contouring (what I am calling "visual induction"), but also that images can derive from prior interpretation and then be used to come to a more particular conclusion (what I call "visual deduction"). In 1937, in one of his first entirely theoretical papers, Simpson directly confronted the issue of how misleading graphical representations of phylogenies can result from incorrect interpretations of true evolutionary history: "In the fine optimistic years of early evolutionary paleontology, it became fashionable to draw up genealogical trees . . . by representing the living animals as the ends of twigs and their ancestors . . . as . . . stems, branches, and trunks of the tree of life. . . . The use of such graphic means of indicating descent is still common. . . . However, a cynical note has crept in, and phylogeny has been subjected to severe criticism . . . [a]s almost every family tree proposed . . . proved, beyond much doubt, to be erroneous in essential features."[12] Simpson further noted that this "old picture of phylogeny" was replaced by a more noncommittal one instead by "diagrams suggesting many different lines [of descent] that are parallel or that diverge but are not connected. . . . These new pictures with their multitudes of independent lines are unquestionably nearer the truth than were the naïve trees formerly in vogue."[13]

Simpson's purpose in this paper was to present new methods for phylogenetic research that would obviate the shortcomings of such older representational approaches in depicting evolutionary history. Rather than come up with yet another configuration of twigs, stems, or branches, Simpson used graphical means to illustrate his views of how to depict morphological characters of "similar specimens from a limited horizon and locality." And, of course, such specimens are the raw data of the paleontologist.

Simpson was questioning the phylogenetic interpretation of graphical representations, and not the use of such representations themselves. He wanted to show that more than one interpretation of a given graphic presentation of data was possible, and he used a series of graphical illustrations to argue his position. He first presented two views of how a scatter of measurements of a morphologic character of individual fossil specimens within a single assemblage might be interpreted as either coming from more than a single species or as variants within a single species. He next discussed how to decide between the two alternative interpretations by advising the collection and measurement of additional specimens so that they clustered either into single or multiple unimodal frequency distributions. If one unimodal distribution, then a single, more widely variable species; if more than one unimodal distribution, then more than one species, which were less variable. Simpson was thus emphasizing the populational, rather than the typological, approach to the recognition of species — "recognition" that is both visual and conceptual.[14]

Simpson explored further this deductive means of visual imaging when he considered how one might interpret twenty-two hypothetical measurements of a given morphologic character that increases in value through geologic time (refer back to fig. 6.3). He displayed graphically three possible evolutionary interpretations of these data — phyletic size increase of a single species, species replacement, or species splitting — and how one could test, statistically, which interpretation is correct. By allowing the reader to grasp visually exactly what was meant, Simpson inverted the more usual practice of incorporating images to complement text by making the text supplemental to the images. Thus, Simpson's images, rather than the text, carried the burden of the discussion. One can clearly follow Simpson's argument by "reading" the images alone without referring to the text.

In that same year (1937), Simpson published a paper — on the nature of taxonomic categories above the species level — that made this rather abstract discussion more concrete. Simpson adopted an illustration from Alfred C. Kinsey, whose monograph (1936) was serving as Simpson's foil, and he gave a diametrically opposed interpretation (fig. 10.4).[15] Whereas Kinsey used the figure to demonstrate how "[h]igher categories are not groups of species with a common origin," Simpson used the same figure to demonstrate precisely the opposite conclusion, namely that "supra-specific groups do have an objective reality . . . [t]he status given these groups, species, genera, families, or whatever it may be, is arbitrary but the groups themselves are natural."[16] Simpson thus showed that, although a diagram can be used to visualize an otherwise rather abstract concept by itself, it is no proof of a given proposition.

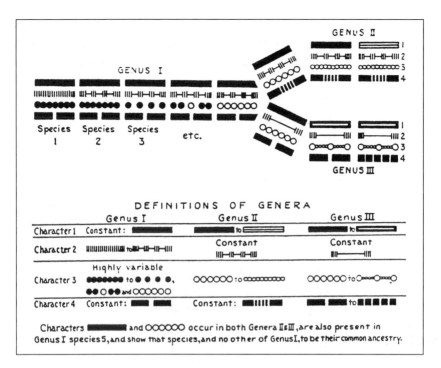

FIGURE 10.4 The relationships of species characters (1–4) and the higher categories they comprise (Genera I, II, III). The original diagram had been previously published by A. C. Kinsey to support his claim that higher categories like the species are "not groups of similar species." (Simpson 1937d, "Supra-specific Variation," 56; © University of Chicago Press)

In that same paper, Simpson made use of another of Kinsey's images, this time by modifying rather than copying it exactly (refer back to fig. 6.7). In this illustration Simpson showed dichotomous branching of species within a lineage and how, over the time represented by the space within the figure, higher categories "appeared" (genera in the example), and how subsequent extinction of intermediate taxa would result in extant taxa lacking living, connecting links.[17] Given that one of the two major themes of Simpson's paper was to rebut Kinsey's views on the nature of higher,"supra-specific," categories, Simpson obviously found Kinsey's illustrations doubly useful, rhetorically as well as conceptually: while using the same visual language of Kinsey, Simpson was able to depict an opposing interpretation.

Simpson apparently found such visualizations of abstract concepts essential, because there are many such graphic images in his later writings. These visualizations contain crucial aspects of his arguments and are not merely incidental to them. But just as the accompanying diagrams in Euclidian geometry are "no proof," so too the visual images in Simpson's paleontology are no proof. Yet, in both cases, the graphics help carry the burden of the claims being made.

Organism-Environment Interactions

Simpson's contribution to the "consolidation of the evolutionary synthesis" in the 1940s was based on his understanding of the dynamic interaction of organisms with their environment, whereby the inherited variation found within specific populations resulted in adaptation through the mediation of natural selection exerted by particular environments.[18] The fossil record of life, as interpreted by Simpson, was therefore the outcome of long-term, geological accumulation of short-term, microevolutionary interactions. Simpson thus countered the interpretations of many of his paleontological contemporaries that explained macroevolution by inherent forces such as inertia, momentum, and orthogenesis. Simpson's understanding and expression of this critical relationship between organisms and their environments depended, for him, upon the visual expression of several key images.

THE ADAPTIVE LANDSCAPE

This image was first introduced by Sewall Wright, who according to Simpson, "has suggested a figure of speech and a pictorial representation that graphically portray the relationship between [natural] selection, [morphologic] structure, and [environmental] adaptation. The field of possible structural variations is pictured as a landscape with hills and valleys."[19]

Following Wright, in spirit if not to the letter,[20] Simpson in *Tempo and Mode* introduced a series of images depicting natural selection as a vector moving in space, creating degrees of adaptation of organisms as represented by hills ("more adapted") and valleys ("less adapted") (fig. 10.5). In this way Simpson could readily portray linear, centripetal, and centrifugal selection by means of a contoured surface with varying adaptive topography. On such a contoured surface, he could then show the splitting of populations as either a single population that divides and moves up two different, nearby adaptive peaks, or just part of one well-adapted population, which moves down off its peak, across a val-

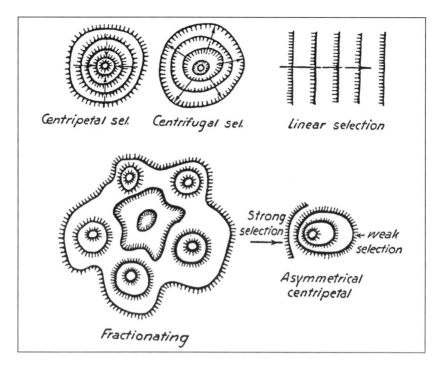

FIGURE 10.5 Simpson's visualization of natural selection as a vector moving populations across an adaptive landscape (hachures on contour lines point downhill). (Simpson 1944b, *Tempo and Mode in Evolution*, 90. Courtesy Anne Roe Simpson Trust)

ley, and up another peak near to the one where the rest of the original population remains (fig. 10.6).[21]

Of course, whether in fact these images accurately portray reality in nature is not the point here. What interests us is the manner in which Simpson used these visual images to explain and clarify how he thought nature operates.

FUNCTIONS OF ORGANISMS AND OF ENVIRONMENTS

Further on in *Tempo and Mode* Simpson introduced the concepts of "prospective and realized functions of organisms and of environments" and converted them to simple images in the form of Venn diagrams (fig. 10.7, *upper*).[22] Simpson wanted to show that only part of the full range of an organism's "prospective," or potential, adaptation is actually "realized" within a given environment, which itself has wider adaptational possibilities than that being realized by the specific organism. Be-

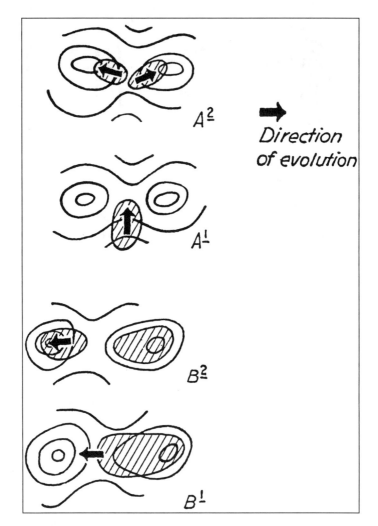

FIGURE 10.6 Selection either subdividing a population and driving it to new adaptive peaks (A^1–A^2) or moving part of a population from its ancestral adaptive peak to a new, adjacent peak (B^1–B^2). (Simpson 1944b, *Tempo and Mode in Evolution*, 91. Courtesy Anne Roe Simpson Trust)

cause the concept is more easily grasped in pictures than in words, Simpson showed two circles, each suggesting respectively the prospective function of the organism and of the environment, with an area of overlap representing the realized functions of each. He then provided an example illustrating the individual prospective function of whales

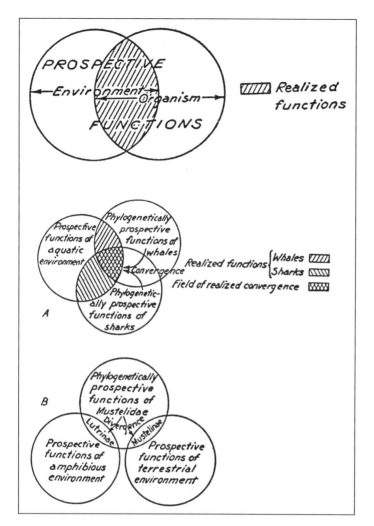

FIGURE 10.7 Venn diagrams showing the "prospective" and "realized" functions of organism and environment (*upper*) with the particular examples of whales and sharks (*middle*) and of otters and weasels (*lower*). (Simpson 1944b, *Tempo and Mode in Evolution*, 184, 185. Courtesy Anne Roe Simpson Trust)

and sharks within the prospective functions of the aquatic environment (fig. 10.7, *middle*). The small amount of overlap between the realized functions of sharks and whales represents convergent evolution. A second example for otters and weasels illustrated the prospective functions of the mustelid family within amphibious and terrestrial environments (fig. 10.7, *lower*). In this case, the lack of overlap between the two realized functions represents evolutionary divergence within the mustelid family.

Besides indicating these realized functions of organisms and their environments, Simpson went on to explain how organisms may be "preadapted" to a future environment in that a prospective function of an organism, not realized in one environment, may be subsequently realized in another environment. Thus the crossopterygian lobed-fin was adapted to an aquatic environment but also had the prospective, "preadapted" function of maneuvering a primitive amphibian descendant on dry land.

THE ADAPTIVE GRID

The most creative visualization of the interaction of organisms with their environment is Simpson's "adaptive grid," also first presented in *Tempo and Mode in Evolution*. Simpson claimed that "the course of adaptive history may be pictorialized as a mobile series of ecological zones with time as one dimension. Within the limits of a flat page, the basic picture resembles a grid, with its major bands made up of discrete smaller bands and these ultimately divided into a multitude of contiguous tracts . . . the lower bands or those first occupied tended to be wider. The zone of tolerance of fluctuating conditions is broader; adaptation is less specific; adaptability, in a phylogenetic sense, is greater. Successively higher bands . . . tend to be narrower . . . demanding [more] exact and specific adaptation."[23]

The adaptive grid allows the evolutionary history of a group to be depicted in terms of environmental adaptation to stable or changing ecology over time. Different evolutionary histories may be recorded and compared at a glance, with particular emphasis given to what Simpson considered critically important in determining evolution: the specific relationship of organisms to their environment (refer back to fig. 7.3). The initial width of a given adaptive zone within the grid and whether it expands, contracts, or remains the same determine if a group of organisms entering the zone will expand, contract (perhaps even becoming extinct), or remain constant in diversity. Some few organisms might succeed in jumping the gap from one adaptive zone to another, say from aerial to submarine flight, and in so doing experience very high rates of evolution. The characteristic rapid bridging of such gaps between adaptive zones explained for Simpson what he called the "systematic discontinuities" in the record of fossilized organisms that fail to connect taxonomically disparate groups.

Simpson's personal contribution to the "consolidation of the evolutionary synthesis" was well encapsulated by this image: macroevolutionary results, as seen in the fossil record of animals, are brought about from microevolutionary forces, as inferred from laboratory experiments and field observations, of natural selection operating within the context of particular environments upon the inherited genetic variation of local populations. The adaptive grid was thus meant to "synthesize" the paleontological with the genetical and the ecological, and Simpson was well aware of the significance of this image: "Darwin said in his sketch autobiography that he could remember exactly the situation, the place where he was, when he suddenly thought of . . . the idea which later came to be known as adaptive radiation. I'm not about to compare myself with Darwin, but the fact is that I can remember the exact situation when I suddenly thought of a particular way of representing in visual terms and of discussing the adaptation, the progressive adaptation, of animals to their environment. As a matter of fact the situation was rather banal, but at that time I was living in Stamford, Connecticut, and commuting to [the American Museum of Natural History in] New York. . . . The idea was in fact a rather simple one — it can be complicated and I did later complicate it. It was simply the idea of an adaptive grid, one of the things which is gone into in some detail in the book *Tempo and mode in Evolution* and which is, I think, a contribution to the then developing synthetic theory [of evolution]."[24]

This quote from Simpson might suggest that the image of the adaptive grid merely solved a problem of text illustration. However, the detailed way in which Simpson recalls how the image occurred to him, and the importance he accords to it by invoking Darwin, indicates that the adaptive grid was a crucial conceptual visualization of organism-environment adaptation in geological time. Simpson used the adaptive grid image in eight subsequent illustrations in *Tempo and Mode* to portray differing modes of evolution with their accompanying rates — bradytelic, horotelic, and tachytelic evolution — and one specific example of felid evolutionary history.[25]

In *Major Features of Evolution*, the revised and expanded version of *Tempo and Mode*, Simpson used the adaptive grid a dozen more times in similar ways, with several additional specific examples from the fossil record. Curiously, however, in later publications Simpson rarely used the adaptive grid image again, although he continued to write extensively about organism-environment adaptation and adaptive zones.[26] Why so? I believe this further indicates that the adaptive grid was not simply an image for convenient, incidental illustration. If it had been, Simpson would have used it over and over in subsequent discussions as was his habit with favorite illustrations like that, say, of horse evolution. Rather, it was a heuristic mental construct that clarified for him, personally, his original attempts to understand how different sorts of organism-environment interactions would play out over geological time. Once he achieved the understanding, Simpson could discard the image that gave birth to it.

The Visual as Effect and Affect

As I have attempted to demonstrate, Simpson used images or visual language as an "effect," that is, something to support a general design or intention — whether for objective demonstration or for inductive and deductive argument. I wish now to suggest that Simpson's proclivity for visual language sprang fundamentally from his personality and temperament; that images also had for him an "affective" component, that is, an emotional response as distinguished from rational thought.

The published record shows that Simpson delighted in the visual. For example, in the opening two pages of his autobiography, *Concession to the Improbable,* Simpson evinced his interests in the visual by the words he used and by the subjects he chose to mention — images conveying a painter's eye and a widespread variety of interests not often found in the public expression of a scientist: "noon sun . . . last crusts of snow . . . ugly long black stockings . . . yellow johnny-jump-ups . . . white sand lilies . . . religious iconography of the Middle Ages . . . Gothic and Romanesque architecture . . . Egyptian hieroglyphs . . . Makonde wood carving . . . mosques of Cairo . . . Picasso's paintings."[27]

Was this mere affectation, a self-conscious desire to show a sensitive and artistic soul? I think not, for even Simpson's letters to his sister Martha, written in his teens and twenties, are replete with doodles and cartoons and verbal portraits; for example, at age twenty, writing to Martha from Colorado: "This is country such as a colorist might dream of. The atmosphere is heavy & hazy, so the lavenders are here, and of course the sky is unoriginally blue. But the canyons are deep gashes . . . cut thru varying beds of intensely red sandstone. The walls are precipitous, castellated & buttressed. In the bottom grow the pines & spruces, silvery to deep green & aspens & willows supplying a lighter green which contrasts vividly with the red where they climb up little gulches in it."[28]

Simpson's youthful susceptibility for the visual was not restricted just to nature, for his visual imagination was also fired by human monuments. Seeing the cathedral of Chartres at night, he wrote, "the glory of the cathedral in full moonlight, all golden, the night, the centuries, the old houses, deep shadows, ghosts speaking from an epoch where one constructed this dream in stone, which seems to spring right up before us, from the fertile soil. . . . A solemn thing! The stones of the flying buttresses seem to melt, to blend, only the golden lines outlined on the mass remained — a mass that is the symbol of an idea that is completely ineffable. The clocks strike midnight. Oh my God, if one could tell of it!" (33).

Later, in his early thirties, Simpson was quite particular about the bookplate his artist sister Martha was designing for him, telling her what the images were meant to signify, and included his own rough sketches. Simpson intended the bookplate to be an icon for who he was (fig. 10.8). "I've thought a lot about what kind of design to put in my books. . . . I want a single one for my whole life. . . . First, it has to be me. My name, yes, but also the design . . . it has to be me. . . . I am a student of ancient beasts — some beasts then . . . it must be a design for all my books. There are art books, Egyptian books, books on travel, on languages, some novels, poetry . . . a design that embraces the whole world, all of existence" (124–25). Simpson also had Egyptian hieroglyphs ranged vertically along the two sides of the bookplate, which read: "I will make thee love literature; I will make its beauty enter before thee" (148). Once again, pictures to express thoughts and feelings.

Simpson also believed that artistic images should have informational content of some sort — intellectual as well as emotional: "What rubbish it is when our rising young artists insist that a work of art should not tell a story. . . . The incomparably gorgeous temples of Asia are not *sculptured* simply, but *storied* — and exactly the same is true of all non-decorative . . . medieval art. The artist expected to be *understood* — he expressed something for everyone to understand. . . .

FIGURE 10.8 Simpson's bookplate as designed by his sister Martha according to his specifications. (Courtesy Anne Roe Simpson Trust)

I find it absurd for an artist to paint something which means nothing. . . . Most illustrators were not geniuses, but most geniuses in art were illustrators. I have just seen much of Holbein, Dürer, & the older Cranach and they never produced a simple picture which was not explicitly meant to tell a story or portray a known individual (or place, or object, of course — one must give much latitude to the idea)" (100; emphasis in the original).

* * *

As Martin Rudwick has indicated, graphic presentation is yet another language for communicating, along with numerical and verbal presentation.[29] In this regard, in his publications Simpson was certainly "multilingual" as he was thoroughly fluent in all three modes of communication: writing, biostatistics and (simple) mathematics, and, as we have seen here, visual images. Thus, along with his generally recognized verbal and numeric skills, one must also acknowledge visual language to be an essential and expert part of his abilities — all of which enabled him to communicate his scientific conclusions more effectively.

CHAPTER 11

THE AWKWARD EMBRACE

I have a debt, a loyalty to the [American] museum; the best place for me to do what I wanted to do.
— G. G. Simpson[1]

George G. Simpson spent virtually his whole professional life as an employee of natural history museums. In 1927, at age twenty-five, Simpson joined the American Museum of Natural History as assistant curator of vertebrate paleontology. He resigned that position in 1959 to accept appointment as Alexander Agassiz Professor at the Museum of Comparative Zoology (MCZ) at Harvard University where he remained until 1970. Professionally very accomplished from the outset, Simpson expected — and usually received — from both institutions all the prerogatives and privileges he thought due him as the leading American, if not world, student of fossil mammals in particular and, more generally, of ancient-life history and evolution. But each appointment was clouded by Simpson's resentment to presumed personal affronts from colleagues at Yale University and the American Museum, respectively. Similarly, his departure from both institutions was marked by misunderstandings and professional grievances.

Yet despite these mostly self-inflicted difficulties and an ongoing inflated sense of what was due him, Simpson thrived scientifically to a degree not otherwise possible anywhere else. This was certainly true, at least, at the American Museum, which Simpson was quick to acknowledge in old age: "I have a debt, a loyalty to the [American] museum; the best place for me to do what I wanted to do."[2] This was somewhat less true at the MCZ where his second five-year, extended appointment as Agassiz Professor was complicated by his declining health and that of his wife, as well as by the absence of the magnificent fossil mammal collection of the American Museum.

Prelude to the American Museum Appointment

After receiving his doctorate from Yale University, Simpson continued his study of Mesozoic mammals at the British Museum (Natural History). There was an informal understanding that upon his return to the United States he would take up an appointment in vertebrate paleontology at Yale's Peabody Museum. However, Simpson's ongoing marital situation came to the attention of his Yale colleagues, and formal application for the appointment was suddenly in doubt because his estranged wife had complained to the spouse of a senior administrator about Simpson's shortcomings as husband and father. During a visit to London, R. S. Lull, professor and curator of vertebrate paleontology at the Peabody Museum and Simpson's dissertation adviser, raised the issue with Simpson. Simpson satisfied Lull with his version of his marital problems and, soon after, Yale made a formal offer to Simpson to join the Peabody Museum. But Simpson remained resentful because he felt that any personal domestic troubles he was having were none of Yale's business.[3]

W. D. Matthew, vertebrate paleontologist at the American Museum, also visited Simpson in London and discussed the possibility of Simpson's coming to the American Museum. Matthew was well acquainted with Simpson, who had been his field assistant in Texas several years before and, as a paleomammalogist himself, was familiar with Simpson's important Mesozoic mammal work. Soon after, Simpson had a second job offer, this time from the American Museum. He weighed both equally attractive offers and decided on the American Museum: "I declined to go [to Yale] where my prospective associates had been so willing to believe me a scoundrel."[4] As he wrote years later, his going to the American Museum "was a crucial point, indeed the most crucial point in my professional development."[5]

Simpson Joins the American Museum

During the early part of 1927, Matthew was discussing with the University of California the offer of an appointment as head of the newly re-created department of paleontology.[6] His discussions with Simpson, therefore, about the latter's joining the American Museum were undoubtedly intended to prepare the way for Simpson to succeed him, although this was not made known to Simpson at the time. Years later, Simpson claimed that it was Matthew's recommendation that brought him "in at the bottom of the department" when Matthew "went out at the top."[7] Matthew departed for Berkeley in the summer of 1927, and Simpson began his appointment as assistant curator of vertebrate paleontology on November 1, shortly after his return from his postdoctoral year in London, at a salary of $2,500 (or about $22,500 today) with "further increase in salary & position to be the reward of effort."[8]

Simpson's initial museum colleagues included H. F. Osborn, president of the museum and department chairman; Barnum Brown (1873–1963), curator of reptiles and acting department chairman; Walter Granger (1872–1941), curator of fossil mammals; and W. K. Gregory, research associate in paleontology. Edwin ("Ned") H. Colbert (b. 1905), a graduate student at Columbia University, joined the department as Osborn's research assistant several years later in 1930, became assistant curator in 1933, and was awarded his Ph.D. under Gregory's tutelage in 1935. Later, following World War II, other close associates of Simpson's included Bobb Schaeffer (b. 1913), curator of fossil fishes, and Norman D. Newell (b. 1909), curator of fossil invertebrates. Barnum Brown immediately put Simpson to work: "He thought all young squirts should start at the bottom, so he started me at preparation in the laboratory. I was twenty-five, father of a family [of three daughters], and had been working at a professional level for several years, but he was fifty-four and incomparably more experienced. I quite enjoyed my initiation in the laboratory. I liked and admired the men there, and it was interesting to prepare some relatively primitive mammals, which I was later allowed to study and publish."[9]

Young squirt or not, and despite the affable tone of the above recollection, Simpson must have felt a little put down by this initial assignment. For at the time of his appointment, Simpson had published four abstracts, eleven articles (one of which was with his soon-to-be museum colleague W. K. Gregory as coauthor), five reviews, and two popular pieces; he had in press at least seven more articles and two major monographs, and a number of other works in progress. Despite his relative youth, Simpson had already established himself as the world's authority on Mesozoic mammals and through his writings and travel was known to virtually all the leading vertebrate paleontologists of the world. Within a year, however, his past achievements and obvious potential were recognized by his promotion to associate curator of vertebrate paleontology.

American Museum Research and Expeditions

Employment by one of the world's leading museums of natural history that imposed no formal teaching or administrative responsibilities provided Simpson with unusual opportunities for research, some to his liking, some not. Among the specimens that Barnum Brown had Simpson prepare were some Paleocene mammals Brown had collected in Montana, which Simpson then studied and published on. Having finished his study of Mesozoic mammals, Simpson was propelled by this new work into a major research effort on the Paleocene and Eocene mammals of North America that was to last, on and off, throughout his tenure at the museum. His attention was also turning to the rich Tertiary geology and paleontology of Patagonia, whose mammal faunas bore uncertain relationships to those of the rest of the world. Of less importance to these long-term research interests were other museum collections of various Tertiary ages and locales within the United States that he worked up as part of his curatorial responsibilities.[10]

The stock market crash of 1929 had a major financial impact on the museum. The private financial contributions of wealthy New York City patrons that H. F. Osborn had so assiduously cultivated for more than two decades now declined sharply.[11] Ned Colbert, Simpson's colleague, has referred to the Depression era as "the restricted years" at the museum because, until then, "any activities beyond the usual ones depended upon help from interested citizens who had money to contribute."[12] Because Simpson's research program required collecting expeditions of some sustained duration, he took personal initiative to seek out support from one of the museum's wealthy benefactors, Horace Scarritt, a New York City investment banker and broker.[13] As Simpson later recalled, "This took some persuasion and more than a few drinking bouts."[14] Scarritt's money flowed as freely as his liquor for he ended up supporting three of Simpson's major fossil-collecting expeditions: two to Patagonia in 1930–31 and 1933–34 and one to Montana in 1935. Simpson recognized Scarritt's crucial support by naming an extinct South American hoofed-herbivore of early Tertiary age *Scarrittia canquelensis*.[15]

The Patagonian expedition had the full encouragement if not financial support of Osborn and grew out of an earlier cooperative plan between the museum and the University of Tübingen (Germany), which subsequently had to back out owing to a lack of funds. Much of the American Museum's reputation and renown had been built upon such foreign scientific research expeditions, and Simpson's plans were thus in keeping with a museum tradition so carefully nurtured by Osborn. Simpson left for Patagonia in the summer of 1930, returning the following fall to New York. Seven months were spent in the field working out the stratigraphy and age of the mammal-bearing deposits as well as collecting specimens, some three hundred in all. After his assistant returned to the museum with the fossils, where they began to be prepared for detailed study, Simpson passed the next five months visiting related fossil collections in the museums of Buenos Aires and La Plata.[16]

This sojourn to South America resulted in Simpson's first book, *Attending Marvels*, which not only continued the long tradition of the American Museum, strongly encouraged by Osborn, to bring attention of its research to the public, but it also made Simpson himself known to the nonpaleontological world.[17] A radio interview in New York City and front-page coverage in the *New York Times Book Review* immediately followed publication of the book. Simpson clearly had the intention of bringing his research to a broader, educated public. As he remarked in the foreword of the book, "This is an account of a scientific expedition, but it is more concerned with people and events than with science . . . voyages to the far corners of the earth [have] interest and excitement not dependent on technical accomplishments."[18] Simpson was undoubtedly motivated by a genuine interest in popular scientific education, but possibly he also wanted to keep the museum's benefactors in the fold, given the toll that the Depression was having on continued museum support. Increasingly, thereafter, Simpson and the American Museum were to become closely associated, if not synonymous, in the educated public's mind, as was already the case with his colleague, the anthropologist Margaret Mead.

After the first yearlong Patagonian trip, Simpson returned for another eight months in 1933–34, returning to New York via Paris and Moscow where he tried, unsuccessfully, to arrange a visit to Mongolia. In between his South American expeditions, Simpson spent two long summers in 1932 and 1935 collecting Paleocene mammals in Montana (see chapter 5). And, of course, when not in the field, Simpson was continuing to publish the results of this and other research. During this hectic professional pace Simpson's personal life began to improve. He finally obtained a divorce from his wife in 1938 and immediately married Anne Roe. From September 1938 to May 1939 they went off together to collect fossils in Venezuela where Simpson had been invited by the government to expand the search for Pleistocene mammals. Despite unusually heavy rains, Simpson made a large but not particularly well-preserved collection that was arduously prepared by American Museum technicians and partially described by Simpson and one of his graduate students. Before the study of the fossils had been completed, a subsequent change of government in Venezuela claimed that the fossils had been stolen by Simpson and it demanded that all the fossils be returned, which they were.[19]

By his travels, fossil-collecting expeditions, and resultant prolific publication during the 1930s, Simpson strengthened his reputation as a vertebrate paleontologist and the leading student of fossil mammals. His place within the American Museum certainly seemed secure as one of its best-known scientists. However, the appointment of a new museum director a few years later would change all this.

New Directions with a New Director

In the late 1930s, decreasing private and public support of the museum with resultant increasing deficits, turnover of directors, and lack of agreement between the administration and the scientific staff about policies and programs, led to a crisis of leadership. Eventually, in early 1941, the president of the University of Michigan, Dr. Alexander Ruthven, was asked to make a study of the museum. His single most important recommendation to the trustees the following fall was that the director, Roy Chapman Andrews, be replaced. Within several months Dr. Albert Parr, director of Yale's Peabody Museum, was named the new director, taking office in June 1942.[20]

Parr then began a series of important changes including setting up a formal schedule for salaries and promotions, making the museum more effective in public education, and developing exhibits that instructed more actively than merely the display of objects, however exotic and unfamiliar. Simpson personally soon benefited, receiving a significant raise in salary that was later followed by promotion to curator after having been stalled as associate curator for eighteen years. However, Parr's most provocative initiative was to "de-emphasize evolutionary studies and taxonomic zoology as much as possible." Parr thought that evolution had become an established scientific fact and that further study would add little new knowledge so that the museum's limited resources should be placed elsewhere. Not only was this viewpoint anathema to Simpson, but Parr added insult to injury by recommending further that the department of paleontology be abolished, which was done in summer 1942. The three paleontologists — Simpson, Colbert, and Otto Haas — were assigned to other departments (mammals, reptiles, and invertebrates, respectively).[21]

Perhaps because of the turmoil going on at the American Museum, Simpson received an offer about this time of a Stirling Professorship at Yale University that he provisionally accepted.[22] Provisionally, because he was also considering enlistment in the U.S. Army now that the United States was seriously engaged in World War II. At age forty and a married man with five dependents (his wife and four daughters), Simpson could have avoided military service; but nevertheless he enlisted, attaining the rank of captain in December 1942 and thereby distancing himself from the problems he was facing at the museum. In June 1943, while serving in North Africa in military intelligence under General Eisenhower's command, Simpson reiterated his opposition to Parr's reorganization plan to a museum official: "A whole new chapter in the history of evolutionary theory is just beginning. It is almost incredible that the Museum, with its great tradition of interest in evolutionary studies, is not taking a more leading part in this work."[23] Part of that tradition, of course, was that Simpson's own work was about to be crowned by publication of his book *Tempo and Mode in Evolution*.[24]

By the time Simpson was released from active military service in December 1944, Parr had relented on his scheme of reorganization, so Simpson declined the Yale offer and took up the chairmanship of the new department of geology and paleontology.[25] The rift between Parr and Simpson was apparently healed, and the next decade and a half would become what Simpson years later called his "halcyon period."[26] He gathered up honors, degrees, medals, and prizes from universities and scientific societies from around the world. His list of publications grew and included a number of texts (one of which was translated into some dozen languages) and popular books, many scientific monographs, and innumerable articles, reviews, and letters to the editor. He became formally associated with the departments of geology and of zoology at Columbia University, offering graduate seminars on fossil mammals and on evolution. He served on several governing councils of professional societies and was elected president of the Society of Vertebrate Paleontology and of the Society for the Study of Evolution, both of which he helped found. Besides bringing glory and distinction upon himself, these honors of course also brought honor to the American Museum.

Simpson as Administrator

Simpson's administrative style during his tenure as chairman of the department of geology and paleontology, from 1944 to 1958 at the American Museum, has been well described by his longtime museum colleague N. D. Newell.[27] Newell has said that Simpson was very cordial when he showed up at the American Museum after the war to take up his appointment as curator of fossil invertebrates, but Simpson hadn't at all anticipated what Newell would need in the way of facilities. They walked all over the museum and couldn't find any available space, so Newell was temporarily housed for six months with a comparative anatomist. Newell was "irate — I was a member of Simpson's team and he ignored me." Newell kept prodding and finally went to see Parr, the museum director, and space was found immediately on the fifth floor in a dead-storage area. The fifth floor was then completely renovated for the whole department, starting first with Simpson's area, then that of Colbert, curator of fossil reptiles, and finally down the line to Newell's. All this work took another six months.

With Simpson, there was no socializing among the departmental scientists and staff, so "things got tense and unhappy. Technicians in vertebrate paleontology were often at odds with the technicians in invertebrate paleontology, so that problems got exaggerated and distorted." Newell found it difficult to become acquainted with Simpson, who was "just a name, a ghost. . . . I found him extraordinarily shy."

According to Newell, Simpson never saw himself as a solver of administrative or staff problems nor even as an arbitrator. Only under duress did Simpson take responsibility for the departments of fossil invertebrates and of mineralogy, which were both formally under him. For example Newell, who also had an appointment at Columbia University as a professor of geology, was once accused by another colleague in his department of spending too much museum time with students. They were at an impasse over this, so Newell insisted they go discuss it with Simpson to settle the issue. After ten minutes of total silence following initial explanation of what the problem was, Newell had to pry out of Simpson just what his responsibilities were, and in the end Simpson backed Newell.

Newell observed that with all Simpson's travels, Colbert (who was next in seniority) got stuck with the departmental administrative chores, but his decisions had to be approved by Simpson — several were reversed months later when Simpson returned to the museum — despite Simpson's prior assurance to Colbert that he would back up whatever Colbert decided. But one important decision by Colbert did not get overturned by Simpson. Renovation of the exhibition halls was discussed at one rare departmental meeting, and Simpson wanted straightforward halls of the Paleozoic, Mesozoic, and Cenozoic eras. Colbert, on the other hand, as curator of fossil reptiles, of course wanted the dinosaurs segregated and treated separately. Newell agreed with Colbert because obviously his invertebrate ammonites would be overshadowed if the dinosaurs were displayed with them. Bobb Schaeffer, curator of fossil fishes, remembered this incident too. He and Colbert privately discussed from all sides Simpson's plan to design the displays chronologically, or so they thought. But when they went to talk to Simpson, he immediately turned their logic upside down, and before they knew it they were back outside in the corridor. They had got nowhere. Nevertheless, this was one time that Colbert's opinion carried the day, because while Simpson was away in South America, Colbert went ahead and had the dinosaur halls done as he wished.[28]

Newell became restive over the lack of endowment for his department of fossil invertebrates, and like Simpson he didn't hesitate to try to raise private funds. However, Simpson wasn't keen on Newell's getting oil company support for his research, and when Newell did receive oil money for work in South America, Simpson told him to be very careful that he wasn't consulting for the oil industry by using the museum's prestige. Because Simpson was not at all enthusiastic about this arrangement, Newell had a contract written up between the oil company and the museum that would permit the company to see any written manuscript being sent out for publication, which would, de facto, give the company about one year's lead time before the information became fully public. Simpson also didn't look with favor on Newell's oil company support for his Permian Reef study in West Texas. Newell had to go to Parr for his approval, which he quickly gave because of the financial pinch the museum was in, despite the fact that the trustees wanted all new, private money to go into a general fund to maintain the museum and not to individual research projects.

Colbert was more succinct in his judgment of Simpson's administrative skills. "He was just so-so, average, as an administrator. I've got a theory that every administrator thinks he's a better administrator than he really is. Simpson often didn't tell us what he was going to do. He was a lone wolf in everything."[29]

Accident in Brazil

In the summer of 1956, Simpson participated in a joint Brazilian–American Museum expedition to a remote region in the headwaters of the Amazon River to collect mammalian fossils. Two months into the trip, while a campsite was being cleared, a felled tree struck Simpson, and as he later wrote, "it hit my head, giving me a concussion; my left shoulder, dislocating it; my back, bruising it; and my legs, dislocating my left ankle and shattering my right leg with a compound fracture of tibia and fibula. . . . That event changed my life quite radically."[30] The remoteness of the accident resulted in a week's delay before Simpson received adequate medical treatment. Its seriousness meant that for two years he was unable to pay full attention to his responsibilities as chairman of the department. Or at least that is how it was perceived by the museum director, his former nemesis Parr, and by several colleagues in his own department.

According to Newell, he had urged Simpson not to go to Brazil, arguing that from his experience you couldn't do much real paleontology in heavy jungle: few outcrops, thick vegetative cover, deep soils. Instead, Newell suggested that Simpson work the flanks of the Andes, which had many oil concessions whose companies would be willing to assist his research. Besides, the outcrops were much better and the geology better known. Newell urged Simpson to contact the appropriate oil companies, but Simpson paid no attention to his advice.

Simpson's morale after the Brazil accident was very low; eventually he was up and about, but quite lame. He was mostly working part-time for two years, and when he did return to work full time in 1958, he received an invitation from the Brazilian Academy of Sciences to visit. Simpson told Colbert that he would again have to cover administratively for him. As Mayr recalled, "Colbert rebelled, was beside himself with rage. He refused to run the vertebrate paleontology department without real authority, despite Simpson's assurance that he'd back him up on all decisions." When Parr heard about this, "he said no more overseas trips! Simpson resigned on the spot."[31]

Already in the year before, Newell and some of his colleagues had asked Parr to relieve Simpson of his administrative responsibilities, but Parr said no. Newell had a serious and cordial discussion about this with Simpson, asking why he didn't simply resign the chairmanship, thereby resolving all these difficulties. Simpson said that since Colbert was the next senior man it was up to him to take on the departmental responsibilities while he was away. Simpson added that his not resigning the chairmanship wasn't a question of prestige, but rather "that it wouldn't be fair to Colbert to burden him with that assignment."[32]

Colbert himself recollected that after the Brazil accident Simpson was effectively away from the museum for about two years. Colbert would go up to the hospital or Simpson's nearby apartment to talk about departmental business. "We had to get decisions made, but he didn't want to talk about them. It was a chore, no great honor, to be chairman. For two to three years I was virtually doing all of it, but I lacked a certain amount of authority. Finally, Parr discussed it with the trustees and they suggested that I take over as chairman. I was very dubious. I didn't think Simpson would go along with that. Parr went to tell him and apparently they had one helluva a row."[33]

Schaeffer corroborates the circumstances surrounding this event, although more tartly than Colbert. He suspects "Colbert groused to Parr that he was doing all the work, and Parr said he'd talk to Simpson about it. Parr went up to Simpson's apartment and Simpson got all hot under the collar." He believes that Simpson was convinced that Colbert and Parr "connived and contrived" to get him out of the chairmanship.[34]

Although Mayr had by then already left the American Museum for Harvard and the Museum of Comparative Zoology, his explanation of the blowup at the museum was that it had been long festering. "Colbert worked for Simpson for years, doing all the routine departmental work while Simpson insisted on reserving all the policy decisions for himself. Colbert finally went to Parr to complain that his research was suffering by being acting chairman. Colbert either wanted to be chairman or relieved of the hack work. Parr called Simpson in and said 'Look here. . . .' Simpson felt that Colbert stabbed him in the back and never forgave him."[35]

"Plus Ça Change . . ."

When Parr asked Simpson to resign the chairmanship, Simpson was infuriated — as he later said, "I resigned rather than accept such a humiliating situation."[36] Not long afterward, it must have stuck in Simpson's craw that he had to request permission from Colbert — now chairman — to accept an invitation to attend the 300th anniversary of the Royal Society of London: "I presume that there will be no serious objection from you or from the director [Parr] . . . but in view of the recent discussion of the inadvisability of absenting myself from the museum, I would like reassurance on that point."[37]

Simpson's dissatisfaction with his situation soon led to an offer of appointment as an Alexander Agassiz professorship at Harvard's Museum of Comparative Zoology, offered through the good graces of the director and his colleague, A. S. Romer. Accompanying his formal letter of resignation, Simpson attached a memo to Parr with a copy to Colbert where he announced his MCZ appointment, effective September 1, 1959. In that memo, Simpson expresses "loyalty and affection" for the museum, but he cannot resist a parting shot. "My essential reason for leaving now is that I am offered a position elsewhere that will make my final professional years more useful and productive than they would be likely to be at the Museum. A scientific staff member in a department here active in exhibition and other direct public services, even if not nominally involved in administration, finds that a large proportion of his time is taken up by routine, sometimes necessary and sometimes, frankly, unnecessary or futile and in either case not satisfyingly productive. Thirty-odd years of that is enough, and the Museum even with recent improvement in this respect does not really and adequately provide for relief of its senior staff members from sheer routine. It therefore becomes only sensible to move on to an institution that can and does provide fuller freedom for scientific activity at this advanced stage in a career.

"There are, of course, other and more personal considerations involved, but these are of secondary importance and I will not dwell on them. I will mention only one: the fact that I have become partially but permanently crippled. It has not been suggested that this injury, which occurred in the service of the Museum, makes me incapable of holding a curatorial position, but it has resulted in certain restrictions and suggestions as to future activities that are uncongenial to me."[38]

In a letter to his sister Martha, Simpson described in glowing terms what he thought the new appointment would require — and not require — of him. "I do have an offer of a better job: Agassiz Professor at Harvard. More money & literally *no* duties — just to sit & think if so disposed, & occasionally to say a kind word to students (but no *teaching!*) & other faculty. Free, too, to come & go as I please. The professorship explicitly does *not* require even residence in Cambridge. . . . There are drawbacks of course: leaving the [fossil] collections I'm working on — but Ned [Colbert] & Bert Parr would probably be so glad to see me go that they'd lend collections to Harvard for my work."[39]

In a letter to his mother a few weeks later, Simpson repeated this rosy picture: "I have been offered a good job at Harvard University . . . [that] pays well & has no duties except to write & do such research as I please. (No teaching, which I do not like — I mean, I do not like to teach & do not have to though I will be a professor.)"[40]

Simpson's colleagues at the museum and Columbia University were quite surprised at his resignation. Schaeffer and John Moore, professor of zoology at Columbia, went to see Simpson to try to talk him out of going to Harvard. Simpson listened without comment to their urgings. His only reply was to say that he really did want to watch the ball game that was soon coming up on television.[41] Newell thought that "the MCZ appointment was a pretty sore arrangement. Simpson was the only person in an enormous dead storage area."[42]

Although, strictly speaking, Agassiz professors did not need to be in residence or to teach, in fact there was some pressure to expand their responsibilities. The reason for this was that Agassiz professorships were renewable five-year appointments that did not carry tenure. Agassiz professors were therefore given tenured joint appointments in Harvard departments to ensure continued association with the university if and when the term Agassiz appointments lapsed, and thus Simpson also held appointments in both biology and geology. While departmental appointments were pro forma, some additional duties were implied, such as teaching an occasional course, working with graduate students, sitting on campus committees — the very activities that Simpson had found "not satisfyingly productive" at the American Museum.

Ernest Williams, who held just such a joint appointment as an Agassiz Professor and biology professor, thinks "these distinctions were probably not made clear to Simpson at his time of appointment by the director, Alfred Romer, who liked to get along with people."[43] Mayr, who succeeded Romer as MCZ director, recalled that the MCZ had "fantastic possibilities for Simpson, with all of the benefits and none of the usual obligations; not a single Ph.D. student and away much of the time. He could have taught more than he did, but instead he just gave a few lectures in my course on evolution, and the rest of the time just sat there."[44] Nevertheless, Simpson did continue to thrive at the MCZ, much as he did at the American Museum, at least for the first four years, traveling even more extensively, giving invited lectures, and garnering still more awards and honors, including the National Medal of Science from President Lyndon Johnson in 1966. The MCZ, too, must have been pleased to have Simpson because a year after his initial appointment Simpson turned down the offer of directorship of the MCZ and instead recommended his colleague Ernst Mayr.[45]

His relations with the American Museum, however, remained strained. In his autobiography Simpson mentions, among other grudges, that "I had planned a monograph on the fossil marsupials of North America and many specimens were made available to me, but I was refused access to the crucial collection [of didelphids] at my old institution, the American Museum of Natural History, and I therefore had to abandon that plan."[46] The particular specimens he wanted were already being studied by a Columbia University graduate student working under the direction of Simpson's successor at the museum, Malcolm C. McKenna. As with some other disagreements, this was another one where Simpson overreacted. McKenna recalls that, "He wanted the material 'right now' [but] it was in use by someone with a legitimate claim to keep [it] on loan . . . for a while. There was a hot exchange of correspondence. Simpson thereafter declared he'd never set foot in the museum again."[47]

In 1964, the beginning of his next five-year appointment as Agassiz Professor, both Simpson and his wife suffered "his and her heart attacks." The health of each declined, compounding Simpson's own medical problems from the Brazilian accident. Later, in 1967, Simpson and his wife decided to move to Tucson, Arizona, because the Cambridge winters were getting ever more difficult for them, especially for Anne Roe who was prone to recurrent pneumonia attacks. Simpson wrote MCZ director Ernst Mayr, saying he was ready to resign his Agassiz professorship. Mayr went to the dean and they instead agreed that Simpson was a special case, so he was given the remaining years of his second five-year appointment without residence requirements, at half-pay, and with his fringe benefits continued. When time came for reappointment in 1969, Simpson wrote Mayr and said how he greatly appreciated the current arrangement and that he'd like the next five-year appointment the same way. This time, however, Mayr and the dean "both had a good laugh and said no — which Simpson resented bitterly."[48] It would seem that Simpson was unrealistic with this request, given that he was in his late sixties, in poor health, and some twenty-three hundred miles from his home institution. Moreover, the Museum of Comparative Zoology with its limited Agassiz professorship funds was in no position to offer Simpson such a sinecure, which is what he apparently expected as his due.

Simpson thus ended his ten-year association with the MCZ and accepted a half-time appointment in the department of geosciences at the University of Arizona, which lasted until he finally retired in June 1982 at the age of eighty. Because he was overage when first appointed, Simpson had to be reappointed, each year, by "special action of the [Arizona] State Board of Regents."[49]

*   *   *

Given the magnitude of Simpson's achievements based upon his voluminous publications and their quality as judged from the many honors and awards he received, it is clear that Simpson was one of the leading scientists of the middle half of the twentieth century. Simpson therefore was not exercising illusions of personal grandeur in expecting more than usual consideration and freedom of operation from his museum employers. However, it is also clear that these expectations were at times inflated, going beyond the bounds of what was reasonable in terms of other people's needs and interests and responsibilities. He may have thought himself a completely free agent and able to do as he pleased, but in the real world of institutional employment, he was not and could not.

No doubt many of Simpson's problems with his museum employers were the result of his personal traits of shyness, social awkwardness, and a disinclination to engage in one-on-one debate, however mild. In short, he not only was a difficult person to deal with but he had no interest in dealing with people. His genuine scientific achievements and the very visible rewards that came with them also seemed to encourage an attitude that he was always in the right, justifiably so or not.

Museum research had great appeal for Simpson, whose only interest was in his lifeless fossils and their evolutionary history. He was clearly unsuited for academic employment, his only other professional option. However, even in a museum, Simpson couldn't avoid people: his institutional superiors, professional associates, and graduate students — all of whom he tried to keep at arm's length. Simpson thus often made life for himself and others difficult and unpleasant. Finally, is there a broader lesson to be learned here about ornery individuals and institutional personality? I think not, because Simpson was sui generis in so many ways, scientifically and temperamentally, so that his resultant behavior must also be recognized as simply one-of-a-kind.

CHAPTER 12

CONCESSION TO THE INELUCTABLE

He did, in fact, leave a communication on the wild chance that it might come to other human eyes. It did so, but not until any return communication with him was out of the question.
— The Universal Historian[1]

In a posthumously published work of science fiction, *The Dechronization of Sam Magruder* (Simpson 1996), Simpson tells the story of Sam Magruder, a "chronologist" living in 2162 a.d., who is experimenting on the "quantum theory of time-motion" when he suffers a "time-slip" that puts him back in the late Cretaceous of New Mexico. Helplessly lost in time and with no hope of returning to the present, Magruder ekes out a primitive existence for some years until a fatal accident befalls him. Before his death Magruder manages to chisel out his experience and philosophy of life on eight rock slabs that are recovered many millions of years later, and so his story becomes known and discussed by several Everyman characters.[2]

Simpson spins a reasonably engaging tale, but its main interest is the degree to which Magruder's philosophy of life may reflect Simpson's own feelings toward the end of his life. Always more comfortable in expressing his views in writing than in speaking, Simpson appears to use this work of science fiction to reveal his own, mostly melancholy views about life's meaning and purpose, the importance of adapting to the here and now, and how historical contingency controls subsequent outcomes. Is Simpson speaking for himself when Magruder declares, "My real purpose in engraving these slabs is a search for comprehension. . . . I am exploring my own nature"?

When Simpson died in 1984, his professional papers went to the

American Philosophical Society Archives in Philadelphia. However, his wife, Anne Roe Simpson, kept a number of private miscellaneous papers that, after her death, passed to Joan Simpson Burns, Simpson's daughter by previous marriage. Among the latter papers was discovered an 88-page manuscript entitled "The Dechronization of Sam Magruder"; the science fiction novella written by Simpson had been carefully typed, edited, and appeared to be ready for press. His daughter felt the manuscript was of sufficient interest to warrant publication and eventually had it published.[3]

Given that G. G. Simpson was a major contributor to twentieth-century evolutionary theory and vertebrate paleontology, there are three aspects of the *Magruder* work that arouse biographical interest: Simpson's preoccupation with the notion of time; his well-informed reconstruction of the Age of Dinosaurs; and, especially, the views of his protagonist Sam Magruder as to the "meaning of life." One wonders if the writing of this novella was simply a private amusement for Simpson that he had no intention of ever making public or whether it was an oblique and covert way of stating a strong emotional view about his own life that he did indeed hope would see the light of day after he was long gone. This chapter will argue that the latter situation is indeed the case.

Magruder's Misadventure

In the year 2162 a.d., Sam Magruder, a distinguished "chronologist" (a researcher on the nature of time), disappears from his laboratory while investigating his theory that time proceeds not continuously but in minute incremental jumps. Using extreme time magnification, Magruder is beginning to perceive the discrete, particulate flow of time when he suddenly "slips" between time-quanta and instantly finds himself back in a late Cretaceous swamp in what will later become the state of New Mexico. For the next two decades he manages to survive by his wits amidst a varied dinosaurian fauna. Aware of his hopeless situation, Magruder nevertheless accepts his fate and laboriously inscribes eight sandstone slabs with an account of his Cretaceous experiences and his philosophic reflections thereon, "on the wild chance that it might come to other human eyes" (7). Many years of unspecified duration after 2162 a.d., a geological expedition discovers the slabs, and eventually Magruder's misadventure does indeed become known, as told by the Universal Historian to several of his skeptical friends.

After an introductory chapter, Simpson's tale of Magruder divides itself into three general commentaries — on the nature of time (chapters 2–5), on important aspects of life history (chapters 6–9), and on his personal philosophy of life (chapters 10–13).

Background

Although the Magruder manuscript is not explicitly dated, the typewriter's font is distinctive and, judging from contemporary letters and the manuscript itself, identical to that used by Simpson's secretaries from 1974 to 1979. This six-year interval was a time in Simpson's life when he was in poorer health than usual and was doing a great deal of autobiographical writing, both private and public. Among other biographical vicissitudes during this time were the Harvard conferences on the evolutionary synthesis in May and October 1974, organized by Ernst Mayr (Simpson could not attend but did complete a lengthy autobiographical questionnaire, in advance, vis-Ö-vis the evolutionary synthesis).[4] In November of that same year, Simpson's wife Anne was in a Tucson hospital following open-heart surgery, and his daughter Elizabeth lay unconscious in a Los Angeles hospital with tubercular meningitis. As Elizabeth noted in a later memoir, "My father was dreadfully beset. Informed that I was dying, he had come to say good-by, had left my mother in intensive care . . . where she had just had an elaborate operation. . . . When he kissed me as he left to return to her side, one of my children told him that my head had moved, that I had responded. But, for him, I was already and permanently labeled 'Out of Reach.'"[5]

In August 1975, Simpson started to dictate notes about his research and publications — in part to elaborate on his perceived role in the foundation of the evolutionary synthesis, because he wanted to set the record straight (at least according to his own lights). "Questionnaires are almost inevitably slanted by the people who compose them and I'm afraid that was quite clearly true of the questionnaire that was circulated regarding the [Harvard] conference."[6] Later that same year, Simpson learned that his "great & old friend Dobzhansky [had] died. Quite a few friends have died in recent years, but none has affected me more deeply."[7] Beginning in 1976 and continuing through most of 1977,

Simpson wrote his autobiography *Concession to the Improbable*.[8] In late 1979 Simpson wrote that he was "terribly saddened to learn that my old and dear friend, Bryan Patterson, . . . died while I was away."[9] And finally, in the late 1970s and into 1980, Simpson prepared the lead article for a book on the paleontology of the terrestrial deposits spanning the Cretaceous-Tertiary boundary in New Mexico where Sam Magruder's misadventure takes place.[10]

Clearly, then, this period of Simpson's life was a time when he was acutely aware of his own mortality and was anxious about posterity's judgment of his scientific contributions. Consequently, he set about to put on public record his reflections on his professional life and scientific work. However, being an emotionally private person, it is not surprising that Simpson preferred to express his last will and testament of personal philosophy more indirectly in the didactic, if science fictional, form that we find in "The Dechronization of Sam Magruder." Simpson's autobiography paints a brighter picture than what is found in *Magruder*: "It is part of my good fortune that the pain and sorrow are more than overbalanced by pleasure and delight. Here in the closing years of my life I have a loving and fascinating wife and family and many good friends."[11] The *Magruder* novella, however, presents a darker and more emotionally intense perspective.

It is also suggestive that throughout his life Simpson was fond of puzzles and codes. An example of the latter was the code he developed in his teens and used periodically thereafter, especially when communicating with his future second wife during protracted marital difficulties with his first wife.[12] So, to repeat, perhaps the Magruder manuscript can be viewed as an encoded version of Simpson's personal philosophy. As one of the story's characters notes: "He did, in fact, leave a communication on the wild chance that it might come to other human eyes. It did so, but not until any return communication with him was out of the question."[13]

What then are the messages that Simpson wished to communicate — and communicate them safely — "from beyond the grave"?

The Nature of Time

At the beginning of *The Dechronization of Sam Magruder*, Simpson distinguishes two time universes: the single point "time-motion universe" consisting simply of the present moment that is in constant movement (11) and the linear "time-dimension" universe that has only two directions, away from and toward the present moment (12). The nature of time was a recurring preoccupation of Simpson's throughout his life. Even as a teenager he wrote in a letter from college to his friend Anne Roe: "Time is a figment of man's imagination. Only in *time* must things have a beginning or end. Actually *nothing* is in time. Time itself is in our *perception* of things, not *in the things* or *governing* them."[14] And a few years later he wrote in a doleful letter to his sister Martha from Yale graduate school where he was studying the fossils of primitive Mesozoic mammals, some 100 to 200 million years old: "The spectacle at which I attend is vastly moving. There is an almost painfully epic sweep to the vastness of geologic ages which pass between my fingers in tattered fragments [i.e., fossils]. The commonplace room is always filled with the mute cries of ages impossible to contemplate in which life has blindly toiled upward, or at least to further complication and further ability to realize that it cannot realize anything at all. It is all very strange and thrilling in a way which is, I am afraid, incommunicable. All of which can only serve as a deeper contrast to the utter futility of it all. The reconstruction of the past, even so great a past as that which lies before me here, can add only a melancholy significance to the fact which we know but dare not realize that the present must become as truly past and perhaps even more irrevocably."[15] And more than sixty years later, in a book published just before he died, Simpson wrote: "That some people are frightened by geological time and more find it difficult to grasp its immensity may well be related to a sense of human mortality, to a fear of personal aging, and to the comparative brevity of our own lives. One of the great values of paleontology is that it enables us to live in our own complex minds not just a few score years but more than three billion years [of life history]."[16]

Simpson's professional specialty, of course, was paleontology, and elsewhere he defined the broader intellectual activity within which that specialty lay: "I am trying to pursue a science that . . . has no name: the science of four-dimensional biology or of time and life."[17] The key word in the title of the *Magrunder* work — "dechronization" — immediately suggests an active reworking of the concept of time, rather than just a narrative taking place in past time. The Universal Historian claims that, "What cut [Sam Magruder] off irrevocably from his fellow man was time, simply time. . . . He was the victim of what is technically known as dechronization. In vulgar terms, he suffered a time-slip. . . . What interests me in this connection is the reaction of a man who lived with the conviction that he would surely never return to the human world or receive any word from it" (8, 5).

Did Simpson suffer his own time-slip? Did he feel cut off from the rest of humanity owing to long years of intense paleontological research, cut off both because of the subject matter of the research itself and his single-minded dedication to it? Or was it simply the normal awareness of an aging human being facing his own mortality? In short, was Simpson experiencing a deep melancholy at the end of a long life to which he gave vent in this story? Consider the title of the first chapter: "How to Be Alone."

Past Life's Lessons

The Universal Historian observes that, "The connection with universal history and chronology must already be obvious, and that with paleontology will become evident" (11). And indeed there is another aspect of time — historical contingency — that Simpson explores in this story that emerges directly from his paleontological specialty. "The past has a reality, an objective and external existence, even more truly than does the present. It *produces* the present. The growth of the past is conditioned and determined by the whole of that past as existent up to any given present" (12; emphasis in the original).

From a paleontological viewpoint, of course, given the nature of genetic inheritance with its material ancestor-descendant physicality, life at any given moment during its billions of years of existence necessarily depends greatly upon what came before: life of the past does indeed produce the life of the present. From a more immediate human perspective, Simpson is acknowledging that where he finds himself at the end of his life is, for better or worse, also the result of historical contingency rather than, say, either a predetermined end or one that was simply the outcome of blind forces.

The story that Simpson spins about Magruder's "time-slip" is a seemingly commonplace tale of science fiction with the usual sort of characters, namely the "lost explorer" surrounded by an alien dinosaurian fauna and flora. Nevertheless, the geographic locales (e.g., San Juan Basin, Ojo Alamo), geologic strata (e.g., Kirtland and Fruitland formations), and dinosaurs (e.g., *Pentaceratops* and *Alamosaurus*) are all real. As a paleontologist who spent many years working the fossil beds of the New Mexico region, it is natural that he would choose the familiar and factual rather than the fictional. By sticking to as much authentic detail as possible, Simpson strove for a verisimilitude that would, however, be lost on most nonspecialist readers.

And yet an interesting puzzle emerges. Simpson was one of the world's foremost authorities on the early Tertiary mammals of this region, and he could have just as well used those animals for his supporting cast of characters instead of Cretaceous dinosaurs.[18] Readers would have found such mammals far more unfamiliar and therefore more exotic than dinosaurs. Nevertheless, Simpson made this choice favoring recognizable dinosaurs over obscure mammals quite intentionally. But why? Could it be that he wanted to attract as wide an audience as possible to hear his tale, and to that end he believed meat- and plant-eating dinosaurs would appear more familiarly archaic to most readers than, say, the early insectivores and rodents, weasel-like carnivores, and sheeplike herbivores that he himself studied? Apparently to make what he wanted to say more accessible to the general reader, Simpson framed the story within the Mesozoic dinosaurian world rather than within the primitive mammalian world of the Cenozoic Era of which he was the acknowledged expert. Unusually, then, Simpson was setting aside his special expertise for the more important goal of attracting an audience who might receive his message that transcends any specific fauna and flora in any particular time and place.

In the beginning of his narrative, Magruder emphasizes that human survival depends critically upon "social organization, on cooperation, on division of labor, on the building up and passing on of knowledge, of tools and methods. . . . The sole individual hasn't a prayer without his group" (45). And although Magruder in his predicament is cut off from the rest of humanity and thus ultimately doomed, he nevertheless reminds himself that humans are also the most adaptable of animals and so he must try to survive in the threatening world of the Cretaceous. He reminds himself that, "You've still got [brains] and they must be some use to you" (46). But then — perhaps resonating with how Simpson himself was feeling at the time of writing — Magruder has another mood swing and laments: "I miss my colleagues. I miss my friends. I miss them terribly and the ache persists as the years pass, but I no longer think much about them. The daily hazards of life demand most of my attention, and I have carefully schooled myself to give as little time as possible to regrets that are vain" (50). But then as Magruder proceeds to tell of his Mesozoic trials and tribulations, he becomes more didactic as he dispenses standard paleontological and evolutionary fare. For example, he remarks on the uncertainty of correct interpretation of functional morphology from fossils, how variations in environment are reflected in the changing composition of animal and plant communities, and the enduring enigma of dinosaurian extinctions. He points out especially that adaptations so successful, like that of the ancient crocodile, may appear misleadingly as evolutionary dead ends: "Is it perhaps not the success but the failure of adaptation that has forced life onward to what we, at least, consider higher levels?" (55). To survive his time-slip, Magruder realizes that he too must adapt to the new circumstances in which he finds himself, however desperate it may be. So Sam Magruder deals as best he can with his new world, observing and coping with the fierce animals about him.

Melancholia

This brings us to the last question that this story raises for us. What is the real message that Simpson wishes to deliver? Surely this tale is more than a playful excursion into the nature of time developed within the framework of the tried-and-true fantasy world of dinosaurs. The descriptive narrative of Magruder's life among the Cretaceous biota provides the occasion for the remainder of the book that is introduced with the apposite chapter title "Brooding" (83) wherein Magruder confesses, "My real purpose in engraving these slabs is a search for comprehension . . . the search is for my own sake. . . . I am exploring my own nature. . . . I cannot entirely abandon hope that these words will sometime be read by other humans. . . . I am never to know whether this message reaches others . . . that my desperate voice will be heard, that someone, sometime will be aware of Sam Magruder and will feel interest in, perhaps sympathy for, his fate" (83).

Magruder expresses a sentiment about his "adventure" that could well be Simpson's feelings about his own life: "What I have done and what I have seen are essential parts of the story, of course, but the only really important thing is what I have felt and thought. This is much harder to convey, but it is the essence of my story" (83–84). Magruder describes a "complex emotional state that has settled on me permanently . . . a melancholy, a dully aching sadness, for which there is no remedy but death. A melancholy tinged with wonder, with attempts to find some meaning in my life . . . there are episodes when I give way to acute despair and horror. . . . Melancholy! It lies too near the surface for any of my few diversions to subdue it wholly or for long" (84, 86).

To allay these bouts of melancholy, Magruder amuses himself with homemade musical instruments and putting himself to sleep by mentally recalling whole symphonies. He also enjoys "looking at scenery. I have always been a sensitive admirer of nature" (85). Simpson, a self-taught mandolin player, enjoyed playing music and listening to it. And if travel off the beaten track is one medium for observing nature, then Simpson surely had his share of that.[19]

There is one curious passage in *Magruder* that challenges facile interpretation. During his first winter in the Cretaceous, Magruder comes down with a "great delirium" during which he experiences a vivid interchange with his father, who reminds Magruder of their "cabin . . . in the mountains, where we used to go fishing together" and of the "little girl with yellow curls next door."[20] His father also asks him to find his cap and gown because it's "almost time to go to commencement exercises" (87). When Magruder wakes, he realizes that his father was not really there, that he was lost to him, as completely, irrevocably lost as all other human beings (88). Simpson seems to have had an ambiguous relationship with his father, one that he did not fully resolve. He was deeply attached to his father but, despite his father's apparent approval in later life, Simpson was somewhat resentful that his father did not feel he was a "real boy" during his childhood.[21] Except for this delirium, Magruder remains rational throughout his experience, having "rigidly rejected the childish expedients of a dream world. So I remain sane, and I pay for sanity the horrible price of unmitigated loneliness" (90).

Simpson as Sam Magruder

The Universal Historian warns that while there is pathos and tragedy in what befell Magruder, what in fact happened "was a complex, unparalleled experience undergone by an extremely intricate personality" (92). This is a claim that can also be made for Simpson, who himself had complex, unparalleled experiences, both in his private and professional life, and who could be accurately described as having an extremely intricate personality, such that few people, including some even within his own family, could ever get emotionally close to him.

So what then "was Magruder's appreciation of his experience?" asks the Ethnologist. The Universal Historian replies, "Did you ever stop to think, really think, what you are? And why? . . . The questions are the answer, and no one has ever been in a position to think so clearly about what and why as Sam Magruder. That is what he appreciated in his experience" (94).

As Magruder's tale unfolds, in slab 7 he eventually goads himself to start "getting to the point . . . [and so] what is this point I should be getting to?" (98). Apparently, the point is his encounter with small, furry, rodentlike mammals that he captures and cages. Magruder reflects that their "descendants would inherit some spark that would keep them fighting successfully in the long struggle for survival. They would come into their own long after the great dinosaurs had failed. Is it not, indeed, that drive that animates me now? Here is the real reason for my own survival. . . . To you, reader . . . the important thing is not *my* survival but the survival of that savage little furry creature I am looking at. You, with your culture and civilization, are the outcome of his struggles and the realization of his potentialities." Magruder resists the impulse of "selectively breeding the little animals. But for what good? . . . They are going to make it. Their descendants will be men, and they'll get there under their own power. . . . What is holy in mankind is that mankind, through this little beast and so many others, has created itself" (101–102; emphasis in the original).

As the Universal Historian remarks, "Here is where Magruder refuses to play God, and where he sees the what and why of himself, of you, and of me." And so Magruder opens the cage and says, "Go with God, Great-grandpa!" (102). Simpson is, of course, making the essential point (for him) that humans are the historical result of an evolutionary process that requires no intervention — supernatural or otherwise.

Magruder's story then ends with a coda of personal insight that may perhaps serve as Simpson's own summing up toward the end of his life: "There isn't much more to say. I've had no joy, but a little satisfaction, from this long ordeal. I have often wondered why I kept going. That, at least, I have learned and I know now at the end. There could be no hope and no reward. I always recognized that bitter truth. But I am a man, and a man is responsible for himself" (104).

Simpson's temperament was such that he did not — probably

could not — easily express the darker and more pessimistic side of his emotions. Outwardly, he seemed coldly rational, aloof, preoccupied, always intently focused on his scholarly work, with little obvious interest in introspection, whether personal or much less that of others. Yet he obviously bore the universal human anxieties regarding the meaning and purpose of existence. It would seem that as he approached the end of his life he needed to set down for himself and others what it all added up to. Typically, however, he could only do this indirectly and in the language of science with which he felt most comfortable. Hence, his is the tale of Sam Magruder.

NOTES

1. Biographical Introduction

This chapter was first published in slightly different form as the "Introduction" in G. G. Simpson, *Simple Curiosity: Letters from George Gaylord Simpson to His Family, 1921–1970*, edited by L. F. Laporte (Berkeley: University of California Press, 1987), 3–12. Copyright © 1987 by The Regents of the University of California. Used by permission of the University of California Press.

1. Further biographical sources for G. G. Simpson include the following: **Simpson's autobiography**, *Concession to the Improbable: An Unconventional Autobiography* (Simpson 1978); **family letters**, *Simple Curiosity: Letters from George Gaylord Simpson to His Family, 1921–1970* (Simpson 1987); a brief **self-assessment of his life's work**, "The Compleat Paleontologist?" (Simpson 1976); Simpson's **journal of his first Patagonian expedition**, *Attending Marvels* (Simpson 1934).

Simpson's **personal papers** can be found in the American Philosophical Society's archives in Philadelphia, Manuscript Collection no. 31, George G. Simpson Papers (hereafter, GGS/APS), which include **professional correspondence**; **wartime letters** written to his wife Anne Roe (Simpson 1942–1944), selected and edited by her (272 pp.); transcription of **comments on his bibliography** (Simpson 1975) that review chronologically his published writings (135 pp.); series of **autobiographical notes** (Simpson 1933), updated by him in 1954 and 1970 (85 pp.); and a **response to Ernst Mayr's questionnaire** (Simpson 1974b) on the "evolutionary synthesis" and Simpson's self-perceived role in its formulation (13 pp.). Simpson's edited response and Ernst Mayr's commentary on it can be found in "G. G. Simpson," in E. Mayr and W. Provine, eds., *The Evolutionary Synthesis*, 452–63.

An **interview** was done just a few months before Simpson's death and published in the University of Arizona campus paper: Jon Marks, "George Simpson: A Scientist's Scientist, a Writer's Writer," *Arizona Daily Wildcat*, February 2, 1984, 8. (Jonathan Marks was a part-time secretary to Simpson while a graduate student in anthropology at the University of Arizona. After receiving his Ph.D., he has taught at UC Berkeley and at Yale University.)

An **introduction to a "festschrift" volume** for Simpson on his seventieth birthday has a biographical sketch and a fairly complete **bibliography** of Simpson's publications up through 1971 (Hecht et al. 1972).

Obituaries and memorials include those by A. W. Crompton, S. J. Gould, and E. Mayr, "George Gaylord Simpson," *Harvard University Gazette* 81.36 (May 16, 1986); Stephen J. Gould, "Recording Marvels: The Life and Work of George Gaylord Simpson," *Evolution* 39 (1985): 229–32; Léo F. Laporte, "George Gaylord Simpson," American Philosophical Yearbook for 1984 (June 1985), 6 pp.; Robert D. McFadden, "George Simpson, Authority of Vertebrate Paleontology," *New York Times*, October 8, 1984; Everett C. Olson, "Memorial to George Gaylord Simpson," *Geological Society of America* Memorial Reprint, 6 pp.; Bobb Schaeffer and Malcolm McKenna, "George Gaylord Simpson," *Society of Vertebrate Paleontology News Bulletin* (February 1985): 62–63; H. B. Whittington, "George Gaylord Simpson," *Biographical Memoirs of Fellows of the Royal Society* 32 (1986): 525–39.

2. Paleontology and the Expansion of Biology

This chapter was first published in slightly different form as "George G. Simpson, Paleontology, and the Expansion of Biology," in Keith Benson, Jane Maienschein, and Ronald Rainger, eds., *The Expansion of American Biology* (New Brunswick, N.J.: Rutgers University Press, 1991), 80–106. Used by permission of Rutgers University Press.

1. Simpson 1946 ("Tempo and Mode in Evolution"). This paper was first given as a lecture in November 1945 in the New York Academy of Science's Section of Biology, and its title was the same as the book Simpson published the previous year. Simpson noted in his introduction to this lecture that the book attempted "to take some of the data available to paleontologists and not to geneticists, and to relate these to recent developments in evolutionary theory," whereas the lecture focused on the "historical

background of modern evolutionary theories" and the divergence and subsequent convergence of paleontology and genetics on this subject.

2. Rainger 1988 ("Paleontology as Biology"). For a full development of the position of paleontology vis-à-vis biology around the turn of the century, see also Rainger 1981 ("Morphological Tradition"); Rainger 1985 ("Paleontology and Philosophy"); Rainger 1986 ("Just Before Simpson"); and Rainger 1989 ("What's the Use"). For a review of the state of paleontology just before the evolutionary synthesis, see Gould 1980 ("Paleontology and the Modern Synthesis"), esp. 153–57, and Simpson 1946 ("Tempo and Mode in Evolution").

3. Simpson 1933 ("autobiographical notes").

4. Simpson 1978 (*Concession to the Improbable*, 16).

5. University of Colorado undergraduate transcript.

6. See Ernst Mayr's abridged and edited answers that Simpson gave to Mayr's questionnaire dealing with Simpson's role in the evolutionary synthesis, in Mayr and Provine 1980 (*Evolutionary Synthesis*, 453); see also Woodruff 1922 (*Foundations of Biology*, 378).

7. Woodruff 1923 (*Development of the Sciences*, 258).

8. Author interview with Simpson, February 3, 1979, Tucson, Arizona.

9. Yale graduate school transcript.

10. Mayr and Provine 1980 (*Evolutionary Synthesis*, 453).

11. Yale graduate school transcript; Lull 1917 (*Organic Evolution*).

12. Simpson 1958a ("Lull Memorial," 130).

13. Mayr and Provine 1980 (*Evolutionary Synthesis*, 453).

14. Simpson 1928a (*Catalogue of Mesozoic Mammalia*) and Simpson 1929a ("American Mesozoic Mammalia").

15. Simpson 1926c ("Mesozoic Mammalia. IV," 228).

16. Simpson 1926b ("Fauna of Quarry Nine").

17. Simpson 1980c (*Why and How*, 86).

18. Simpson 1978 (*Concession*, 39).

19. Ibid., 34.

20. Simpson 1986 ("Reflections on Matthew," 202).

21. Matthew 1928 (*History of Life*, 6).

22. Matthew 1925 ("Recent Progress and Trends").

23. Matthew 1915 ("Climate and Evolution"). See also chapter 9, this volume, as to why Simpson's biogeographic theory led him astray with respect to Wegener's hypothesis for continental drift.

24. Simpson 1971 ("W. K. Gregory tribute").

25. Ibid., 158–60. The notion is that a given kind of organism "is a mosaic of primitive and advanced characters, of generalized and special features," owing to differing evolution of its various components (Mayr 1963, *Animal Species and Evolution*, 598).

26. Simpson 1937d ("Supra-specific Variation," 236). The venue for Simpson's "door-opener" was most appropriate. As noted on the title page, the paper was read December 30, 1936, in Atlantic City, N.J., at a symposium of the American Society of Naturalists in joint session with the American Society of Zoologists, the Botanical Society of America, the Genetics Society of America, the American Phytopathological Society, and the Ecological Society of America. This was precisely the audience for "presenting the paleontological view-point on the zoological problem of higher [taxonomic] categories" (ibid., 236).

27. Simpson to William K. Gregory, November 16, 1936 (folder G, GGS/APS).

28. See chapter 7 in this volume.

29. Simpson 1944a ("Henry Fairfield Osborn," 585).

30. Rainger 1988 ("Paleontology as Biology," 221).

31. Simpson 1946 ("Tempo and Mode," 52). "Aristogenesis" was the name Osborn gave to the inherent drive within organisms toward future perfection. Simpson's use of "philosophical" here was to indicate that Osborn was often seeking to address in his scientific writings the larger issues of "how and why" with respect to organic evolution, not just merely the "what." In this colloquial sense Simpson, too, was "philosophical." See also chapter 4 in this volume.

32. Simpson 1944a ("Henry Fairfield Osborn," 586).

33. Simpson 1978 (*Concession*, 81). Why the ten-year gap (1926–1936) between graduate school and Simpson's first "theoretical" papers? Recall that it wasn't until the early 1930s that the work of the population geneticists, like Fisher, Haldane, and Wright, was beginning to have an impact, and not until 1937 did Simpson discover them through Dobzhansky's classic (see n. 37, this chapter).

34. Simpson 1937a ("Fort Union of the Crazy Mountain Field").

35. Simpson 1937c ("Phyletic Evolution") and 1937d ("Supra-specific Variation").

36. Simpson and Roe 1939 (*Quantitative Zoology*, vii).

37. Key writings of this time that Simpson cited in his *Tempo and Mode* were Fisher 1930 (*Genetical Theory of Natural Selection*), Haldane 1932 (*Causes of Evolution*), various articles by Sewall Wright, esp. 1931 ("Evolution in Mendelian Populations"), and Dobzhansky 1937 (*Genetics and the Origin of Species*). In tape-recorded comments Simpson made in 1975, providing background on his various published works, he specifically acknowledged the important influence of these writings on his own research. See Simpson 1975 ("Comments on His Bibliography"). There is no evidence, however, that Dobzhansky personally influenced Simpson in one-on-one conversations during this period. Both did later overlap in time in New York City during 1940–1942 when Dobzhansky came to Columbia

from the California Institute of Technology in 1940, and before Simpson went to war in 1942. Even so, Simpson did not yet have an adjunct appointment at Columbia — that wasn't until 1945 — and he had a reputation for not engaging in conversation on these issues, even at the museum. As Mayr has remarked, "We [Simpson and Mayr] never talked about these things. We sat at the same lunch table at the museum and never once discussed evolutionary theory." (Author interview with Ernst Mayr, October 22, 1980, Cambridge, Mass.)

38. Simpson 1975 ("Comments," 45). See also Simpson's comments in Mayr and Provine 1980 (*Evolutionary Synthesis*, 456). How accurate are such post hoc memories, some forty years later? It does seem that Dobzhansky's book was the key turning point in Simpson's appreciation of genetics for his paleontological research, but he probably was not as explicitly aware of the work of the geneticists until then, because he does not refer to any of that research in his 1937 "theoretical" papers (two delivered orally in 1936). However, thereafter in many of his publications, besides *Tempo and Mode*, he does cite not only Dobzhansky's 1937 book but also the earlier genetic work on which it was based and to which it referred.

39. Simpson 1937a ("Fort Union," 63).

40. Ibid., 64; Matthew 1930 ("Range and Limitations of Species").

41. The hypothesis is that organisms closely similar in morphology are for that reason inferred to be closely similar in ways of life as well, so that interspecific competition would eventually drive one species out, but by living in separate geographic areas they would not compete with one another.

42. Simpson 1937c ("Phyletic Evolution," 308–309).

43. Simpson 1980c (*Why and How*, 112). "Typological" because the species is defined upon a single type specimen to which all subsequent specimens are referred for inclusion or exclusion with respect to the named species. Species named according to a "population" concept are based upon a sample of several specimens, usually not identical, and therefore the species represents some integrated combination of all the specimens. However, by the rules of zoological nomenclature, one specimen must still be designated the "type specimen," which becomes the physical entity bearing the new species name. The important difference is that "typological thinking" views the type specimen as a manifestation of some idealized, unvarying species; whereas "population thinking" allows for — even expects — variation among individuals drawn from an interbreeding population, which is, of course, the basic unit of evolution.

44. Simpson 1937d ("Supra-specific Variation," 250). Kinsey did extensive research on the life history and evolution of gall wasps in Mexico and Central America; later in his career he came well known to the general public for his studies on human sexual behavior.

45. Simpson and Roe 1939 (*Quantitative Zoology*, 23).
46. Author interview with Simpson, February 2, 1979, Tucson.
47. Simpson 1931 ("A New Classification of Mammals"); Simpson 1945 ("Principles of Classification"); and Simpson 1961c (*Principles of Animal Taxonomy*).
48. Simpson 1945 ("Principles of Classification," 2, 7).
49. Simpson 1937d ("Supra-specific Variation," 250).
50. Simpson 1944b (*Tempo and Mode*, xviii).
51. Gould 1980 ("Modern Synthesis," 36ff.); Mayr 1982 (*Growth of Biological Thought*, 568, 607ff.). See also chapter 7 of the present volume. *Tempo and Mode* was "written at intervals between the spring of 1938 and the summer of 1942" (Simpson 1953c, *Major Features of Evolution*, ix).
52. Rainger 1986 ("Just before Simpson") and Rainger 1988 ("Paleontology as Biology").
53. See Gould 1980 ("Modern Synthesis," 153–72), for a discussion of what was unique about Simpson's treatise and the role it played in the evolutionary synthesis, and chapter 7 in this volume for a more detailed discussion of the specific content of *Tempo and Mode* as well as the state of paleontology just before its publication, how the book was received, and its subsequent status.
54. Simpson 1953c (*Major Features*, xii).
55. Simpson 1944b (*Tempo and Mode*, 92–93 and 8ff.).
56. Although Simpson made these three distinctions, he abandoned them in his later discussions, preferring to follow geneticist Richard Goldschmidt's usage "that historical changes within species be called microevolution while those from species upward are to be called macroevolution." Simpson made this distinction for convenience only, and noted that it coincided with a difference in the domain of materials available for direct study to the experimental biologist and paleontologist, respectively, and not to qualitative differences between the two kinds of evolution. See Simpson 1953c (*Major Features*, 338–40).
57. Simpson 1976 ("Compleat Paleontologist?" 5).
58. Dobzhansky 1945 ("Review of *Tempo and Mode*," 114); Glass 1945 ("Review of *Tempo and Mode*," 261); Hutchinson 1944 ("Review of *Tempo and Mode*," 356); Huxley 1945 ("Review of *Tempo and Mode*," 3); Wright 1945 ("A Critical Review," 415).
59. Jepsen 1946 ("Review of *Tempo and Mode*," 538). Glenn L. Jepsen (1903–1974) was a contemporary of Simpson's, a Princeton professor of vertebrate paleontology, and like him a student of early Cenozoic mammals. Because of their closeness in age and field of research, Jepsen was impressed by, yet somewhat envious of, Simpson's greater professional renown.

60. Rainger 1981 ("Morphological Tradition").
61. Simpson 1950a ("Trends in Research").
62. Ibid., 498 and 499.
63. Simpson 1978 (*Concession*, 129).
64. Several such American contemporaries of Simpson's who were consistently theoretically inclined were Preston E. Cloud (1912–1991) and Norman D. Newell (b. 1909) among the invertebrate paleontologists, and Everett C. Olson (1910–1993) and Glenn Jepsen among vertebrate paleontologists; all were sufficiently prominent to have been elected to the National Academy of Sciences.
65. Olson 1974 ("Workshop on the evolutionary synthesis," 27), MES/APS.
66. Simpson, Roe, and Lewontin 1960 (*Quantitative Zoology*, v). Simpson defined systematics as "the scientific study of the kinds and diversity of organisms and of any and all relationships among them," whereas "taxonomy is the theoretical study of classification, including its bases, principles, procedures, and rules." Simpson 1961c (*Animal Taxonomy*, 7, 11).
67. Mayr 1982 (*Biological Thought*, 570).
68. Simpson 1940d ("Types in Modern Taxonomy," 418).
69. Romer 1959 ("Vertebrate Paleontology," 919). Romer at that very time was about to make Simpson the offer of an Alexander Agassiz professorship at Harvard's Museum of Comparative Zoology, of which he was the director.
70. Dunbar 1959 ("Half Century of Paleontology," 911). Dunbar (1891–1970) was professor of invertebrate paleontology and stratigraphy at Yale and director of its Peabody Museum of Natural History.
71. Simpson 1946 ("Tempo and Mode," 49, 57; emphasis in the original).
72. Mayr 1963 (*Animal Species*, 11).
73. Simpson 1961c (*Animal Taxonomy*, 153, 150).
74. Simpson 1961d ("Some Problems of Vertebrate Paleontology," 1679).
75. Ibid., 1679 and 1680.
76. This summary of the history of the SVP follows Simpson 1941a ("History of the Society," 1–3).
77. Simpson 1941d ("Some Recent Trends and Problems," 2–4 [part 1], and 10–11 [part 2]).
78. Gingerich 1986 ("Empirical Theoretician," 3–9).
79. Simpson 1978 (*Concession*, 268).
80. Simpson was mostly ill when he was president of the ASZ. He and his wife Anne suffered "his and her heart failures and were hospitalized" over the summer, and they spent most of the rest of the year on a conva-

lescent South Sea cruise (Simpson 1978, *Concession*, 214–15). President-elect Theodore Bullock became de facto ASZ president during 1964, before assuming his own formal presidency in 1965.

3. The Summer of 1924

This chapter was first published in slightly different form as "George G. Simpson (1902–1986): Getting Started in the Summer of 1924," *Earth Science History* 9 (1990): 62–73. This article was printed in and is used by permission of the journal of the History of the Earth Sciences Society, *Earth Sciences History*.

1. University of Colorado and Yale University transcripts.
2. Author interview with Simpson, February 3, 1979, Tucson, Arizona.
3. Simpson 1975 ("Comments," 8); Simpson 1978 (*Concession*, 17). "In later years [Lull] used to remark that the best collecting he knew was in the basement of the Peabody Museum, where reposed much of the great Marsh Collection, including innumerable undescribed specimens that Marsh himself had never laid eyes on" (Simpson 1958a, "Lull Memorial," 129).
4. Lull to Osborn, January 18, 1924 (R. S. Lull correspondence, folder III, DVP/AMNH).
5. Osborn to Matthew, January 21, 1924 (R. S. Lull correspondence, folder III, DVP/AMNH).
6. Matthew to Lull, January 24, 1924 (R. S. Lull correspondence, folder III, DVP/AMNH).
7. The meeting was suggested in Lull to Matthew, January 25, 1924 (R. S. Lull correspondence, folder III, DVP/AMNH).
8. For a recent appraisal of Matthew's importance see Rainger 1986 ("Just Before Simpson") and also Matthew's biography by E. H. Colbert, *William Diller Matthew: Paleontologist* (1992).
9. Owing to the rate of inflation since 1924, $100 would be worth about $900 at the end of the century.
10. Matthew to Simpson, April 8 and 25, 1924 (GGS/APS).
11. Matthew 1924c ("A New Link") and Matthew 1926 ("Evolution of the Horse"). The latter article became a classic in the literature of fossil evidence demonstrating Darwinian evolution.
12. Simpson 1987 (*Simple Curiosity*, 37).
13. Simpson 1986 ("Reflections on Matthew," 201). Some years later one of Simpson's field assistants, George Whitaker of the American Museum, had a similar experience when he broke a delicate jaw that Simpson had just handed him. Seeing Whitaker's embarrassment, Simpson immediately reassured him, "Look, I know how these things hurt, because that happened to me once." (Author interview with Whitaker, November 7, 1980, Avenel, N.J.)

14. Simpson 1978 (*Concession*, 34).

15. Simpson 1986 ("Reflections on Matthew," 203).

16. Matthew 1924b ("List of species"). At the bottom of the third and final page is a note by Matthew: "Please have copied & send copy to Mr. Frick & one to Prof. Osborn" — to Osborn, president of the American Museum, and to Childs Frick because of his support of fieldwork.

17. Matthew 1924a ("Fossil Horses from Texas," 630).

18. Matthew 1924c ("A New Link," 2). Subsequent work revised Matthew's generic assignment, keeping *simplicidens* within the broad genus *Equus*, but recognizing it as *Dolichohippus*, one of many subgenera within *Equus*. One of Matthew's admonitions to Simpson that summer was not to be disappointed that their discovery did not warrant a new species name: "Dr. Matthew cooled me down by emphasizing the fact that mere naming of species is nothing" (Simpson 1986, "Reflections on Matthew," 203).

19. DVP/AMNH (Simpson, "Field Diary" [1924]). Fifty-nine pages unpaginated, Field Diaries, #5GGS: WDM, 1925 (*sic*), Llano Estacado; Santa Fe Beds, DVP/AMNH.

20. Simpson 1951a (*Horses*, 94–95).

21. Matthew to Simpson, July 19, 1924 (GGS/APS).

22. *New York Times* obituary, May 10, 1965, 33; and "Childs Frick and the Frick Collection of Fossil Mammals," *Curator* 18 (1975): 3–15.

23. Frick to Simpson, April 4, 1946 (DVP/AMNH). This letter was written to remind Simpson that none of the fund could be used to support prospecting or collecting of Mesozoic reptiles by Edwin Colbert, then curator of fossil reptiles and amphibians in the department of vertebrate paleontology.

24. Frick 1926b ("Prehistoric Evidence," 446–47).

25. Simpson to Matthew, July 14, 1924 (DVP/AMNH).

26. Matthew to Simpson, July 23, 1924 (GGS/APS).

27. Falkenbach to Frick, July 21, 1924 (DVP/AMNH).

28. Simpson to Matthew, July 22, 1924 (Field Correspondence [one-page letter], DVP/AMNH).

29. Simpson to Matthew, July 22, 1924 (Field Correspondence [three-page letter], DVP/AMNH).

30. Matthew to Simpson, July 28, 1924 (W. D. Matthew folder, GGS/APS).

31. Falkenbach to Matthew, July 27, 1924 (DVP/AMNH).

32. Matthew to Falkenbach, July 31, 1924 (DVP/AMNH).

33. Matthew to Osborn, August 1, 1924 (DVP/AMNH).

34. Osborn to Matthew, August 4, 1924 (DVP/AMNH).

35. Simpson to Matthew, August 4, 1924 (DVP/AMNH). For a while, in his twenties, Simpson experimented with the contemporary fashion of

"modernizing" spelling by dropping the "ugh" in words like although, thorough, and brought.

36. DVP/AMNH (Simpson, "Field Diary" [1924]). (See n. 19, this chapter, for full citation.)

37. Falkenbach to Matthew, August 23, 1924 (Field Correspondence, 1923–26, box 5, folder 4, DVP/AMNH).

38. Matthew to Simpson, September 24, 1924 (W. D. Matthew folder, GGS/APS).

39. Simpson to Matthew, October 27, 1924 (Field Correspondence, 1923–26, box 5, folder 4, DVP/AMNH).

40. Matthew to Simpson, November 18, 1924 (W. D. Matthew folder, GGS/APS).

41. Frick 1926b ("Prehistoric Evidence," 447).

42. Frick 1926a ("The Hemicyoninae," 19).

43. Prof. Ronald Rainger, Texas Tech University, has informed me that the department of vertebrate paleontology personnel files indicate that, while Falkenbach had worked for the museum from 1916 to 1919, by 1924 he was working exclusively for Frick (Rainger to Laporte, January 31, 1990).

44. Simpson to Matthew, November 24, 1924 (Field Correspondence, 1923–26, box 5, folder 4, DVP/AMNH). The abstract was published the next year by Simpson, "Reconnaissance of the Santa Fé Formation," but the paper on which the abstract was based was never published (as explained later in the chapter). Because formal stratigraphic terminology has changed in the years since, I use the terminology then current when Simpson studied these rocks, namely the Santa Fe formation, although the rock unit today is referred to as the Santa Fe Group

45. Matthew to Simpson, December 3, 1924 (W. D. Matthew folder, GGS/APS).

46. "A Reconnaissance of part of the Santa Fé formation of the Rio Grande Valley," unpublished manuscript, 14 pp. (W. D. Matthew folder, GGS/APS). The photographs and their captions are in separate archives (Field Correspondence, 1923–26, box 5, folder 4, DVP/AMNH) from the typed manuscript.

47. Simpson to Matthew, January 17, 1925 (Field Correspondence, 1923–26, box 5, folder 4, DVP/AMNH). No such map was ever drawn, no doubt because the project was aborted by Frick's opposition.

48. Matthew to Simpson, April 21, 1925 (W. D. Matthew folder, GGS/APS).

49. Simpson to Matthew, April 22, 1925 (Field Correspondence, 1923–26, box 5, folder 4, DVP/AMNH).

50. Simpson to Matthew, May 1, 1925 (Field Correspondence, 1923–26, box 5, folder 4, DVP/AMNH).

51. Simpson 1975 ("Comments," 8).

52. Simpson 1978 (*Concession*, 35). This opinion has been challenged by Dr. Malcolm C. McKenna, curator of vertebrate paleontology at the American Museum, who asserts that there was no museum policy — formal or informal — that prevented Simpson's access to the Frick Collections. (Author interview with M. C. McKenna, Toronto, October 28, 1998.)

53. Simpson to Frick, April 1, 1946 (Childs Frick Correspondence, box 15, folder 12, DVP/AMNH).

54. Frick to Simpson, April 4, 1946 (Childs Frick Correspondence, box 15, folder 12, DVP/AMNH).

55. Simpson to Frick, April 4, 1946 (Childs Frick Correspondence, box 15, folder 12, DVP/AMNH).

56. Simpson to Frick, July 24, 1949 (Childs Frick Correspondence, box 15, folder 12, DVP/AMNH).

57. Frick to Simpson, August 19, 1949 (Childs Frick Correspondence, box 15, folder 12, DVP/AMNH).

58. Simpson to Colbert, September 16, 1949 (G. G. Simpson Correspondence, folder II, 1948–51, DVP/AMNH). To my knowledge there is no written acknowledgment by Simpson to Frick's rejection of the offer to participate in the 1950 field conference.

59. Simpson 1950b ("The Upper Cenozoic of the Rio Grande Depression," 83–84).

60. Simpson 1929a ("American Mesozoic Mammalia").

61. The Department of Paleontology at Berkeley was initially established in 1909, but merged again with geology in 1921 when its chairman, J. C. Merriam, became president of the Carnegie Institution. The department was reestablished as an independent academic unit in 1927 under Matthew.

62. When the Society of Vertebrate Paleontology was formally organized in 1941, it counted an initial membership of some 150 practitioners, and so from this statistic we may get some idea of the proportionally still smaller size of the discipline two decades earlier (Simpson 1941a, "History of the Society").

4. Darwin's World

This chapter was first published in slightly different form as "The World into Which Darwin Led Simpson," *Journal of the History of Biology* 23 (1990): 499–516. Used with kind permission from Kluwer Academic Publishers.

1. MES/APS, pp. 2, 3, 5, 9.

2. MES/APS, pp. 67, 72, 74, 75.

3. Letter from Simpson to the author, April 27, 1981 (GGS/APS).

4. Simpson 1964b (*This View of Life*, 37). The essay in which this quote appears ("One Hundred Years Without Darwin Are Enough") is a reprint of an article of the same title published in the *Teachers College Record* 62.8 (May 1961): 617–26.

5. "[Simpson] once told me that he had read everything Darwin had ever written" (Lawrence Gould 1984, "Memorial Tribute").

6. Unlike Ernst Mayr, who has also written extensively on Darwin but more formally, for historians and philosophers (e.g., *Toward a New Philosophy of Biology* [1988], 161–264).

7. Simpson's religious background and quotes from Simpson 1978 (*Concession*, 10 and 25–27).

8. Simpson 1933 ("autobiographical notes").

9. Ibid.

10. Kingsley 1925 (*Madam How and Lady Why*, vii, xi).

11. Simpson 1980c (*Why and How*, vii).

12. Undated letter from G. G. Simpson to Anne Roe from the University of Colorado in Boulder. This and the following letter seen in photocopy among several sent to me by Dr. Anne Roe Simpson, July 9, 1985. According to Dr. Simpson, these two letters were written about 1920 — 21, when she was sixteen and Simpson was eighteen.

13. Second undated letter to Dr. Anne Roe Simpson (emphasis in the original of both letters).

14. Simpson 1964b (*This View of Life*, 27–28).

15. Ibid., 37.

16. Simpson 1960 ("The World into which Darwin Led Us," 967 and 969).

17. Simpson 1977a ("New Heaven and a New Earth," 69).

18. Simpson 1947c ("Problem of Plan and Purpose," 481). Simpson remarked later in life that the Vanuxem Lecture at Princeton on which this article was based was his "first public expression of a growing interest in evolutionary philosophy" (Simpson 1964b, *This View of Life*, 297).

19. Simpson 1947c ("Problem of Plan and Purpose," 487–88).

20. Simpson 1933 ("autobiographical notes" with 1954 addition, p. 17); in his 1970 updating of these notes, Simpson crossed out "throw at you" and substituted "say."

21. Simpson 1969 ("Present Status of the Theory of Evolution," 153).

22. Simpson 1944b (*Tempo and Mode*) and Simpson 1953c (*Major Features*).

23. Simpson 1959a ("Darwin Led Us into This Modern World," 271).

24. Ibid., 272.

25. Simpson 1933 ("autobiographical notes," 1970 addition, p. 17).

26. Simpson 1953b (*Life of the Past*, 142).

27. Simpson 1970a ("Darwin's Philosophy and Methods," 1362).
28. Simpson 1933 ("autobigographical notes").
29. Simpson 1980c (*Why and How*).
30. Simpson 1961c (*Animal Taxonomy*, 53) and Simpson 1980c (*Why and How*, 121).
31. Simpson 1961c (*Animal Taxonomy*, 54).
32. Simpson 1983 (*Fossils*, 120).
33. Simpson 1980c (*Why and How*, 123).
34. Simpson 1961b ("Lamarck, Darwin, and Butler," 245).
35. Mayr 1988 (*Toward a New Philosphy*, 440–41).
36. I follow the definition of positivism given by N. C. Gillespie (1979): "Positivism signifies that attitude toward nature . . . which saw the purpose of science to be the discovery of laws which reflected the operation of purely natural or 'secondary' causes. It typically used mechanistic or materialistic models of causality, rejected supernatural, teleological, or other factors which were in principle beyond scientific inquiry" (*Charles Darwin and the Problem of Creation*, 8).
37. Simpson 1964b (*This View of Life*, 38).
38. Simpson 1960 ("The World into which Darwin Led Us," 966).

5. Paleocene Mammals of Montana

This chapter was first published in slightly different form as "G. G. Simpson and the Paleocene Mammals of the Fort Union Group of Montana," *Journal of Vertebrate Paleontology* 15 (1995): 187–194. Used by permission of the *Journal of Vertebrate Paleontology*.

1. Mayr and Provine 1980 (*Evolutionary Synthesis*).
2. Everett C. Olson (1910–1993), a distinguished American vertebrate paleontologist and somewhat younger contemporary of Simpson's, claimed that "Simpson's papers in the 1930s were mostly standard and not harbingers of *Tempo and Mode* or of his later theoretical approach." (Author interview with Olson, March 6, 1979, Santa Cruz, Calif.)
3. Simpson 1937a ("Fort Union").
4. Simpson 1937d ("Supra-specific Variation") and 1937c ("Phyletic Evolution"); and Simpson 1978 (*Concession*, 80–81).
5. Simpson and Roe 1939 (*Quantitative Zoology*).
6. Simpson 1975 ("Comments," 3).
7. Simpson 1942–1944 ("Observations," 167).
8. SI/USNM, Gilmore: 10/6/31.
9. SI/USNM, Wetmore: 11/9/31.
10. SI/USNM, Simpson: 11/18/31.
11. Simpson 1937a ("Fort Union," 6).

12. See Gidley for 1909, 1915, 1919, and 1923.
13. Simpson 1928a (*Catalogue of Mesozoic Mammalia*) and Simpson 1929a ("American Mesozoic Mammalia").
14. Simpson 1928b ("A New Mammalian Fauna") and 1928c ("Third Contribution") and 1929b ("A Collection of Paleocene Mammals").
15. SI/DVP, Simpson: 12/19/31.
16. SI/DVP, Simpson: 1/6/32.
17. SI/USNM, Wetmore: 5/2/32.
18. Simpson 1978 (*Concession*, 52).
19. Simpson 1987 (*Simple Curiosity*, 177–78; emphasis in the original).
20. Ibid., 179.
21. SI/USNM, Simpson: 7/5/32.
22. DVP/AMNH (Simpson, "Field Diaries," [1932]: box 7, folder 7).
23. SI/USNM, Simpson: 8/4/32.
24. SI/USNM, Simpson: 2/3/33.
25. SI/USNM, Wetmore: 2/14/33.
26. SI/USNM, Simpson: 3/15/33.
27. SI/USNM, Prentice: 6/1/33.
28. SI/USNM, Simpson: 7/7/33.
29. Simpson 1933 ("autobiographical notes").
30. SI/USNM, Wetmore: 7/11/33.
31. Simpson 1936c ("Third Scarritt Expedition," 13) and 1937a ("Fort Union," 10). The Silberling and Gidley quarries are of middle Paleocene age, with Gidley five miles southwest of Silberling. The Scarritt Quarry is late Paleocene in age and lies five and a half miles to the northwest of Gidley.
32. Simpson 1937a ("Fort Union," 14).
33. Simpson 1936b ("A New Fauna," 2).
34. Simpson 1978 (*Concession*, 79).
35. Ibid., 199–206.
36. Simpson 1987 (*Simple Curiosity*, 202, 203, 205).
37. DVP/AMNH, Simpson: 3/10/36.
38. Simpson 1937a ("Fort Union," iii).
39. Simpson 1975 ("Comments," 23).
40. Simpson 1937d ("Supra-specific Variation") and 1937c ("Phyletic Evolution").
41. Simpson 1937a ("Fort Union," 71, 73). Subsequent page numbers are cited in the text.
42. Mayr 1982 (*Biological Thought*, 567).
43. Simpson 1937a ("Fort Union," 65).
44. Ibid., 2. Subsequent page numbers are cited in the text.

45. Simpson and Roe 1939 (*Quantitative Zoology*).
46. Simpson 1937a ("Fort Union," 21).
47. Ibid., 21–22. Subsequent page numbers are cited in the text.
48. Matthew 1930 ("Range and Limitations of Species"); Cabrera 1932 ("Incompatibilidad ecológica"); and Simpson 1937a ("Fort Union," 64).
49. Simpson 1937a ("Fort Union," caption to his fig. 4, p. 66). Subsequent page numbers are cited in the text.
50. Simpson 1944b (*Tempo and Mode*, 99ff.).
51. Simpson 1937d ("Supra-specific Variation") and 1937c (Phyletic Evolution").
52. Simpson 1978 (*Concession*, 80–81).

6. On Species

This chapter was first published in slightly different form as "Simpson on Species," *Journal of the History of Biology* 27 (1994): 141–59. Used with kind permission from Kluwer Academic Publishers.

1. Simpson 1944b (*Tempo and Mode*). See chapter 7 of the present volume.
2. Simpson 1976 ("The Compleat Paleontologist?" 5). Readers suspecting historical whiggery are encouraged to read the introduction to *Tempo and Mode*, esp. xvi–xvii, where Simpson made the same claim in almost precisely the same language. For different views on what *Tempo and Mode* did (or did not) accomplish, see Eldredge 1985 (*Unfinished Synthesis*); Gould 1980 ("Modern Synthesis," esp. 157–65); and chapter 7 in this volume. Provine has called the evolutionary synthesis the "evolutionary constriction," claiming that it "was not so much a synthesis as it was a vast cut-down of variables considered important in the evolutionary process" (Provine 1988, "Progress in Evolution," 61). Simpson participated in this constrictionary aspect of the synthesis as well by devoting chapter 5 of *Tempo and Mode* ("Inertia, Trend, and Momentum") to a debunking of notions of inherent evolutionary forces like orthogenesis, aristogenesis, and racial senescence that were favored by contemporary paleontologists.
3. Among several key characteristics of the evolutionary synthesis, Ernst Mayr identified the "overwhelming importance . . . of the populational structure . . . [and the] evolutionary role of species" (Mayr 1982, *Biological Thought*, 370). For the role of speciation in the development of the synthesis, see Eldredge 1982 ("Introduction," xv–xxxvii).
4. Ernst Mayr formally introduced these terms. "Alpha taxonomy is the definition of units, while beta taxonomy is the combining of these units

into a system of classification. Gamma taxonomy [is] the study of the biological meaning of all these units" (Mayr 1958, *Concepts of Biology*, 118).

5. Simpson 1925a ("American triconodonts," 145–65, 334–58) and 1925b ("Preliminary Comparison," 559–69). In fig. 3, p. 567 of the latter work, Simpson published his first phylogeny, "Diagram suggesting the probable relationships of three chief types of Jurassic molars."

6. Eldredge 1985 (*Unfinished Synthesis*, 62).

7. Gould 1980 ("Modern Synthesis," 162).

8. Mayr 1988 (*Toward a New Philosophy*, 440).

9. Simpson 1928a (*Catalogue of Mesozoic Mammalia*) and 1929a ("American Mesozoic Mammalia"). Simpson not only created new species but also higher taxonomic categories. Thus, of the total of 72 genera, 10 families, and 6 orders with which he worked, 27 genera, 3 families, 2 suborders, and 1 order were defined by him as new.

10. A "type" in zoological nomenclature is a particular, designated specimen that bears the name of the specific category which it defines. Taxonomic practice requires that any other specimen being described must be explicitly referred to the type specimen to determine if it is the same or different. The basis by which one judges the degree of sameness, of course, is the raw material of taxonomic interpretation and debate.

11. Roe 1985 ("Tyler Address," 313); Simpson and Roe 1939 (*Quantitative Zoology*). Although published in 1939, the manuscript for the book was completed in 1937, the year before their marriage.

12. For example, in eight semesters of college mathematics, including three of calculus, Simpson had an overall average of 99.3 and never earned a grade less than 97 (University of Colorado transcript).

13. Simpson 1937c ("Phyletic Evolution," 308–309). Subsequent page numbers are cited in the text.

14. Simpson 1980c (*Why and How*, 112).

15. Simpson 1937b ("Clark Fork," 1–24). Subsequent page numbers are cited in the text.

16. Simpson 1937a ("Fort Union"). Simpson alerts the reader immediately in the introduction: "Throughout this work, wherever they proved useful, statistical methods have been employed. These are summed up in Fisher (1925) and also in a paper soon to be published" (2). The "paper" expanded into a full-blown text, *Quantitative Zoology*; of particular interest to paleontologists in that book was chapter 11, "Small Samples and Single Specimens."

17. Simpson 1937a ("Fort Union," 73–76).

18. Simpson 1940d ("Types in Modern Taxonomy," 413–31).

19. Ibid., 414–15. Years later, at the invitation of physical anthropologists, Simpson elaborated further on this subject of the logical relations

between specimens as samples of a species and the species itself as a biological, not just taxonomic, entity (Simpson 1963a, "The Meaning of Taxonomic Statements"). (See chapter 8 of the present volume.)

20. Simpson 1937a ("Fort Union," 66).

21. Simpson credits W. D. Matthew for articulating this principle (ibid., 64). See Matthew 1930 (" Range and Limitations of Species," 271–74).

22. More precisely, Simpson introduced yet another term, "megaevolution," for evolution at the levels of family and higher, whereas strictly macroevolution was only for levels above species to genera. However, he subsequently discarded these latter distinctions, collapsing mega-evolution into macroevolution in *Major Features of Evolution* (1953).

23. Simpson 1944b (*Tempo and Mode*, 191–96, and esp. figures 26–30). (See figure 7.5, this volume.) Subsequent page numbers will be cited in the text.

24. I use formal publication dates to mark off these various periods in Simpson's thinking, but obviously he must have been working along these lines earlier. For example, he started the Fort Union study in 1932, published a preliminary note in 1935, and had the monograph manuscript completed in early 1936, with publication occurring in summer 1937.

25. Simpson 1937d ("Supra-specific Variation," 246). Simpson gave force to his argument by reducing to a single species of Eocene hoofed-herbivore what the South American paleontologist Florentino Ameghino had previously divided into three families, seven genera, and seventeen species.

26. In a letter to paleontologist Preston Cloud, March 14, 1978, Simpson asserted that "*all* evolution is phyletic if one defines 'phyletic' as 'reproductive continuity in a continuous sequence of populations.' That is true of any evolved or evolving species, and it makes no difference whether another species has branched off from the same ancestry or an ancestral species has continued unchanged" (GGS/APS, Cloud Folder; emphasis in the original).

27. Simpson 1937d ("Supra-specific Variation," 259)

28. Simpson 1959d ("Nature of Supraspecific Taxa," 255).

29. Simpson 1961c (*Animal Taxonomy*, 7).

30. Simpson 1953 (*Evolution and Geography*). See chapter 9 for further discussion of Simpson's approach to historical biogeography.

31. Simpson 1951c ("The Species Concept") and 1961c (*Animal Taxonomy*).

32. Simpson 1961c (*Animal Taxonomy*, 153).

33. Ibid., 150.

34. Mayr 1942 (*Systematics*, 210).

35. This sort of distinction is common to plate tectonics, where one can

more easily describe the movement of crustal plates (their kinematics) than determine the precise forces involved (their dynamics).

36. If one may be allowed a "presentist" viewpoint, to many students of evolution both definitions seem complementary rather than mutually exclusive concepts, and therefore to champion one definition misses important insights provided by the other.

37. Smocovitis 1992 ("Unifying Biology," 20).

38. For Simpson's positivistic philosophy, see chapter 4 in this volume; for Simpson's quantification of evolutionary phenomena, see *Tempo and Mode*; and for Simpson's quantitative style of argumentation more generally, see, for example, Simpson and Roe 1939 (*Quantitative Zoology*) and Simpson 1980c (*Why and How*).

7. Tempo and Mode in Evolution

This chapter is adapted from "Simpson's *Tempo and Mode in Evolution* Revisited," first published in *Proceedings of the American Philosophical Society* 127 (1983): 365–417. Used by permission of the American Philosophical Society.

1. See Hecht et al. 1972 ("George Gaylord Simpson").

2. The paper was first presented orally at a conference celebrating the tercentenary of Harvard College in September 1936. And although Simpson included both "Phyletic Evolution" and "Supra-specific Variation" in the Works Cited section of *Tempo and Mode*, he does not explicitly refer to the first, and his reference to the second is relatively unimportant.

3. Read in December 1936 at a symposium of the American Association for the Advancement of Science in Atlantic City, N.J.

4. Kinsey 1936 ("Origin of Higher Categories").

5. Simpson 1937d ("Supra-specific Variation," 240). The fourth conclusion found in *Tempo and Mode* is that the evolution of one character may be the function in fact of another, as with his example of changes in horse skull proportions being a result of overall phyletic size increase (Simpson 1944b, *Tempo and Mode*, 6, 12).

6. Simpson 1937d ("Supra-specific Variation," 250). Subsequent page numbers will be cited in text. See chapter 10 of the present volume for more regarding Simpson's use of "visual language."

7. Stebbins and Ayala 1981 ("Is a New Evolutionary Synthesis Necessary?").

8. Simpson 1975 ("Comments," 44–47), 1976 ("Compleat Paleontologist?" 4–5), and 1978 (*Concession*, 114–15). Five years after *Tempo and Mode*, Simpson remarked, "I think it fair to say that no discussion of evolutionary theory need now be taken seriously if it does not reflect knowledge of

these studies [of Fisher, Haldane, and Wright] and does not take them strongly into account" (Simpson 1949, *Meaning of Evolution*, 266).

9. Simpson 1978 (*Concession*, 115).

10. Ibid.

11. Simpson 1975 ("Comments," 29).

12. The geneticist Hampton Carson was overly impressed by Haldane's discussion of the fossil record when he claimed that "the paleontological record is woven [by Haldane] into the argument: the synthesis had begun in earnest" (Carson 1980, "Cytogenetics," 89). In fact, Haldane was uncertain what the fossil record meant in terms of evolutionary mechanisms, although he clearly read the record as an unequivocal demonstration of evolution.

13. Mayr 1982 (*Biological Thought*, 569). Mayr 1980 ("Prologue," 1) approximately dates the beginning of the synthesis in 1936 with Dobzhansky's Jesup Lectures at Columbia University that formed the basis of his *Genetics and the Origin of Species* published the following year, and its essential consolidation in 1947 at the Princeton conference subsequently published as *Genetics, Paleontology, and Evolution* (Jepsen, Simpson, and Mayr 1949).

14. Simpson 1980a ("Biographical Essay," 456).

15. Dobzhansky 1980 ("The Birth of the Genetic Theory," 229–42).

16. Ibid., 239.

17. Mayr 1982 (*Biological Thought*, 569). One clear indication of this bridge-building is the innumerable variety of field studies Dobzhansky cited to support his genetic arguments, ranging from Jimson weed, snapdragons, silkworms, and land snails to herring, storks, deer mice, and sea urchins, to mention but a few of several dozen examples given by him. Obviously, this book was not written by "a person who shut himself in a room, pulled down the shades, watched small flies disporting themselves in milk bottles, and thought that he was studying nature" (Simpson 1944b, *Tempo and Mode*, xv).

18. Dobzhansky 1937 (*Genetics*, xv).

19. Dobzhansky 1937 (*Genetics*, 11–12).

20. Gould 1982 ("Introduction").

21. Dobzhansky 1937 (*Genetics*, 13).

22. Ibid., 176–85.

23. Goldschmidt 1940 (*Material Basis of Evolution*).

24. Ibid., 390.

25. Simpson 1978 (*Concession*, 115).

26. This series of quotes is from Goldschmidt 1937 (*Material Basis*, 321).

27. Ibid., 322.

28. Admittedly, Goldschmidt cannot be faulted for the paleontological concept of orthogenesis which, as we will see in the next section, was in fact widely promoted by paleontologists until buried by Simpson in *Tempo and Mode*.

29. Simpson himself has provided a review of the state of relations between paleontologists and geneticists in the early part of this century, prior to his writing of *Tempo and Mode* (Simpson 1946, "Tempo and Mode").

30. Rainger 1991 (*Agenda for Antiquity*).

31. Osborn 1925–1927 ("Origin of Species," parts 1–5).

32. Osborn 1932 ("Biological Inductions") and 1934 ("Aristogenesis").

33. Simpson was in residence at Yale when Osborn gave his address at the dedication ceremonies for the Peabody Museum. In fact, he was first introduced to Osborn at that time by Professor R. S. Lull on the steps of the new museum (author interview with Simpson, June 17, 1982, Tucson, Arizona).

34. Osborn 1926b ("Origin of Species [Part 4]," 341; emphasis in the original).

35. Ibid., 341.

36. Osborn 1934 ("Aristogenesis," 210). Subsequent pages will be cited in the text.

37. Osborn 1932 ("Biological Inductions," 503, 502). Simpson suggested that Osborn supported his "nonmechanistic . . . explanation . . . as an escape from the dilemma" that the contemporary schools of August Weismann, Hugo de Vries, and the neo-Lamarckians had created. "It was almost inevitable that such an escape from the dilemma would be sought by a deeply philosophical paleontologist of [this] period" (Simpson 1946, "Tempo and Mode," 52).

38. Lull 1929 (*Organic Evolution*, vii).

39. Simpson 1978 (*Concession*, 114). In the preface to the revised 1929 edition, Lull acknowledged the valuable criticism Simpson (among others) had offered for the revision. He also cited Simpson's recently published monographs on Mesozoic mammals in his chapter on early archaic mammals.

40. Lull 1929 (*Organic Evolution*, 15). Subsequent page numbers are cited in the text.

41. Although making no reference to Lull's stance on evolutionary materialism, Simpson notes in another context that Lull "was deeply religious. For him science and religion were in different spheres, rarely touching and never conflicting" (Simpson 1958a, "Lull Memorial," 131). One wonders.

42. By choosing Osborn, Lull, and Schindewolf to sketch the contemporary state of paleontology in the 1930s, I am deliberately focusing on people who obviously influenced Simpson and who were as theoretically

inclined and attempted to read fossil history in light of the current biological knowledge. Therefore, I am intentionally and necessarily omitting other paleontologists and their viewpoints that are even farther apart from the materials, methods, and conclusions that we find in *Tempo and Mode*. Yet even with this minimalist position that thus avoids setting up "straw man" arguments, we can discover how radical *Tempo and Mode* was as compared to the contemporary paleontology.

43. Simpson 1975 ("Comments," 47).

44. Simpson 1978 (*Concession*, 115).

45. Simpson 1975 ("Comments," 47).

46. As David Kitts has observed: "The arguments in defense of synthetic theory against the attacks of Schindewolf, particularly those of Simpson, greatly enriched the theoretical content of paleontology" (Kitts 1974, "Paleontology," 168).

47. Simpson 1976 ("Compleat Paleontologist?" 4–5).

48. There was perhaps another important personal factor at play here. Just before starting this work, Simpson had finally resolved a very trying marital situation by divorcing his first wife, marrying Dr. Anne Roe, and regaining the permanent custody of four young daughters from his previous marriage. Simpson has referred to these "happy events and outcomes" as contributing to what was "a vintage year" for him (Simpson 1978, *Concession*, 81).

49. Hecht et al. 1972 ("George Gaylord Simpson," 13–15).

50. Simpson 1944b (*Tempo and Mode*, xv–xviii). Subsequent page numbers are cited in the text.

51. Simpson contrasts phenogenetics with "cryptogenetics," which is the study of the transmission of inherited characters. Paleontology can say nothing about how heredity occurs (i.e., cryptogenetics), but it can contribute to the observable expression of heredity in ancient organisms (i.e., phenogenetics). (See *Tempo and Mode*, xvi.)

52. As noted in a previous section ("Two Door-Openers"), conclusions 2, 3, and 4 were also arrived at in an earlier paper on South American notoungulates (Simpson 1937, "Supra-specific Variation").

53. Years later, after Simpson introduced the idea, survivorship curves continued to enjoy a great vogue in paleobiology and often fueled the debate on both sides of the issue of "phyletic gradualism vs. punctuated equilibrium." (See, for example, Stanley 1978, *Macroevolution*.)

54. In fact, Simpson's reformulation of the new genetics is so thorough that he gave relatively few citations to the genetic literature. When he did, he referred chiefly to Fisher (1930), Wright (1931), Haldane (1932), and Dobzhansky (1937), some half-dozen times each.

55. Simpson 1937d ("Supra-specific Variation").

56. Mayr 1963 (*Animal Species*, 34).

57. Simpson also remarked: "If, as is now generally admitted, the heredity of an animal is fixed permanently before the first cleavage of the developing embryo, no further evolutionary change . . . occurs until the next generation appears" (62). Simpson, like the other major synthesists, thus discounted any evolutionary importance of what Mayr has termed "soft inheritance . . . [t]he almost universal belief in [which] was a major stumbling block in the path of a neo-Darwinian interpretation" (Mayr 1980, "Prologue," 15). According to Mayr, soft inheritance includes genetic changes effected by use and disuse, by internal progressive tendencies, or by direct action of the environment.

58. Wright (1945) reviewed *Tempo and Mode* and pointed out that Simpson had incorrectly reproduced his own earlier notation (Wright 1931), in that Simpson had used $1/2\ N$ for $1/(2\ N)$, $1/2\ s$ for $1/(2\ s)$, and $1/4\ u$ for $1/(4\ u)$. These were typographical errors obviously introduced between manuscript and printed page, because Simpson's actual calculations indicate his proper use of Wright's mathematical terms.

59. Depending upon the different values of other variables, a population of "intermediate size" might usually be between 250 and 25,000 interbreeding individuals, "very large" and "small" populations being, respectively, greater or less (Simpson 1944b, *Tempo and Mode*, 68). Simpson clearly suggested these numbers to provide orders of magnitude only.

60. Mayr has referred to the first two as examples of the belief in "soft inheritance" (Mayr 1980, "Prologue," 15). About the third alternative, he has stated that the "thoroughly unbiological assumption that structure precedes function was curiously revived by the mutationists after 1900: Cuénot, de Vries, and Bateson claimed . . . that organisms are exposed to the mercy of their mutations but that some mutations 'preadapt' them to new behaviors and adaptive shifts" (Mayr 1982, *Biological Thought*, 463).

61. Brinkmann 1929 ("Ammoniten"). Later reanalysis of his original data "shows that the correspondence between morphological jumps and physical breaks [in the record] is not as consistent as was thought by Brinkmann" (Raup and Crick 1982, "*Kosmoceras*," 90).

62. A review of such systematic gaps in the fossil record by Everett C. Olson shows that the situation as described by Simpson, four decades earlier, had not changed since (Olson 1981, "Problem of Missing Links").

63. In his review of *Tempo and Mode*, Wright took Simpson to task for asserting that he, Wright, claimed that rapid evolution is most favorable in populations of intermediate size (as on page 67 of *Tempo and Mode*), when in fact Wright believed that "conditions are enormously more favorable in a population which may be large but which is subdivided into many small local populations almost but not quite completely isolated from each

other" (Wright 1945, "Review," 416). Simpson agreed with Wright and even cited him as the authority (e.g., *Tempo and Mode*, 70–71).

64. Subsequent study modified some of the details of this example, but the general interpretation remains the same (Simpson 1980b, *Splendid Isolation*, 70–73).

65. Here Simpson was clearly demonstrating that he was not a "hyperselectionist" adhering to the "adaptationist program," but rather was giving due credit to history and "bauplan" (i.e., basic anatomic blueprint) as well as to selection and occasional random factors in effecting evolutionary results.

66. Simpson's discussion of the adaptive zone undoubtedly was drawn, in part at least, from Wright (1932, "Mutation, Inbreeding, Crossbreeding") — although not so referenced. That paper of Wright's had also supplied Simpson with the metaphor of the adaptive landscape, with its peaks and valleys.

67. In commenting on *Tempo and Mode* years later, Simpson said, "What . . . all this really leads up to, the relationship between organism and environment, the nature of adaptation, and here comes in the adaptive grid . . . " (Simpson 1975, "Comments," 47).

68. Simpson and Roe 1939 (*Quantitative Zoology*, 99).

69. Although subsequent discussion of quantum evolution by others has usually only considered it with respect to higher categories, Simpson clearly saw it as not restricted to them. In *The Major Features of Evolution* (1953), he reiterated this point: "Quantum evolution may lead to a new group at any taxonomic level. It is probable that species, either genetic or phyletic, often arise in this way" (389). Somehow down through the years, Simpson's broader application of quantum evolution has been ignored.

70. Glass 1945 ("Review," 261).

71. Wright 1945 ("Review," 415).

72. Dobzhansky 1945 ("Review," 114).

73. Glass 1945 ("Review," 261).

74. Wright 1945 ("Review," 415). Mayr has emphasized that thinking about species as variable populations rather than as ideal types was one of several cornerstones of the modern synthesis (Mayr 1980, "Prologue," 18, 28; Mayr 1982, *Biological Thought*, 570). However, Mayr considered it to be a gross oversimplification when the conclusion was drawn that evolution was merely temporal shifts in gene frequencies, what he termed "bean bag genetics" (Mayr 1980, "Prologue," 12, 13; Mayr 1982, *Biological Thought*, 558).

75. Wright 1945 ("Review," 418–19).

76. Dobzhansky 1945 ("Review," 115).

77. Simpson 1953c (*Major Features*, xi). Not surprisingly, however, it is

Tempo and Mode that is more frequently consulted and cited today than is *Major Features*, just because it was so refreshingly original and so succinctly expressed. E. C. Olson anticipated this in his review of *Major Features*: "The solidity of the study is welcome, but your reviewer, at least, cannot but feel small twinges of nostalgia for the 'brave new world' approach of the earlier volume" (Olson 1953, "Review," 87).

78. Hubbs 1945 ("Review," 271–72).
79. Hutchinson 1944 ("Review," 356).
80. Hutchinson 1945 ("Marginalia," 120).
81. Zirkle 1947 ("Review," 109).
82. Huxley 1945 ("Review," 3).
83. Ibid., 4. Huxley's point is that while Simpson does speak of very rapid evolutionary rates for the quantum mode — "more rapid than the later recorded evolution, almost surely twice as fast and probably more, quite possibly ten or fifteen times as rapidly in some cases" (*Tempo and Mode*, 121) — the implication is that absolute duration might, in fact, be on the order of hundreds of thousands or even a few million years.
84. Jepsen 1946 ("Review," 538, 540, 541). What Jepsen saw as clarity, other reviewers occasionally viewed as opaqueness: "The chapter on Organism and Environment — the least satisfying to the reviewer — might be revised" (Hubbs 1945, "Review," 275); "[T]he author's style is not always fortunate" (Zirkle 1947, "Review," 110).
85. Bucher 1942 ("Geological Research," 1339–40). Such thoughtful, catholic interest in science, and in geology particularly, was typical of Bucher, who himself was well known for structural geological research that dealt with the deformational mechanisms and history of the earth's crust. Bucher also accurately predicted two other fields of historical geology that were ripe for renewed consideration: the "hypothesis of the drifting of continents" and "the lowly field of . . . tracks, trails, and burrows" (ibid., 1341). Both fields, of course, saw great resurgence and revitalization in the 1960s.
86. My account here is pieced together from Jepsen 1949 ("Foreword") as well as Simpson 1975 ("Comments") and 1978 (*Concession*).
87. Mayr 1980 ("Prologue," 42–43) and 1982 (*Biological Thought*, 568).
88. Simpson 1978 (*Concession*, 130).
89. Davis 1949 ("Comparative Anatomy," 74, 76, 77, 86). See also Kitts (1974, "Paleontology"), who similarly argued that, strictly speaking, the fossil record *by itself* proves nothing about evolutionary processes.
90. Stebbins 1949 ("Rates of Evolution in Plants," 235).
91. Wright 1949 ("Adaptation and Selection," 387–88). Gould has claimed that Wright never meant his adaptive landscape to be applied to hierarchical levels above the species, but rather that he "devised the model

to explain differentiation among demes *within* species" (Gould 1982, "Introduction," xxxvi; emphasis in the original). However, in the passage from which the above is quoted, Wright definitely also used what he calls "our geometric model to explain the rise of higher categories."

92. Stebbins 1950 (*Variation and Evolution in Plants*, 515).
93. Ibid., 551ff.
94. Dobzhansky 1951 (3rd ed., *Genetics and the Origin of Species*, 17, 18, 100).
95. Simpson 1944b (*Tempo and Mode*, xviii).
96. Laudan 1982 ("Tensions in the Concept of Geology").
97. Shapere 1980 ("Meaning of the Evolutionary Synthesis," 389).
98. Simpson 1944b (*Tempo and Mode*, xviii).

8. Mentor for Paleoanthropology

This chapter was first published in slightly different form as "George Gaylord Simpson as Mentor and Apologist for Paleoanthropology," *American Journal of Physical Anthropology* 84 (1991): 1–16. Copyright © 1991 by Wiley-Hiss, Inc. Reprinted by permission of John Wiley & Sons, Inc.

1. Other key works were Dobzhansky 1937 (*Genetics*); Huxley 1942 (*Evolution*); Mayr 1942 (*Systematics*); Jepsen, Simpson, and Mayr 1949 (*Genetics, Paleontology, and Evolution*); and Stebbins 1950 (*Variation*).

2. Birdsell 1987 ("Some Reflections," 1, 4). Prof. Jonathan Marks of Yale University has suggested that another reason may have been that, because some geneticists were promoting eugenics as a social program, anthropologists were unenthusiastic about the lessons population geneticists had to offer during the 1920s and 1930s (personal communication, August 11, 1989).

3. Haldane 1949 ("Human Evolution").

4. Mayr and Provine 1980 (*Evolutionary Synthesis*); MES/APS.

5. As Prof. Joseph Birdsell of UCLA has remarked: "A bunch of anthropologists got together for several summer conferences right after the war to refurbish themselves, get the rust off, by reeducating themselves in the field, including the developments in the evolutionary synthesis. This was a blessing for all those anthropologists of the Hooton school — he had trained about 90 percent of the [American] anthropologists at that time" (author interview with Birdsell, June 11, 1987, Los Angeles, Calif.). See also Washburn 1983 ("Evolution of a Teacher" 12); Haraway 1988 ("Sherwood Washburn," 226).

6. For both broader and more specific discussions of the role that the evolutionary synthesis played within paleoanthropology, see Boaz 1981 ("History of Paleoanthropological Research"); Boaz and Spencer 1981 ("Ju-

bilee Issue"); Spencer 1982 (*History of American Physical Anthropology*); and Bowler 1986 (*Theories of Human Evolution*).

7. Simpson had an early interest in human evolution. In 1927, at the age of twenty-five, he thanked his father for $10 with which he bought "Duckworth's 'Morphology & Anthropology' — a rare out-of-print book of greatest value to me. I'm spending time studying human evolution & I hope your gift will blossom some day" (Simpson 1987, *Simple Curiosity*, 61). That same year, during his postdoctoral stay at the British Museum (Natural History), Simpson socialized with anthropologists Sir Arthur Woodward, Grafton Elliot Smith, and Wilfrid Le Gros Clark. Simpson remarked presciently about the latter, then in his early thirties: "I think he's going to be the great comparative anatomist of my generation" (ibid., 66).

8. See Stocking 1988 (*Bones, Bodies, and Behavior*) and Bowler 1986 (*Theories of Human Evolution*, part 3) with respect to these and other biases in human evolution; Bowler 1983 (*Eclipse of Darwinism*), about biases within evolutionary theory in general; and Simpson 1949 (*Meaning of Evolution*), for the case he made against these various biases. See, too, Haraway 1988 ("Sherwood Washburn") for her argument that the new physical anthropology had its own bias as well.

9. Simpson 1931 ("A New Classification"). W. K. Gregory, Simpson's senior colleague at the American Museum, clearly influenced Simpson both with respect to human evolution and to a neo-Darwinian position. See also Rainger 1989 ("What's the Use") for the intellectual and theoretical ambience that Gregory created in vertebrate paleontology at the museum in general, and in his work on primates in particular.

10. Simpson 1940c ("Earliest Primates").

11. Simpson 1945 ("Principles of Classification," 182).

12. Ibid., 187–88. "Simpson's position is, I think, consistent over the years: as a gradist, he thought in terms of human vs. non-human, and used 'mentality' as a principal criterion for [the] human [grade]. By such a criterion, the australopithecines fall into the 'non-human' category, although they might certainly have a special relationship to humans. When, however, the consensus started to focus on bipedalism as the defining attribute of hominids, the Australopithecinae could simply be lifted from the 'non-human' category to the 'human' category" (Marks, personal communication, August 11, 1989).

13. Both Sherwood Washburn and Joseph Birdsell emphasized this point in their interviews with the author on January 27, 1984 (Berkeley, Calif.), and June 11, 1987, Los Angeles, Calif., respectively.

14. Simpson 1949 (*Meaning of Evolution*, 81–82, 88, 90). Washburn himself anticipated this interpretation: "In the fall of 1938 [at Harvard], Dr. Hooton asked me to teach a course on primates. . . . I stressed that all the

major families of primates could be seen as adaptive radiations" (Washburn 1983, "Evolution of a Teacher," 6).

15. Cartmill 1982 ("Basic Primatology," 151–52). Cartmill makes another interesting point about the combined influence of Le Gros Clark and Simpson in their creating such an atmosphere of extreme skepticism in the "search for phylogenetic affinities between early and later primates and so tend[ing] to discourage research" in prosimians for a generation (ibid., 152)

16. Simpson 1949 (*Meaning of Evolution*; compare first ed., p. 93, with 1967 revised ed., p. 94).

17. Washburn topped off his own recommended reading list for all doctoral candidates in anthropology with Simpson's *Meaning of Evolution* and Dobzhansky's *Genetics, Evolution, and Man* (Haraway 1988, "Sherwood Washburn," 217).

18. Simpson 1951b ("Some Principles of Historical Biology").

19. Howells 1951 ("Concluding Remarks," 79).

20. Simpson 1959d ("Nature of Supraspecific Taxa," 255).

21. Ibid., 268, 270.

22. Simpson 1961c (*Animal Taxonomy*, 187–227, and esp. 125, 126, 214, 216).

23. Author interview with Washburn, January 27, 1984, Berkeley, Calif.

24. Simpson 1962 ("Primate Taxonomy," 497). Subsequent page numbers are cited in the text.

25. Simpson 1945 ("Principles of Classification," 188, re "Eoanthropus"). I thank Prof. Marks for bringing Weidenreich's views on this to my attention.

26. Simpson 1962 ("Primate Taxonomy," 509).

27. Simpson 1975 ("Comments," 93–94).

28. Simpson 1980c (*Why and How*, 164).

29. Simpson 1963a ("Meaning of Taxonomic Statements," 4). Subsequent page numbers are cited in the text.

30. Simpson 1983 (*Fossils*, 212).

31. Simpson 1963 ("Meaning of Taxonomic Statements," 21, 25–26). (See chapter 7 in this volume for further discussion of the Simpsonian concept of the adaptive grid.)

32. Howells 1962 (*Ideas on Human Evolution*, v, 1).

33. Simpson 1966a ("Biological Nature of Man," 475).

34. Ibid., 475.

35. Bronowski 1969 ("Review," 680).

36. Simpson 1972 ("Evolutionary Concept of Man," 35).

37. Simpson 1973 ("Divine Non Sequitur," 100, 102). Simpson has remarked about the conference in San Francisco where this paper was first

given: "The only others there who knew something about evolution, Dobzhansky and Louis Leakey, both dodged the issue — Dobzhansky by praising Teilhard for seeking a synthesis of science and religion without judging whether the synthesis was valid, and Leakey by not talking about Teilhard at all" (Simpson 1978, *Concession*, 138).

38. Simpson 1968 ("What Is Man," 6, 20).

39. Simpson 1963b ("Review of *The Origin of Races*"). Ernst Mayr, too, gave Coon's book a favorable reading, but Theodosius Dobzhansky was strongly opposed (Mayr 1962, "*The Origin of Races*"; and Dobzhansky 1963, "*The Origin of Races*").

40. Simpson 1944b (*Tempo and Mode*, esp. 4–20).

41. Simpson 1959c ("Polyphyletic Origin of Mammals").

42. Simpson 1963b ("Review," 269). When read in full context it is clear that Simpson was making the point that, by definition, geographic subspecies or races could not be identical, and therefore were not equal. He was not, of course, making any reference to equality before the law.

43. Simpson 1974a ("Recent Advances," 16).

44. Luckett and Szalay 1975 (*Phylogeny of the Primates*, 3). At age seventy-two, perhaps Simpson felt he had already said about all there was for him to say. And because of trying personal circumstances — both his wife and his daughter Elizabeth were hospitalized with life-threatening illnesses — he had neither the time nor energy to attempt a substantial revision of the preprint. He may, too, have seen how far outdistanced he had become in keeping up with a rapidly developing field that was not central to his own research.

45. Le Gros Clark 1955 (*Fossil Evidence*, 10).

46. Simpson 1981b ("Prologue").

9. Wrong for the Right Reasons

This chapter was first published in slightly different form as "Wrong for the Right Reasons: G. G. Simpson and Continental Drift," in E. T. Drake and W. Jordan, eds., *Geologists and Ideas: A History of North American Geology*, Centennial Special Publication No. 1 (Boulder, Colo.: Geological Society of America, 1985), 273–85. Used by permission of the publisher, the Geological Society of America, Boulder, Colorado USA. Copyright © 1985 Geological Society of America.

1. See, for example, Frankel 1981 ("Paleobiogeographical Debate"), Hallam 1983 (*Great Geological Controversies*), Marvin 1973 (*Continental Drift*), and references therein. Oreskes 1999 (*The Rejection of Continental Drift*) provides an in-depth treatment, in all its rich complexity, of the debate over Wegener's theory.

2. See, for example, Axelrod 1960 ("Flowering Plants") and Cloud 1961 ("Paleobiogeography").

3. Cracraft 1974 ("Continental Drift," 215); McKenna 1973 ("Sweepstakes," 295) and McKenna 1983 ("Holarctic Landmass," 475).

4. Simpson 1978 (*Concession*, 33–34); and chapter 2 in this volume.

5. Croizat 1981 ("Biogeography," 500, 518); Romer 1973 ("Continental Connections," 345).

6. See, for example, Simpson 1940a ("Antarctica," 765).

7. Marvin 1973 (*Continental Drift*) and Frankel 1981 ("Paleobiogeographical Debate"). Oreskes, too, places Simpson within the overall framework of the debates (Oreskes 1999, *Rejection*).

8. Wegener 1966 (*Origin of Continents and Oceans*). Subsequent page numbers from this translation of Wegener's 1929 edition are cited in the text.

9. Du Toit 1937 (*Our Wandering Continents*).

10. Gregory 1929 ("Atlantic Ocean") and 1930 ("Pacific Ocean"); Ihering 1927 (*Die geschichte*); and Simpson 1940a ("Antarctica"), 1940b ("Mammals and Land Bridges"), 1943 ("Nature of Continents"), 1947b ("Holarctic Mammalian Faunas"), and 1952 ("Probabilities").

11. Simpson 1945 ("Principles of Classification").

12. Simpson 1940a ("Antarctica").

13. See Simpson 1943 ("Nature of Continents," 30) for Joleaud references.

14. Wegener 1966 (*Origin of Continents and Oceans*, 100; emphasis added).

15. Simpson 1943 ("Nature of Continents," 14).

16. Simpson 1944b (*Tempo and Mode*, 191–92).

17. Simpson 1940b ("Mammals and Land Bridges," 138).

18. Simpson 1953a (*Evolution and Geography*, 20).

19. Ibid., 53

20. Simpson 1947a ("Evolution, Interchange, Resemblance"), 1936a ("Data on Relationships") and 1943 ("Nature of Continents").

21. Simpson 1943 ("Nature of Continents," 19).

22. Wegener 1966 (*Origin of Continents and Oceans*, 18–19); Du Toit 1937 (*Our Wandering Continents*, 219ff.).

23. Simpson 1947b ("Holarctic Mammalian Faunas," 616*n*).

24. Ibid., 685–86.

25. Simpson 1952 ("Probabilities").

26. Simpson 1953a (*Evolution and Geography*).

27. Ibid., 61–62.

28. Du Toit 1944 ("Tertiary Mammals," 149, 163).

29. Cloud 1961 ("Paleobiogeography," 194); Axelrod 1960 ("Flowering Plants," 280; emphasis in the original).
30. Opdyke 1962 ("Paleoclimatology").
31. Simpson and Beck 1965 (*Life*, 32).
32. Simpson 1966b ("Southern Continents," 3).
33. Elliott et al. 1970 ("Triassic Tetrapods"); Simpson 1970b ("Drift Theory," 678).
34. Simpson 1983 (*Fossils*, 115).
35. Colbert 1971 ("Tetrapods and Continents," 263).
36. See, for example, Simpson 1976 ("Compleat Paleontologist?" 8ff.), and Simpson 1983 (*Fossils*, 93ff.).
37. Simpson 1975 ("Comments," 96).
38. Simpson 1978 (*Concession*, 272–73).
39. See, for example, Mayr 1982 (*Biological Thought*, 449ff.); McKenna 1973 ("Sweepstakes") and 1983 ("Holarctic Landmass").
40. See, for example, Colbert 1952 ("Mesozoic Tetrapods").
41. Romer 1968 (*Notes and Comments*, 228–29).
42. Westoll 1944 ("Haplolepidae," 109).
43. Simpson 1980b (*Splendid Isolation*).
44. Marshall et al. 1982 ("Great American Interchange").

10. The Mind's Eye

This chapter was first published in slightly different form as "The Mind's Eye: George G. Simpson's Use of Visual Language," *Earth Science History* 14 (1995): 37–46. This article was printed in and is used by permission of the journal of the History of the Earth Sciences Society, *Earth Sciences History*.

1. Author interview with Simpson, December 31, 1980, Tucson, Arizona; letter to author, April 4, 1981.
2. Simpson 1953c (*Major Features*). In the December 31, 1980, interview with the author, Simpson acknowledged that Ernst Mayr in particular found Simpson's visualizations more confusing than helpful. And indeed Mayr, even in his most abstract work, rarely used the original sort of visualizations found so often in Simpson, although of course he did include the more usual graphs, maps, charts, and drawings common to the biological literature. Contrast, for example, Simpson 1953c (*Major Features*) with Mayr 1963 (*Animal Species*).
3. Simpson 1980c (*Why and How*, 38).
4. Simpson 1925a ("American Triconodonts," 147–48).
5. Simpson 1926a ("*Dromatherium* and *Microconodon*," 87–108).
6. Simpson 1928a (*Catalogue of Mesozoic Mammalia*); Simpson 1929a ("American Mesozoic Mammalia"). In the former publication Simpson

used stereophotographs for Figure 4 in Plate I and in Plate II. Of the thirty-two plates in the latter publication, Simpson shows sixty-two pairs of stereophotos (enlarged 2–25 diameters) in twenty-one plates, all taken by himself.

7. Simpson 1976 ("Compleat Paleontologist?" 3).

8. Simpson 1941c ("Pleistocene Felines"). See appendix in this paper for a description of Simpson's methodology for constructing ratio diagrams.

9. Simpson 1980c (*Why and How*, 57; emphasis added).

10. Simpson 1964a ("Species Density," 57–73).

11. Simpson 1980c (*Why and How*, vii).

12. Simpson 1937c ("Phyletic Evolution," 303–304). In particular, Simpson was thinking of the sort of evolutionary trees (all vertical branches with no roots) drawn by his late American Museum colleague, H. F. Osborn.

13. Ibid., 304.

14. Ibid., fig. 1, p. 305.

15. Kinsey 1936 ("Origin of Higher Categories," fig. 14, p. 56); Simpson 1937d ("Supra-specific Variation," fig. 7, p. 256).

16. Kinsey 1936 ("Origin of Higher Categories," caption to fig. 14, p. 56); Simpson 1937d ("Supra-specific Variation," 255, 257). Credit for the visual imagery, of course, must go to Kinsey not only in the figure itself but, more relevantly, to Kinsey's use of graphics to illustrate the points he was making.

17. Kinsey 1936 ("Origin of Higher Categories," fig. 13, p. 55); Simpson 1937d ("Supra-specific Variation," fig. 5, p. 252).

18. Simpson 1944b (*Tempo and Mode*).

19. Wright 1932 ("Mutation, Inbreeding, Crossbreeding," fig. 2, p. 358); Simpson 1944b (*Tempo and Mode*, 89).

20. Provine claims that, in fact, the concept "is one of Wright's most confusing and misunderstood contributions to evolutionary biology," misunderstood not only by Simpson and other evolutionary synthesists but subsequently by Wright himself! (Provine 1986, *Sewall Wright and Evolutionary Biology*, 308ff.).

21. Simpson 1944b (*Tempo and Mode*, fig. 11, p. 90; and fig. 12, p. 91).

22. Ibid., 183–86. Simpson bases his discussion upon concepts first articulated by Parr 1926 ("Adaptiogenese und Phylogenese"); Simpson 1944b (*Tempo and Mode*, fig. 24, p. 184).

23. Ibid. (fig. 26, p. 191; and 192).

24. Simpson 1975 ("Comments," 45–46). Darwin's insight: "I can remember the very spot in the road, whilst in my carriage, when to my joy the solution occurred to me . . . the modified offspring of all dominant and increasing forms tend to become adapted to many and highly

diversified places in the economy of nature" (Darwin, *Autobiography*, 120–21).

25. Simpson 1944b (*Tempo and Mode*, figs. 27a, 27b, 28, 29, 30, 35, 36a, and 36b). "Bradytely," "horotely," and "tachytely" were Simpson's neologisms for very slow, slow-to-fast, and very fast rates of evolution. The last rate was usually also referred to as "quantum evolution" by Simpson.

26. Simpson 1961c also uses a simplified, modified version of the adaptive grid (*Animal Taxonomy*, figs. 23 and 25).

27. Simpson 1978 (*Concession*, 1–2).

28. Simpson 1987 (*Simple Curiosity*, 24). Subsequent page numbers are cited in the text.

29. Rudwick 1976 ("The Emergence of a Visual Language," 149–95).

11. The Awkward Embrace

This chapter was first published in slightly different form as "G. G. Simpson and His Museum Employers," in M. Ghiselin and A. Levinton, eds., *Institutions of Natural History* (San Francisco: California Academy of Sciences, 2000). Used by permission of the California Academy of Sciences.

1. The expression "awkward embrace" was coined by Joan Simpson Burns, Simpson's third daughter, and used for the title of her book that examines what former editor-in-chief of *Time* magazine Hedley Donovan called "the balance between the rights — or possibilities — of the individual, and the necessities — or claims — of organized society." It seems thoroughly appropriate to extend the metaphor to the particular case of the tension between her father's needs as a highly creative scientist and those of his museum employers as institutions of academic and popular education (Burns 1975, *Awkward Embrace*, xv).

2. Author interview with Simpson, August 18, 1981, Tucson, Arizona.

3. The description of Simpson's negotiations with Yale and the American Museum are based upon several sources, including Simpson's autobiography (Simpson 1978, *Concession*, 38ff.); letters to his parents and sister at the time that provide more specific details about the Yale and AMNH offers, informal and formal, and Simpson's protracted consideration thereof (Simpson 1987, *Simple Curiosity*, 60, 67, 70, 73, 80, 87, 89, 91); and unpaginated autobiographical notes (GGS/APS). (See chapter 3 regarding the strong relationship Simpson had established earlier with W. D. Matthew.)

4. Simpson 1978 (*Concession*, 38).

5. Ibid., 38.

6. Colbert 1992 (*William Diller Matthew*).

7. Simpson 1978 (*Concession*, 38).

8. Simpson 1987 (*Simple Curiosity*, 87).

9. Simpson 1978 (*Concession*, 39).

10. Ibid., 41.

11. See Kennedy 1968 ("Philanthropy and Science," esp. 219ff.); and Rainger 1991 (*Agenda for Antiquity*).

12. Colbert 1980 (*Fossil Hunter's Notebook*).

13. *New York Times*, obituary, May 17, 1949, 26.

14. Simpson 1934 (*Attending Marvels*, 302).

15. "These were the very animals of which we had seen one small fragment almost three years earlier. Dozens of them! Hundreds perhaps! Complete skeletons weathering out of the ancient rocks where they had lain for millions of years undisturbed! One of these, crushed flat but perfect down to the last joint, is now on display as a treasure of the American Museum in New York and another is in the National Museum in Washington. As the animals did prove to have been totally unknown to science, I named them *Scarrittia* after the patron of our expeditions" (Simpson 1941b, "How We Knew Where to Dig," 209). At the end of Simpson's article, the editor adds: "*Scarrittia canquelensis* . . . was about the size of a horse, but stockier and heavier built. It was cumbersome . . . herbivorous, browsing on leaves and twigs rather than grazing on grasses. It probably had no tail. It is unrelated to anything . . . today."

16. Although the Patagonian expedition had a sound scientific basis, the yearlong overseas trip had the added benefit for Simpson of relief from his marital woes. His marriage was in trouble practically from the first and went downhill quickly thereafter. By summer 1930, when he departed for South America, Simpson had been estranged from his wife, more or less continuously from the time he left Yale to go to the British Museum, and was finally legally separated from her the year after he returned.

17. Simpson 1934 (*Attending Marvels*). As Kennedy points out: "This kind of outdoor science pleased Professor Osborn's wealthy Wall Street trustees and their friends. They particularly liked the museum's combination of exploration and science with the 'traditional American' entrepreneurial virtues of 'boldness' and 'hard work'" (Kennedy 1968, "Philanthropy and Science," 157). Rainger, too, notes that "Osborn operated as a salesman who promoted new projects to museum administrators, authored popular articles in leading newspapers and magazines, and employed rhetoric, exaggeration, and supreme confidence to acquire economic and political support for major projects in what was otherwise an expensive, non-practical, and peripheral field of inquiry [vertebrate paleontology]" (Rainger 1991, *Agenda for Antiquity*, 3).

18. Simpson 1934 (*Attending Marvels*, xxiv–xxv).

19. Simpson 1978 (*Concession*, 85, 89, 95).

20. Kennedy 1968 ("Philanthropy and Science," 233ff.).

21. Ibid., 243.
22. Simpson 1978 (*Concession*, 38).
23. Quoted in Kennedy 1968 ("Philanthropy and Science," 246).
24. Simpson 1944b (*Tempo and Mode*). Three other works that are considered important in the "consolidation of the evolutionary synthesis" in the late 1930s and early 1940s were based on the Jesup Lectures at Columbia University, a program of public instruction funded by the late Morris K. Jesup, a wealthy American Museum founder and former Columbia University president who had hired the young Henry Fairfield Osborn to start the Department of Paleontology (Preston 1986, *Dinosaurs in the Attic*, 64).
25. Simpson 1978 (*Concession*, 38–39).
26. Ibid., 129.
27. Author interview with Norman D. Newell, May 29–30, 1979, Las Vegas, Nev.
28. Author interview with Bobb Schaeffer, November 6, 1980, New York City.
29. Author interview with E. H. Colbert, July 7, 1982, Flagstaff, Ariz.
30. Simpson 1978 (*Concession*, 170).
31. Author interview with Ernst Mayr, October 22, 1980, Cambridge, Mass. (See also note 48.)
32. Newell interview, May 29–30, 1979.
33. Colbert interview, July 7, 1982. Anne Roe, Simpson's wife, had another perspective. She noted that the relationship between her husband and Parr prior to Simpson's resignation "had suddenly turned antagonistic." Simpson didn't know why Parr would no longer talk to him or why he would leave the staff room when Simpson entered. For a long time Anne thought her husband imagined these slights, until she saw Parr turn away from her husband at a museum Christmas party. Sometime later, Simpson discovered that Parr had found out that Parr's grant proposal to the National Science Foundation had been turned down upon Simpson's recommendation. As for Simpson's wanting to remain chairman of the department, Anne had asked her husband why he didn't resign, letting Colbert take over. "After all it was just extra worry and work for him anyway." Simpson replied that he would, except he was afraid he wouldn't continue to get the support he needed to carry on his research. (Author interview with Anne Roe Simpson, December 17–18, 1985, Tucson, Ariz.)
34. Schaeffer interview, November 6, 1980.
35. Mayr interview, October 22, 1980. (See also note 48).
36. Simpson 1978 (*Concession*, 170).
37. Letter, Simpson to Colbert, September 23, 1958 (GGS/APS).
38. Letter, Simpson to Parr, April 17, 1959 (GGS/APS).
39. Simpson 1987 (*Simple Curiosity*, 294; emphasis in the original).

40. Ibid., 295.

41. Author interview with John Moore, June 14, 1979, Riverside, Calif.

42 Newell interview, May 29–30, 1979.

43. Author interview with Ernest Williams, October 3, 1980, Cambridge, Mass.

44. Mayr interview, October 22, 1980.

45. Simpson 1978 (*Concession*, 195). (See also note 48.)

46. Ibid., 219.

47. E-mail, M. McKenna to the author, September 9, 1998.

48. Mayr interview, October 22, 1980. In his review of this chapter, Mayr wanted to correct the record because, since my interview with him, he had learned about Simpson's resignation directly from Parr. According to Parr, when Simpson was fully recovered from his accident, Parr offered him two options. Either Simpson could return full time to the departmental chairmanship or resign and take on an appointment as senior scientist. Simpson angrily refused both offers and immediately contacted Romer at the MCZ, asking him for an appointment there as Alexander Agassiz Professor, to which Romer agreed. Furthermore, Parr vigorously denied that Colbert had any direct role in Simpson's resignation. Mayr also did not believe that Romer offered Simpson the directorship of the MCZ, given Simpson's personality and previous administrative record at the AMNH. On the contrary, Mayr himself was being groomed by Romer for the directorship. (Mayr review, 27 April 1999).

49. Simpson 1978 (*Concession*, 218).

12. Concession to the Ineluctable

This chapter was first published in slightly different form as "G. G. Simpson as Sam Magruder: Concession to the Ineluctable," *Earth Sciences History* 16 (1997): 44–49. This article was printed in and is used by permission of the journal of the History of the Earth Sciences Society, *Earth Sciences History*.

1. Simpson 1996 (*Sam Magruder*, 7; words spoken by the Universal Historian). Subsequent page numbers are cited in the text.

2. The Universal Historian is one of several Everyman sort of characters who discuss the meaning and significance of Magruder's slab writings; others are the Pragmatist, the Ethnologist, and the Common Man. Additional characters include a "leading geologist," Pierre Précieux (French for "precious stone"), who discovers the slabs; a Cretaceous paleontologist, Saurier ("sauros," Greek for lizard); Magruder's supervisor Kto Znayet (Russian for "who knows"); and the paleographer Schreiben ("to write down, record," German).

3. This science fiction story is the second finished piece of such writing

that Simpson left behind. He and his wife Anne Roe had amused themselves during a rainy field season in Venezuela in 1938 by collaborating on a detective story, "Trouble in the Tropics," which, however, was never published.

4. Simpson 1974b ("Questionnaire"), edited by Mayr and reprinted in Mayr and Provine 1980 (*Evolutionary Synthesis*, 452–63).

5. Simpson [E. L.] 1982 (*Notes on an Emergency*, 32).

6. Simpson 1975 ("Comments," 3).

7. Mayr and Provine 1980 (*Evolutionary Synthesis*, 134). As noted in chapters 2 and 7 of the present volume, Theodosius Dobzhansky initiated the founding of the modern evolutionary synthesis with the publication of his *Genetics and the Origin of Species* in 1937. Dobzhansky was also one of four deceased colleagues to whom Simpson dedicated his penultimate book, *Fossils and the History of Life*, in 1983. (See note 9.)

8. Simpson 1978 (*Concession*).

9. Simpson letter to the author, December 24, 1979. Bryan "Pat" Patterson (1909–1979) was a vertebrate paleontologist, Harvard colleague (1959–1970), and drinking companion of Simpson's for a number of years. See Simpson 1984 (*Discoverers of the Lost World*, 177–183) for more about Patterson and about his relationship to Simpson. Patterson was another dedicatee of Simpson's penultimate book, *Fossils and the History of Life* (see note 7, above).

10. Simpson 1981a ("Vertebrate Paleontology in the San Juan Basin," 3–25).

11. Simpson 1978 (*Concession*, 274–75).

12. Simpson 1987 (*Simple Curiosity*, 193ff.).

13. See note 1, above.

14. Quoted from letter from G. G. Simpson to Anne Roe from the University of Colorado, Boulder, ca. 1920–21 (GGS/APS; emphasis in the original).

15. Simpson 1987 (*Simple Curiosity*, 47).

16. Simpson 1983 (*Fossils*, 207).

17. Simpson 1953c (*Major Features*, xii).

18. See, for example, Simpson 1959b ("Fossil mammals . . . Paleocene of New Mexico").

19. See Simpson 1978 (*Concession*) for a sampling of Simpson's musical and travel diversions.

20. The "little girl . . . next door" might allude to Simpson's childhood playmate, Anne Roe, who lived a few houses down the street from him in Denver and whom he later married.

21. Evidence for this conclusion is based on interviews the author has had with various family members and Simpson himself, and on impres-

sions gained from private correspondence and personal reminiscences, both unpublished and published. For example, in one interview Simpson told the author that his parents always approved and supported his interests, although his father thought maybe he wasn't masculine enough (author interview with Simpson, February 3, 1979, Tucson). In another interview Simpson further remarked that "it annoyed my father that I wasn't athletic enough for him, particularly as regards team games. He got two mitts to play catch so I'd be better at baseball. But [because of nystagmus] I couldn't focus on the ball; I kept missing it, but he'd keep throwing it, and the ball would keeping hitting me. He'd say 'Don't just stand there!' I'd say 'I can't see the ball!' This went on for weeks; it was torture for me" (author interview with Simpson, August 19, 1981, Tucson).

BIBLIOGRAPHY

Archives and Papers

DVP/AMNH: Department of Vertebrate Paleontology Archives, American Museum of Natural History, New York City.
GGS/APS: G. G. Simpson Papers, MS Collection no. 31, American Philosophical Society Archives, Philadelphia.
MES/APS: Transcription of the taped session on genetics and paleontology, October 1974, during the conference on the history of the development of the modern evolutionary synthesis at Harvard University (American Philosophical Society Archives, Philadelphia).
SI/USNM: Smithsonian Institution Archives, Washington, D.C., 1877–1975. Record Unit 192, United States National Museum, Permanent Administrative Files, box 409, folder 10.
SI/DVP: Smithsonian Institution Archives, Washington, D.C., 1889–1957. Record Unit 156, Division of Vertebrate Paleontology Records, 13, folder 17.

Primary and Secondary Sources

Allee, W. C., A. E. Emerson, O. Park, T. Park, and K. P. Schmidt. 1949. *Principles of Animal Ecology*. Philadelphia: W. B. Saunders.
Ayala, F., G. L. Stebbins, and J. W. Valentine. 1977. *Evolution*. San Francisco: W. H. Freeman.

Axelrod, D. 1960. "The Evolution of Flowering Plants." In S. Tax, ed., *Evolution After Darwin* 1:227–305. Chicago: University of Chicago Press.
Birdsell, J. 1987. "Some Reflections on Fifty Years in Biological Anthropology." *Annual Review of Anthropology* 16: 1–12.
Boaz, N. 1981. "History of American Paleoanthropological Research on Early Hominidae, 1925–1980." *American Journal of Physical Anthropology* 56: 397–405.
Boaz, N. and F. Spencer, eds. 1981. "Jubilee Issue." *American Journal of Physical Anthropology* 56, no. 4.
Boucot, A. J. 1978. "Community Evolution and Rates of Cladogenesis." *Evolutionary Biology* 11: 545–655.
Bowler, P. 1983. *The Eclipse of Darwinism*. Baltimore: Johns Hopkins University Press.
———. 1986. *Theories of Human Evolution: A Century of Debate, 1844–1944*. Baltimore: Johns Hopkins University Press.
Brinkmann, R. 1929. "Statistisch-biostratigraphische Untersuchungen an mittel-jurassichen Ammoniten über Artbegriff und Stammesentwicklung." *Abh. Ges. Wiss. Göttingen* 13.3: 1–249.
Bronowski, J. 1969. "On the Uniqueness of Man: Review of *Biology and Man* by G. G. Simpson." *Science* 165:. 680–81.
Bucher, W. 1942. "National Research Council and Cooperation in Geological Research." *Geological Society of America Bulletin* 53: 1331–54.
Burns, J. S. 1975. *The Awkward Embrace: The Creative Artist and the Institution in America*. New York: Knopf.
Cabrera, Á. 1932. "La incompatibilidad ecológica: una ley biológica interesante." *Anales de Sociedad Científica de Argentina* 114: 243–60.
Carson H. L. 1980. "Cytogenetics and the Neo-Darwinian Synthesis." In Mayr and Provine, eds., *The Evolutionary Synthesis*, 86–95.
Cartmill, M. 1982. "Basic Primatology and Prosimian Evolution." In F. Spencer, *History of American Physical Anthropology*, 147–86.
Cloud, P. E. 1961. "Paleobiogeography of the Marine Realm." In M. Sears, ed., *Oceanography*, 151–200. American Association for the Advancement of Science, Publication No. 67.
Colbert, E. H. 1952. "The Mesozoic Tetrapods of South America." *American Museum of Natural History Bulletin* 99: 237–54.
———. 1971. "Tetrapods and Continents." *Quarterly Review of Biology* 46: 250–69.
———. 1980. *A Fossil Hunter's Notebook*. New York: Dutton.
———. 1992. *William Diller Matthew: Paleontologist*. New York: Columbia University Press.
Cracraft, J. 1974. "Continental Drift and Vertebrate Distribution." *Annual Review of Ecology and Systematics* 5: 215–61.

Croizat, L. 1981., "Biogeography: Past, Present, and Future." In G. Nelson and D. E. Rosen, eds., *Vicariance Biogeography*, 500–23. New York: Columbia University Press.

Darlington, P. J., Jr. 1980. *Evolution for Naturalists*. New York: Wiley-Interscience.

Darwin, C. 1969 (paperback ed.). *Autobiography* (1958; New York: Harcourt, Brace and World). Edited by Lady Nora Barlow. New York: Norton.

Davis, D. D. 1949. "Comparative Anatomy and the Evolution of Vertebrates." In Jepsen, Simpson, and Mayr, eds., *Genetics, Paleontology, and Evolution*, 64–89.

Dobzhansky, T. 1937 (3d ed., 1951). *Genetics and the Origin of Species*. New York: Columbia University Press.

———. 1945. "Genetics of Macro-evolution: A Review of *Tempo and Mode in Evolution*." *Journal of Heredity* 36: 113–15.

———. 1963. "*The Origin of Races* by Carleton S. Coon" (review). *Scientific American* 2: 169–72.

———. 1980. "The Birth of the Genetic Theory of Evolution in the Soviet Union in the 1920s." In Mayr and Provine, eds., *The Evolutionary Synthesis*, 229–42.

Dodson, E. O. and P. Dodson. 1976. *Evolution: Process and Product*. New York: Van Nostrand.

Dunbar, C. O. 1959. "A Half Century of Paleontology." *Journal of Paleontology* 33: 909–914.

Du Toit, A. L. 1937. *Our Wandering Continents*. Edinburgh: Oliver and Boyd.

———. 1944. "Tertiary Mammals and Continental Drift." *American Journal of Science* 242: 145–63.

Eldredge, N. 1982. "Introduction." In Ernst Mayr, *Systematics and the Origin of Species* (reprint of 1942). New York: Columbia University Press.

———. 1985. *Unfinished Synthesis*. New York and Oxford: Oxford University Press.

Elliot, D. H., E. H. Colbert, W. J. Breed, J. A. Jensen, and J. S. Powell. 1970. "Triassic Tetrapods from the Antarctic: Evidence for Continental Drift." *Science* 169: 1197–1201.

Fisher, R. A. 1925. "Statistical Methods for Research Workers." *Biological Monographs and Manuals* 5 (entire volume).

———. 1930. *The Genetical Theory of Natural Selection*. Oxford: Clarendon Press.

Frankel, H. 1981. "The Paleobiogeographical Debate Over the Problem of Disjunctively Distributed Life Forms." *Studies in the History and Philosophy of Science* 12: 211–59.

Frick, C. 1926a. "The Hemicyoninae and an American Tertiary Bear." *American Museum of Natural History Bulletin* 56: 1–119.

———. 1926b. "Prehistoric Evidence." *Natural History* 26: 440–48.

Gidley, J. W. 1909. "Notes on the Fossil Mammalian Genus *Ptilodus*, with descriptions of new species." *Proceedings of the United States National Museum* 34: 611–26.

———. 1915. "An Extinct Marsupial from the Fort Union with notes on the Myrmecobidae and other families of this group." *Proceedings of the United States National Museum* 48: 395–402.

———. 1919. "New Species of Claenodonts from the Fort Union (Basal Eocene) of Montana." *American Museum of Natural History Bulletin* 41: 541–56.

———. 1923. "Paleocene Primates of the Fort Union, with discussion of the relationships of the Eocene primates." *Proceedings of the United States National Museum* 63: 1–38.

Gillespie, N. C. 1979. *Charles Darwin and the Problem of Creation*. Chicago: University of Chicago Press.

Gingerich, P. D. 1986. "George Gaylord Simpson: Empirical Theoretician." *Contributions to Geology*, University of Wyoming Special Paper no. 3: 3–9.

Glass, B. 1945. "Review of *Tempo and Mode in Evolution*." *Quarterly Review of Biology* 20: 261–63.

Goldschmidt, R. 1940. *The Material Basis of Evolution*. New Haven: Yale University Press.

Gould, L. M. 1984. "Memorial Tribute to G. G. Simpson" (October 25). University of Arizona, Tucson.

Gould, S. J. 1980. "G. G. Simpson, Paleontology, and the Modern Synthesis." In Mayr and Provine, eds., *The Evolutionary Synthesis*, 153–72.

———. 1982. "Introduction." Facsimile reprint of T. Dobzhansky's 1937 *Genetics and the Origin of Species*, xvii–xli. New York: Columbia University Press.

Gregory, J. W. 1929. "The Geological History of the Atlantic Ocean." *Quarterly Journal of the Geological Society of London* 85: lxvii–cxxii.

———. 1930. "The Geological History of the Pacific Ocean." *Quarterly Journal of the Geological Society of London* 86: lxxii–cxxxvi.

Gregory, W. K. and G. G. Simpson. 1926. "Cretaceous Mammal Skulls from Mongolia." *American Museum Novitates* 225: 1–20.

Haldane, J. B. S. 1932. *Causes of Evolution*. New York: Harper.

———. 1949. "Human Evolution: Past and Future." In Jepsen, Simpson, and Mayr, eds., *Genetics, Paleontology, and Evolution*, 405–18.

Hallam, A. 1983. *Great Geological Controversies*, 110–156. New York and Oxford: Oxford University Press.

Haraway, D. 1988. "Remodeling the Human Way of Life: Sherwood Washburn and the New Physical Anthropology, 1950–1980." In G. Stocking, ed., *Bones, Bodies, and Behavior*, 206–59.

Hecht, M. K., B. Schaeffer, B. Patterson, R. van Frank, and F. D. Wood. 1972. "George Gaylord Simpson: His Life and Works to the Present." *Evolutionary Biology* 6: 1–29.

Howells, W. 1951. "Concluding Remarks of the Chairman." *Cold Spring Harbor Symposium on Quantitative Biology* 15: 79–86.

——. 1962. *Ideas on Human Evolution: Selected Essays, 1949–1961*. Cambridge: Harvard University Press.

Hubbs, C. 1945. "Review of *Tempo and Mode in Evolution*." *American Naturalist* 79: 271–75.

Hutchinson, G. E. 1944. "Review of *Tempo and Mode in Evolution*." *American Journal of Science* 243: 356–58.

——. 1945. "Marginalia." *American Scientist* 33: 120.

Huxley J. 1942. *Evolution: The Modern Synthesis*. London: Allen and Unwin.

——. 1945. "Genetics and Major Evolutionary Change: Review of *Tempo and Mode in Evolution*." *Nature* 156: 3–4.

Ihering, H. von. 1927. *Die geschichte des Atlantischen Ozeans*. Vienna: Gustav Fischer.

Jepsen, G. L. 1946. "Review of *Tempo and Mode in Evolution*." *American Midland Naturalist* 35: 538–41.

——. 1949. "Foreword." In Jepsen, Simpson, and Mayr, eds., *Genetics, Paleontology, and Evolution*, v–x.

Jepsen, G. L., G. G. Simpson, and E. Mayr, eds. 1949. *Genetics, Paleontology, and Evolution*. Princeton: Princeton University Press.

Kennedy, J. M. 1968. "Philanthropy and Science in New York City: The American Museum of Natural History" (Ph.D. diss., Yale University).

Kingsley, C. 1925 (orig. pub. 1869). *Madam How and Lady Why: First Lessons in Earth Lore for Children*. New York: Macmillan.

Kinsey, A. C. 1936. "The Origin of Higher Categories in *Cynips*." *Indiana University Publications*, Science Series, no. 4: 1–334.

Kitts, D. 1974. "Paleontology and Evolutionary Theory." *Evolution* 28: 458–72.

Laporte, L. F. 1994. "Australia's Place in Simpson's Biogeography." In D. F. Branagan and G. H. McNally, eds., *Useful and Curious Geological Enquiries Beyond the World*, 7–13. Sydney: 19th INHIGEO Symposium, Sydney, Australia.

Laudan, R. 1982. "Tensions in the Concept of Geology: Natural History or Natural Philosophy?" *History of Geology* 1: 7–13.

Le Gros Clark, W. 1955 (2d ed.). *Fossil Evidence for Human Evolution*. Chicago: University of Chicago Press.

Levinton, J. S. and C. M. Simon. 1980. "A Critique of the Punctuated Equilibria Model and Implications for the Detection of Speciation in the Fossil Record." *Systematic Zoology* 29: 130–42.

Luckett, W. P. and F. Szalay, eds. 1975. *Phylogeny of the Primates*. New York: Plenum Press.

Lull, R. S. 1917 (rev. ed., 1929). *Organic Evolution*. New York: Macmillan.

Marshall, L. G., S. D. Webb, J. J. Sepkoski, Jr., and D. M. Raup. 1982. "Mammalian Evolution and the Great American Interchange." *Science* 215: 1351–57.

Marvin, U. 1973. *Continental Drift: The Evolution of a Concept*. Washington, D.C.: Smithsonian Press.

Matthew, W. D. 1915. "Climate and Evolution." *Annals of the New York Academy of Sciences* 24: 171–318.

———. 1924a. "Fossil Horses from the Texas Pliocene."*Natural History* 24: 629–31.

———. 1924b. "List of species, Amer. Mus. expedition, May-June 1924, Llano Estacado Tex." Handwritten notes, Annual Reports, 1924 (DVP/AMNH).

———. 1924c. "A New Link in the Ancestry of the Horse." *American Museum Novitates* 131: 1–2.

———. 1925. "Recent Progress and Trends in Vertebrate Paleontology." *Annual Report of the Smithsonian Institution for 1923*, 273–89. Washington, D.C.: Government Printing Office.

———. 1926. "The Evolution of the Horse: A Record and Its Interpretation." *Quarterly Review of Biology* 1: 139–85.

———. 1928. *Outline and General Principles of the History of Life*. Synopsis of Lectures in Paleontology I, Syllabus Series, no. 213. Berkeley: University of California Press.

———. 1930. "Range and Limitations of Species as Seen in Fossil Mammal Faunas." *Geological Society of America Bulletin* 41: 271–74.

———. 1939. *Climate and Evolution*. New York: New York Academy of Sciences, Special Publication No. 1 (223 pp.; slightly expanded, posthumous reprint of 1915 ed.).

Mayr, E. 1942. *Systematics and the Origin of Species*. New York: Columbia University Press.

———. 1944. "Wallace's Line in Light of Recent Zoogeographic Discoveries." *Quarterly Review of Biology* 19: 1–14.

———. 1958. *Concepts of Biology*. Washington, D.C.: National Academy of Sciences, National Research Council, Publication 560.

———. 1962. "*The Origin of Races* by Carleton S. Coon" (review). *Science* 138: 420–22.

———. 1963. *Animal Species and Evolution*. Cambridge: Harvard University Press.

———. 1980. "Prologue: Some Thoughts on the History of the Evolutionary Synthesis." In Mayr and Provine, eds., *The Evolutionary Synthesis*, 1–48.

———. 1982. *The Growth of Biological Thought*. Cambridge: Harvard University Press.

———. 1988. *Toward a New Philosophy of Biology*. Cambridge: Harvard University Press.

Mayr, E. and W. Provine, eds. 1980. *The Evolutionary Synthesis*. Cambridge: Harvard University Press.

McKenna, M. 1973. "Sweepstakes, Filters, Corridors, Noah's Arks, and Viking Funeral Ships in Palaeogeography." In *Implications of Continental Drift to the Earth Sciences* 1:295–307. New York and London: Academic Press.

———. 1983. "Holarctic Landmass Rearrangement, Cosmic Events, and Cenozoic Terrestrial Organisms." *Annals of the Missouri Botanical Garden* 70: 459–89.

Morgan, T. H. 1932. *The Scientific Basis of Evolution*. New York: Norton.

Olson, E. C. 1953. "Review of *The Major Features of Evolution*." *Evolution* 8: 87–88.

———. 1974. "Transcription of comments made at the second workshop on the founding of the evolutionary synthesis, October 11–12, 1974, Cambridge, Mass.," MES/APS.

———. 1981. "The Problem of Missing Links: Today and Yesterday." *Quarterly Review of Biology* 56: 405–442.

Opdyke, N. D. 1962. "Palaeoclimatology and Continental Drift." In S. K. Runcorn, *Continental Drift*, 41–65. New York and London: Academic Press.

Oreskes, N. 1999. *The Rejection of Continental Drift*. New York and Oxford: Oxford University Press.

Osborn, H. F. 1925a. "The Origin of Species as Revealed by Vertebrate Paleontology [Part 1]." *Nature* 115: 925–26 and 961–63.

———. 1925b. "The Origin of Species II." *Proceedings of the National Academy of Science* 11: 749–52.

———. 1926a. "The Origin of Species, 1859–1925 [Part 3]." *Scientific Monthly* 22: 185–92.

———. 1926b. "The Problem of the Origin of Species as it appeared to Darwin in 1859 and as it appears to us today [Part 4]." *Science* 44: 337–41.

———. 1927. "The Origin of Species V: Speciation and Mutation." *American Naturalist* 61: 5–42.

———. 1932. "Biological Inductions from the Evolution of the Proboscidea [Part 8]." *Science* 76: 501–504.

———. 1934. "Aristogenesis, the Creative Principle in the Origin of Species [Part 11]." *American Naturalist* 68: 193–235.

Parr, A. E. 1926. "Adaptiogenese und Phylogenese; zur Analyse der An-

passungerscheinungen und ihre Entstehung." *Abh. theor. organ. Ent.* 1: 1–60.
Preston, D. J. 1986. *Dinosaurs in the Attic*. St. Martin's.
Provine, W. B. 1986. *Sewall Wright and Evolutionary Biology*. Chicago: University of Chicago Press.
——. 1988. "Progress in Evolution and Meaning in Life." In M. H. Nitecki, ed., *Evolutionary Progress*. Chicago: University of Chicago Press.
Rainger, R. 1981. "The Continuation of the Morphological Tradition: American Paleontology, 1880–1910." *Journal of the History of Biology* 14: 129–58.
——. 1985. "Paleontology and Philosophy: A Critique." *Journal of the History of Biology* 18: 267–87.
——. 1986. "Just Before Simpson: William Diller Matthew's Understanding of Evolution." *American Philosophical Society Proceedings* 130: 453–74.
——. 1988. "Vertebrate Paleontology as Biology: Henry Fairfield Osborn and the American Museum of Natural History." In Keith Benson, Jane Maienschein, and Ronald Rainger, eds., *American Development of Biology*, 219–56. Philadelphia: University of Pennsylvania Press.
——. 1989. "What's the Use: William King Gregory and the Functional Morphology of Fossil Vertebrates." *Journal of the History of Biology* 22: 103–139.
——. 1991. *Agenda for Antiquity*. Tuscaloosa and London: University of Alabama Press.
Raup, D. M. and R. E. Crick. 1982. "*Kosmoceras*: Evolutionary Jumps and Sedimentary Breaks." *Paleobiology* 8: 90–100.
Roe, A. 1985. "Leona Tyler Award Address." *Counseling Psychologist* (April): 311–26.
Roe, A. and G. G. Simpson. 1958. *Behavior and Evolution*. New Haven: Yale University Press.
Romer, A. S. 1959. "Vertebrate Paleontology, 1908–1958." *Journal of Paleontology* 33: 915–25.
——. 1968. *Notes and Comments on Vertebrate Paleontology*. Chicago: University of Chicago Press.
——. 1973. "Vertebrates and Continental Connections: An Introduction." In D. H. Tarling and S. R. Runcorn, eds., *Implications of Continental Drift to the Earth Sciences* 1:345–49. New York and London: Academic Press.
Rudwick, M. 1976. "The Emergence of a Visual Language for Geological Science, 1760–1840." *History of Science* 14: 149–95.
Schindewolf, O. H. 1936. *Paläontologie, Entwicklungslehre, und Genetik*. Berlin: Borntraeger.
Shapere, D. 1980. "The Meaning of the Evolutionary Synthesis." In Mayr and Provine, eds., *The Evolutionary Synthesis*, 388–98.

Simpson, E. L. 1982. *Notes on an Emergency*. New York: Norton.
Simpson, G. G. 1925a. "Mesozoic mammalia I. American Triconodonts." *American Journal of Science* 210: 145–65, 334–58.
———. 1925b. "Mesozoic mammalia III. Preliminary Comparison of Jurassic Mammals except Multituberculates." *American Journal of Science* 210: 559–69.
———. 1925c. "Reconnaissance of the Santa Fé Formation." *Geological Society of America Bulletin* 36: 230.
———. 1926a. "Are *Dromatherium* and *Microconodon* Mammals?" *Science* 63: 548–49.
———. 1926b. "The Fauna of Quarry Nine." *American Journal of Science*, 5th ser., 12: 1–16.
———. 1926c. "Mesozoic Mammalia. IV. The Multituberculates as Living Animals." *American Journal of Science*, 5th ser., 11: 228–50.
———. 1928a. *A Catalogue of the Mesozoic Mammalia in the Geological Department of the British Museum*. London: British Museum of Natural History.
———. 1928b. "A New Mammalian Fauna from the Fort Union of Southern Montana." *American Museum Novitates* 297: 1–15.
———. 1928c. "Third Contribution to the Fort Union Fauna at Bear Creek, Montana." *American Museum Novitates* 345: 1–12.
———. 1929a. "American Mesozoic Mammalia." *Memoirs of the Peabody Museum of Natural History of Yale University* 3 (part 1): 1–235.
———. 1929b. "A Collection of Paleocene Mammals from Bear Creek, Montana." *Carnegie Museum Annals* 19: 115–22.
———. 1931. "A New Classification of Mammals." *American Museum of Natural History Bulletin* 59: 259–93.
———. 1933. Handwritten autobiographical notes, unpaginated; typed and revised in 1954; handwritten additions to typescript in 1970. GGS/APS.
———. 1934. *Attending Marvels*. With an Introduction by Larry G. Marshall and an Afterword by G. G. Simpson. New York and London: Macmillan. Facsimile reprint, Chicago: University of Chicago Press, 1982.
———. 1936a. "Data on the Relationships of Local and Continental Mammalian Land Faunas." *Journal of Paleontology* 10: 410–14.
———. 1936b. "A New Fauna from the Fort Union of Montana." *American Museum Novitates* 873: 1–16.
———. 1936c. "Third Scarritt Expedition of the American Museum of Natural History." *Science* 83: 13–14.
———. 1937a. "The Fort Union of the Crazy Mountain Field, Montana and Its Mammalian Faunas." *United States National Museum Bulletin* 169: 1–287.

———. 1937b. "Notes on the Clark Fork, Upper Paleocene, Fauna." *American Museum Novitates* 954: 1–24.

———. 1937c. "Patterns of Phyletic Evolution." *Geological Society of America Bulletin* 48: 303–314.

———. 1937d. "Supra-specific Variation in Nature and in Classification: from the Viewpoint of Paleontology." *American Naturalist* 71: 236–67.

———. 1940a. "Antarctica as a Faunal Migration Route." *6th Pacific Congress Proceedings*: 755–68.

———. 1940b. "Mammals and Land Bridges." *Journal of the Washington (D.C.) Academy of Science* 30: 137–63.

———. 1940c. "Studies on the Earliest Primates." *American Museum of Natural History Bulletin* 77: 185–212.

———. 1940d. "Types in Modern Taxonomy." *American Journal of Science* 238: 413–31.

———. 1941a. "History of the Society and Its Predecessors." *Society of Vertebrate Paleontology News Bulletin* 1 (March): 1–3.

———. 1941b. "How We Knew Where to Dig." In S. S. Cramer, ed., *Through Hell and High Water*. New York: McBride.

———. 1941c. "Large Pleistocene Felines of North America." *American Museum Novitates* 1136: 1–27.

———. 1941d. "Some Recent Trends and Problems of Vertebrate Paleontological Research." *Society of Vertebrate Paleontology News Bulletin* 3 (part 1): 2–4; and vol. 4 (part 2): 10–11.

———. 1942–1944. "Observations During a War: Excerpts from Letters of G. G. Simpson to his Wife [Anne Roe]." American Philosophical Society Archives, 272 pp. (GGS/APS).

———. 1943. "Mammals and the Nature of Continents." *American Journal of Science* 241: 1–31.

———. 1944a. "Henry Fairfield Osborn." *Dictionary of American Biography* 11 (Supplement 1): 584–87.

———. 1944b. *Tempo and Mode in Evolution*. New York: Columbia University Press.

———. 1945. "The Principles of Classification and a Classification of Mammals." *American Museum of Natural History Bulletin* 85: 1–350.

———. 1946. "Tempo and Mode in Evolution." *New York Academy of Sciences Transactions*, 2d ser., 8: 45–60.

———. 1947a. "Evolution, Interchange, Resemblance of the North American and Eurasian Cenozoic Mammalian Faunas." *Evolution* 1: 218–20.

———. 1947b. "Holarctic Mammalian Faunas and Continental Relationships During the Cenozoic." *Geological Society of America Bulletin* 58: 613–88.

———. 1947c. "The Problem of Plan and Purpose in Nature." *Scientific Monthly* 64: 481–95.

———. 1948. "The Beginning of the Age of Mammals in South America, part 1." *American Museum of Natural History Bulletin* 91: 1–232.

———. 1949 (rev. ed., 1967). *The Meaning of Evolution*. New Haven: Yale University Press.

———. 1950a. "Trends in Research and the *Journal of Paleontology*." *Journal of Paleontology* 24: 498–99.

———. 1950b. "The Upper Cenozoic of the Rio Grande Depression." In E. H. Colbert and S. A. Northrop, eds., *Guidebook for the Fourth Field Conference of the Society of Vertebrate Paleontology*, 81–85. New York: American Museum of Natural History.

———. 1951a. *Horses*. New York and Oxford: Oxford University Press.

———. 1951b. "Some Principles of Historical Biology Bearing on Human Origins." *Cold Spring Harbor Symposium on Quantitative Biology* 15: 55–66.

———. 1951c. "The Species Concept." *Evolution* 5: 285–98.

———. 1952. "Probabilities of Dispersal in Geologic Time." *American Museum of Natural History Bulletin* 99: 163–76.

———. 1953a. *Evolution and Geography: An Essay on Historical Biogeography with Special Reference to Mammals*. Congdon Lectures. Eugene: Oregon State System of Higher Education.

———. 1953b. *Life of the Past*. New Haven: Yale University Press.

———. 1953c. *The Major Features of Evolution*. New York: Columbia University Press.

———. 1958a. "Memorial to Richard Swann Lull." *Geological Society of America Proceedings (Annual Report for 1957)*, 127–34.

———. 1958b. "The Study of Evolution: Methods and Present Status of Theory." In Roe and Simpson, *Behavior and Evolution*, 7–26.

———. 1959a. "Darwin Led Us into This Modern World." *The Humanist* 5: 267–75.

———. 1959b. "Fossil Mammals from the Type Area of the Puerco and Nacimiento Strata, Paleocene of New Mexico."*American Museum Novitates* 1957: 1–22.

———. 1959c. "Mesozoic Mammals and the Polyphyletic Origin of Mammals." *Evolution*. 13: 405–14.

———. 1959d. "The Nature and Origin of Supraspecific Taxa." *Cold Spring Harbor Symposium on Quantitative Biology* 24: 255–71.

———. 1960. "The World into which Darwin Led Us." *Science* 131: 966–74.

———. 1961a. "Historical Zoogeography of Australian Mammals." *Evolution* 15: 431–46.

———. 1961b. "Lamarck, Darwin, and Butler: Three Approaches to Evolution." *American Scholar* 30: 238–49.

———. 1961c. *The Principles of Animal Taxonomy*. New York: Columbia University Press.

———. 1961d. "Some Problems of Vertebrate Paleontology." *Science* 133: 1679–89.

———. 1962. "Primate Taxonomy and Recent Studies of Nonhuman Primates." *Annals of the New York Academy of Sciences* 102: 497–514.

———. 1963a. "The Meaning of Taxonomic Statements." In *Viking Fund Publications in Anthropology*, vol. 37, *Classification and Human Evolution*, ed. S. Washburn, 1–31. New York: Viking Fund in Anthropology.

———. 1963b. "Review of *The Origin of Races* by Carleton S. Coon." *Perspectives on Biology and Medicine* 6: 268–72.

———. 1964a. "Species Density of North American Recent Mammals." *Systematic Zoology* 13: 57–73.

———. 1964b. *This View of Life*. New York: Harcourt, Brace and World.

———. 1965. *The Geography of Evolution*. New York: Capricorn.

———. 1966a. "The Biological Nature of Man." *Science* 152 (1966): 472–78.

———. 1966b. "Mammalian Evolution of the Southern Continents." *Neue Jahrbuch für Geologie und Paläontologie Abhandlungen* 12: 1–18.

———. 1967. "The Beginning of the Age of Mammals in South America, part 2." *American Museum of Natural History Bulletin* 137: 1–259.

———. 1968. "What Is Man? A Review of *The Naked Ape* by Desmond Morris." *New York Times Book Review*, February 4: 16, 18, 20.

———. 1969. "The Present Status of the Theory of Evolution." *Royal Society of Victoria Proceedings* 82: 149–60.

———. 1970a. "Darwin's Philosophy and Methods: A Review of *The Triumph of the Darwinian Method* by Michael T. Ghiselin." *Science* 167: 1362–63.

———. 1970b. "Drift Theory: Antarctica and Central Asia." *Science* 170: 678.

———. 1971. "William King Gregory, 1876–1970." *American Journal of Physical Anthropology* 35: 155–73.

———. 1972. "The Evolutionary Concept of Man." In Bernard G. Campbell, ed., *Sexual Selection and the Descent of Man*, 17–39. Chicago: Aldine.

———. 1973. "The Divine Non Sequitur." In G. Browning, J. Alioto, and S. Farber, eds., *Teilhard de Chardin in Quest of the Perfection of Man*. Rutherford, N.J.: Fairleigh Dickinson University Press.

———. 1974a. "Recent Advances in Methods of Phylogenetic Inference." In Luckett and Szalay, eds., *Phylogeny of the Primates*, 3–19.

———. 1974b. "Response to Ernst Mayr's 1974 questionnaire on the history of the development of the modern evolutionary synthesis." GGS/APS.

———. 1975. "Transcription of Comments on His Bibliography." GGS/APS.

———. 1976. "The Compleat Paleontologist?" *Annual Review of Earth and Planetary Sciences* 4: 1–13.

———. 1977a. "A New Heaven and a New Earth and a New Man." In D. W. Corson, ed., *Man's Place in the Universe*, 53–75. Tucson: University of Arizona Press.

———. 1977b. "Too Many Lines: The Limits of the Oriental and Australian Zoogeographic Regions." *Proceedings of the American Philosophical Society* 121: 102–120.

———. 1978. *Concession to the Improbable: An Unconventional Autobiography.* New Haven: Yale University Press.

———. 1980a. "Biographical Essay." Edited and abridged by E. Mayr in Mayr and Provine, eds., *The Evolutionary Synthesis*, 452–63.

———. 1980b. *Splendid Isolation.* New Haven: Yale University Press.

———. 1980c. *Why and How: Some Problems and Methods in Historical Biology.* New York: Pergamon Press.

———. 1981a. "History of Vertebrate Paleontology in the San Juan Basin." In S. G. Lucas, J. K. Rigby, Jr., and B. S. Kues, eds., *Advances in San Juan Paleontology.* Albuquerque: University of New Mexico Press.

———. 1981b. "Prologue: Historical Biology and Physical Anthropology." *American Journal of Physical Anthropology* 56: 335–38.

———. 1983. *Fossils and the History of Life.* San Francisco: W. H. Freeman.

———. 1984. *Discoverers of the Lost World.* New Haven: Yale University Press.

———. 1986 [posthumous]. "G. G. Simpson's Reflections on W. D. Matthew." *Palaios* 1: 200–204.

———. 1987. *Simple Curiosity: Letters from George Gaylord Simpson to His Family, 1921–1970.* Edited by L. F. Laporte. Berkeley: University of California Press.

———. 1996 [posthumous]. *The Dechronization of Sam Magruder.* New York: St. Martin's.

Simpson, G. G. and A. Roe. 1939. *Quantitative Zoology.* New York: McGraw-Hill.

Simpson, G. G., A. Roe, and R. Lewontin. 1960. *Quantitative Zoology.* 2d ed. New York: McGraw-Hill.

Simpson, G. G. and W. S. Beck. 1965. 2d ed. *Life: An Introduction to Biology* (1957). New York: Harcourt, Brace, and World.

Smocovitis, V. B. 1992. "Unifying Biology: The Evolutionary Synthesis and Evolutionary Biology." *Journal of the History of Biology* 25: 1–65.

Spencer, F., ed. 1982. *A History of American Physical Anthropology, 1930–1980.* New York and London: Academic Press.

Stanley, S. M. 1978. *Macroevolution: Pattern and Process.* San Francisco: W. H. Freeman.

Stebbins, G. L. 1949. "Rates of Evolution in Plants." In Jepsen, Simpson, and Mayr, eds., *Genetics, Paleontology, and Evolution*, 229–42.

———. 1950. *Variation and Evolution in Plants.* New York: Columbia University Press.

Stebbins, G. L. and F. Ayala. 1981. "Is a New Evolutionary Synthesis Necessary?" *Science* 213: 967–71.

Stocking, G. 1988. "Bones, Bodies, Behavior." In Stocking, ed., *Bones, Bodies, and Behavior*, 3–17. History of Anthropology Series, vol. 5. Madison: University of Wisconsin Press.

Washburn, S. 1983. "Evolution of a Teacher." *Annual Review of Anthropology* 12: 1–24.

Wegener, A. 1929 (4th ed.). *Die Entstehung der Kontinente und Ozeane*. Berlin: Friedrich Vieweg.

———. 1966 (English translation of Wegener's 1929 ed.). *The Origin of Continents and Oceans*. Translated by John Biram. New York: Dover.

Weidenreich, F. 1940. "Some Problems Dealing with Ancient Man." *American Anthropologist* 42: 375–83.

Westoll, T. S. 1944. "The Haplolepidae, a New Family of Late Carboniferous Bony Fishes." *American Museum of Natural History Bulletin* 83: 1–121.

Woodruff, L. L. 1922. *Foundations of Biology*. New York: Macmillan.

———. 1923. "Biology." In Woodruff, ed., *The Development of the Sciences*, 247–59. New Haven: Yale University Press.

Wright, S. 1931. "Evolution in Mendelian Populations." *Genetics* 16: 97–159.

———. 1932. "The Roles of Mutation, Inbreeding, Crossbreeding, and Selection in Evolution." *Proceedings of the 6th International Congress of Genetics* 1: 356–66.

———. 1945. "A Critical Review." *Ecology* 26: 415–19.

———. 1949. "Adaptation and Selection." In Jepsen, Simpson, and Mayr, eds., *Genetics, Paleontology, and Evolution*, 365–89.

Zirkle, C.. 1947. "Review of *Tempo and Mode in Evolution*." *Isis* 37: 109–110.

INDEX

Locators for figures are in **bold**. Locators for tables are indicated by t.

acquired characteristics, 2
adaptation: defined, 151; organism–environment interactions, 151–52; phases, 152, 234; preadaptation (*see* preadaptation); and survival, 151
adaptive grid, 30; defined, 153; in evolution, 106–107, **158**, 289*n*67; in evolutionary synthesis, 235; phyletic evolution, 159; and rates of evolution, 153, **154**, 155; as visualization, 234–36
adaptive landscape: equid evolution, **138**; in natural selection, 137, 139, 230–31, **231**, **232**; Wright on, 290*n*91
adaptive radiation, 124–25, 178, 292*n*14
adaptive zones: bridging of gaps, 234–35; fossil record, 153, 155; invasion of, 161; multidimensional nature, 152–53; preadaptation and, 153; rates of evolution, 153, 155; Wright on, 289*n*66
Allee, W. C., 168
Allen, Garland, 68
alloiometry, 121
American Association for the Advancement of Science, 121
American Association of Physical Anthropologists, 175
American Museum of Natural History: abolition of department of paleontology, 246; Albert Parr as director, 245–46; biological focus on fossils, 17, 21–25; finances during Depression, 243; Henry Fairfield Osborn and, 120; Paleontology Field Fund, 52–53, 64–65, 275*n*23; Silberling fossil collection, 82

— Simpson and: administrative style, 247–49; as assistant curator of vertebrate paleontology, 21–25, 241–43; attitude of, 240; as chairman, department of geology and paleontology, 10, 246–49; complications with initial appointment, 6, 241; fundraising, 243–44; initial assignments, 242; part-time work following injury, 249–51; research and expeditions, 243–45; resignation, 249–51, 301n48; strained relations with, 253

American Society of Mammalogists, 40

American Society of Zoologists, 40, 273n80

anagenesis, 108, 158

Andrews, Roy Chapman, 245

Antarctic fossils, 215–16

anthropology, physical. *See* physical anthropology

arctocyonid carnivores, 91

aristogenesis, 2, 24, 121–22, 270n31

Arldt, Theodore, 198

Axelrod, Daniel, 214

Baitsell, G. A., 20

Balfour, Francis M., 24

Bassler, R. P., 81

biogeography, historical: Colbert on, 218; continued acceptance of, 217; Matthew on, 196; migration patterns (*see* migration)

— Simpson on: arguments for stable continents, 199; development of theory, 109; dispersal of organisms, 203–205; *Evolution and Geography*, 199; explaining paleobiogeographic data, 195; incorporation of plate tectonics, 216–17; quantitative insights, 205–11; reliance on Cenozoic mammals, 199, 208–209, 212–13, 218–19; theory unaffected by continental drift, 219–20

biology: Darwinian impact on, 73–74; Richard Swann Lull on, 20; role of paleontology in, 17–18; Simpson studies at Yale, 19–21

biometry, 92

Birdsell, Joseph, 291n2, 291n5

bookplate, 237, **238**

bradytely, 146–47, 153, 235, 298n25

Brinkmann, Roland, 140, 288n61

Bronowski, Jacob, 190

Brown, Barnum, 242, 243

Bucher, Walter, 165, 290n85

Bullock, Theodore, 273n80

Burns, Joan Simpson, 257, 298n1

Butler, Samuel, 76

Cabrera, Ángel, 93

Caenolestoidea, 147

Carnegie Museum (Pittsburgh), 82

carnivores, arctocyonid, 91

Carson, Hampton, 285n12

Cartmill, Matt, 178, 293n15

cladogenesis, 108, 156, 158

Cloud, Preston, 214, 273n64, 283n26

codes and puzzles, 259

coefficient of faunal similarity, 195, 205–11, 207t

Colbert, Edwin H.: on biogeography, 218; on *Lystrosaurus* migration, 216; work on *Tempo and Mode in Evolution*, 127

— and American Museum of Natural History: as acting administrator for Simpson, 247–51; on finances during Depression,

243; Paleontology Field Fund, 64, 275n23; reassignment during reorganization, 246; role in Simpson's resignation, 249–51; as Simpson colleague, 242; on Simpson's administrative style, 249
Columbia University, 10, 247
Committee of Common Problems of Genetics, Paleontology, and Systematics, 166
competitive exclusion, 26, 271n41
continental drift: Du Toit on, 213–14; fossil data and, 196, 214; Wegener on, 196, 197–98
— attitudes toward theory: basis of objections, 196; geologists, 196; paleontologists, 214
— Simpson opposition to, 195–220; acceptance of plate tectonics as cause for, 215–17; basis of opposition, 11, 109, 196–97, 200, 202; *Evolution and Geography*, 198, 211, 216; fallacies in Simpson's arguments, 218–19; identification of proponents' flawed arguments, 200–202; a priori stabilist assumptions, 196; zoological test of paleogeographic theory, **212**, 212–13. See also plate tectonics
Coon, Carlton, 191–92
corridors, as migration bridges, 203–204
Cracraft, Joel, 196
cryptogenetics, 287n51
Ctenacodon, **100**

Darwin, Charles: on adaptive radiation, 235, 297n24; on convergent evolution, 76; Darwinian themes in Simpson's work, 73–78; eclectic methodology, 76; influence on Simpson, 67–70, 278n5; on notoungulates, 113; Simpson on, 76–77, 78–79
Davis, Dwight, 167
de Beer, Gavin, 23
De Vore, Irven, 175
Dechronization of Sam Magruder, The, 256–66, 301n2
dentition, 89–92
Dobzhansky, Theodosius: career, 117; as Cold Spring Harbor symposium organizer, 178–79; and evolutionary synthesis, 117–19, 168, 285n13, 285n17, 302n7; incorporation of Simpson's views, 170; influence on Simpson, 25–26, 115–16, 270n37, 270n38; on mutation, 133; mysticism and evolution, 293n37; review of *Tempo and Mode in Evolution*, 31–32, 163; Simpson on death of, 258, 302n7
Docodon, **99**
Du Toit, Alexander: on land bridges, 198, 201; on Laurasia, 208; rebuttal of Simpson's arguments, 213–14
Dunbar, C. O., 273n70

ecological incompatibility, 26, 93, 106, 283n21
ecology: evolutionary, 143–44, 202–203; paleoecology, 25, 80–95
Ectocion, **103**
Eldredge, Niles, 97, 281n2
Elliott, David, 215
Emerson, Alfred, 168–69
environment: and adaptive grid, 234; and evolution, 74, 106–107,

environment: and adaptive grid (*continued*) 151–55; function of organisms, 231–34, **233**; interactions with, 230–36; Lull on change in, 125; preadaptation, 234

evolution: acquired characteristics, 2; adaptive grid (*see* adaptive grid); alloiometry, 121; aristogenesis, 2, 121; bradytelic (*see* bradytely); branching, 149; convergent, 76; Darwinian themes, 73–78; determinants, 131–39; Dobzhansky on, 118–19; and environment, 74, 106–107, 151–55; and experimental genetics, 118–19; fossil record, 2, 36, 96; function in, 284*n*5; generation length, 134–35; horotelic (*see* horotely); "how" of nature, 75; inertia and momentum, 2, 148–50; long-term trends, 148; and mutation, 122, 123, 133–34, 286*n*37; natural selection, 136–37, 139; orthogenesis, 2, 120; and population size, 135–36, **144**; purposefulness of organic adaptation, 74; quantification of, 111; quantum (*see* quantum evolution); racial senescence, 2; rectilinear, 148–50; soft inheritance, 288*n*57, 288*n*60; species in, 108–11; survivorship in, 130–31, 287*n*53; synthesis (*see* evolutionary synthesis); tachytelic (*see* tachytely); variability in, 132–33; watchmaker argument, 74

— discontinuities: morphological, 139–44; systemic, 160–61, 234–35, 288*n*62

— and ecology, 143–44, 202–203 (*see also* ecological incompatibility)

— human: man as natural product, 75; mentality in, 177, 292*n*12; mystical approach, 191; Simpson qualifications in, 176, 292*n*7

— modes, 155–62; and adaptive grid, **158**; defined, 155; instability, 160; phyletic, 158–59, 283*n*26, 284*n*2; speciation, 156

— rates: adaptive grid, 153, **154**, 155, 235, 298*n*25; adaptive radiation, 167–68; defined, 272*n*56; high versus low rate lines, 144–47; and mode, **157**, 159; palimpsest theory, 23; in plants, 169–70; Simpson on, 128–31; variation in, 2, 27, 30–31. *See also* macroevolution; megaevolution; microevolution

Evolution (journal), 34, 166, 172

evolutionary synthesis: adaptive grid in, 235; botanical contributions, 169–70; contributors, 168; development, 2, 17, 165, 300*n*24; Dobzhansky and (*see* Dobzhansky, Theodosius); Jesup Lectures and, 300*n*24; Mayr on, 35, 67, 258; paleoanthropology and, 175–76, 182, 193, 291*n*2; paleontology and, 2, 34–35; population concept, 88–89; primates, 180–81; Princeton conference, 165; *Principles of Animal Ecology*, 168–69; Romer on, 34; Stebbins on, 167, 169–70; views of, 281*n*2; Wright on, 167–68

— Simpson and, 1–2, 17–18, 96; *Meaning of Evolution*, 177–78, 292*n*13; *Tempo and Mode in Evolution*, 29, 34–35

extinction, 148, 160, 184

Falkenbach, Charles, 50, 54–55, 60, 276*n*43
filter bridges, 205
Fisher, R. A., 115–16, 168
Fort Union group of mammals. *See* Paleocene mammals
fossils: adaptive zones, 153, 155; and continental drift theory, 196, 197–98, 201–202; as evolutionary record, 2, 25, 96, 202; interpretation, 94–95; as once-living creatures, 21; in population genetics, 107, 172, 177; rate of evolution, 128–31; statistical analysis, 27–28, 99–101, **102**, 103–104, 282*n*16; validating biological theory, 36
— discontinuities: horizontal versus vertical, 140–41; systematic, 141–42, 288*n*62
founder principle, 132
Frankel, Henry, 197
Frick, Childs: animosity toward Simpson, 64–66; and Paleontology Field Fund, 52, 64–65, 275*n*23; paleontology research, 53; relationship with American Museum of Natural History, 51–53
— Santa Fe formation: claim on *Hemicyon* discovery and results, 59–60; funding of 1924 field work, 51–53, 60; interest in, 57; suppression of Simpson paper on, 43–44, 62–64
Frick, Henry Clay, 52

Gause, G. F., 26
generations, length of, 134–35
genetics: and evolution, 118–19; heredity, 149; phenogenetics, 127, 287*n*51; phylogenetic patterns, 227–30, 297*n*12; and rectilinear evolution, 149, 289*n*65
— paleontology: conflict with, 126; consistency with new genetics, 131, 287*n*54; synthesis, 125–26, 287*n*46
— population: and diversity of life, 119; and fossil record, 107, 172, 177; mathematical analysis, 116–17
Geological Society of America, 38
Gidley, James W., 81, 82, 200
Gillespie, N. C., 279*n*36
Gilmore, Charles W., 81, 82
Glass, Bentley, 32, 163
Goldschmidt, Richard: on orthogenesis, 120, 286*n*28; on rates of evolution, 272*n*56; saltation and evolution, 116, 119, 143
Gondwana, 198, 199, 215, 216
Gould, Stephen, 68, 97, 118, 281*n*2
Granger, Walter, 59, 86, 101, 242
Great AEPPS, 168
Greene, John, 67, 68
Gregory, John W., 198, 201
Gregory, William King: on biology in paleontology, 17; classification of mammals, 176, 177; influence on Simpson, 22, 23–24, 270*n*26, 292*n*9; palimpsest theory, 23; as Simpson colleague, 113, 242

Haas, Otto, 246
Haldane, J. B. S., 115–16, 168, 175, 285*n*12
Harrison, Ross, 20
Hemicyon: discovery and excavation by Simpson and Falkenbach, 54–57, 60; Frick claim on discovery and results, 59–60; restored skeleton, **58**; Simpson field notes, 57

historical biogeography. *See* biogeography, historical
Hominidae: and evolutionary synthesis, 180; grade recognition, 180–81; Simpson classification, 176, 183–89, **188**
Hooton, Earnest A., 175–76, 291*n*5
horotely, 146–47, 153, 235, 298*n*25
horses: evolution, 30, 45, **49**, 129–30, 149–50; mutation, 133–34; natural selection, 137, **138**, 139; species names, 47, 275*n*18
Howells, William, 178–79, 189
human characteristics, 190
Hunter, Fenley (Mr. and Mrs.), 86
Hutchinson, G. E., 32
Huxley, J., 32, 164, 168, 290*n*83
Huxley, Thomas H., 24
hypodigm, versus typology, 35–36

incompatibility, ecological. *See* ecological incompatibility
inertia and momentum, 2, 148–50
interdisciplinary communication, 165

Jepson, Glenn: attitude toward Simpson, 272*n*59; Princeton conference papers, 166; review of *Tempo and Mode in Evolution*, 32, 164–65, 290*n*84; theoretical focus, 273*n*64
Jesup, Morris K., 300*n*24
Johnson, Lyndon B., 14
Joleaud, Léonce, 200–201
Jurassic Morrison Formation of Wyoming, 21

Kennedy, J. M., 299*n*17
kinetogenesis, 124

Kingsley, Charles, 69, 70–71
Kinney, Helen J. *See* Simpson, Helen J. (nee Kinney) (mother)
Kinsey, Alfred C.: images, 228–29, 297*n*16; Simpson rebuttal of, 228–29, **229**; on taxonomy, 27, 113–14, 271*n*44
Kitts, David, 287*n*46

Lamarckian inheritance, 122, 123–24, 148
land bridges, transoceanic: continental drift theory, 197; Gregory on, 201; Joleaud on, 200–201; paleobiogeographic data, 195; Simpson on, 200–202, 209
Laudan, Rachel, 171–72
Laurasia, 199, 208, 209
Le Gros Clark, Wilfrid, 178, 193, 292*n*7
Leakey, Louis, 185, 293*n*37
Lull, Richard Swann: on adaptive radiation, 124–25; education and career, 122; on evolutionary inertia and momentum, 148; on kinetogenesis, 124; on Lamarckian inheritance, 122, 123–24; on natural selection, 123, 124; on orthogenesis, 123, 148; on paleobiology, 20, 122–23; on pulse of life, 125; religion and science, 286*n*41; on response to environmental change, 125
— and Simpson: influence on Lull's work, 286*n*39; Lull as dissertation adviser, 6, 43, 44–45; recommendation to Matthew, 44–45; Simpson on work of, 20; Yale job offer, 241
Lyell, Charles, 171
Lystrosaurus, 215, 216

macroevolution: defined, 139; explained by microevolution, 1, 29–30, 106–109, 115, 179; Schindewolf on, 125–26
mammals: classification, 25–26; coefficient of faunal similarity, 206–209; contouring of species density, 224, **226**; primitive, as Simpson dissertation topic, 6; taxonomy, 282n9
Marks, Jonathan, 291n2
Marsh Collection of Mesozoic Mammals, 6, 43, 44, 59, 66, 274n3
Marvin, Ursula, 197
Matthew, William Diller: on biology in paleontology, 17, 22–23; correspondence with Richard Swann Lull, 44–45; on historical biogeography, 196; and horse evolution, 45; influence on Simpson, 22–23, 43; law of ecological incompatibility, 283n21; on mutation, 134; portrait, **43**; recommendation of Simpson to American Museum of Natural History, 241–42; research areas, 45; at University of California, Berkeley, 6, 241–42
— field expedition (1924), 45–50; discovery of new horse genus, 47; Simpson as field assistant, 7, 41–66; sketch by Simpson, **46**
Mayr, Ernst, 168: biological species concept, 110–11, 284n36; on evolution and paleontology, 36; and evolutionary synthesis, 35, 67, 258; as leader of Society for the Study of Speciation, 166; on migration routes, 208; nature of writings, 278n6; on population structure, 281n3; on Princeton conference, 166; on soft inheritance, 288n57, 288n60; on taxonomy, 281n4; on visualization, 296n2
— on Simpson: appointment to Harvard, 253; citations of Simpson's work, 170–71; extension of Harvard appointment after relocation to Arizona, 254; view of species, 97; worldview, 67
McKenna, Malcolm, 196
Mead, Margaret, 244
megaevolution, 139, 283n22
melancholia, 263–64
Merriam, J. C., 45–46, 277n61
Metachriacus, **105**, 105–106
microevolution: defined, 139; explaining macroevolution, 1, 29–30, 106–109, 115, 179; rates and styles, 106; Schindewolf on, 125–26
migration: barriers, 203–205; climate and, 209; coefficient of faunal similarity, 195, 205–11, 207t
— transoceanic dispersal: continental drift theory, 198; dispersal across water, 210; dispersal probability, 210–11, 211t; distribution of species, 109; environmental effects, 211; expanding terrestrial vertebrate populations, **204**; geological time spans, 195
mode, defined, 155
momentum. *See* inertia and momentum
Moore, John, 252
Morgan, Thomas Hunt, 117
morphology, 129, 223–24, **225**

Morris, Desmond, 191
mosaic evolution, 23, 269*n*25
multituberculates: *Ctenacodon* species differentiation, **100**; dental characters, 88–89; dentition scatter diagram, **104**; paleobiology, 21; statistical analysis, 89–92, 103–104; taxonomy, 88–89
Museum of Comparative Zoology, Harvard University, 12, 240, 251–54
mutation: evolution and, 122, 123, 133–34, 286*n*37; and population, 119; probability, 149; soft inheritance, 288*n*60
mysticism, 191

naked ape, 191
National Academy of Sciences, 112, 120, 165
natural history, 171–72
natural selection: adaptation, 152; adaptive landscape, 137, **138**, 139, **231**, **232**; aristogenesis, 121–22; in evolution, 136–39; Lull on, 123, 124; Simpson on, 74
neo-Darwinism, 116
New York Academy of Sciences, 182–83
Newell, Norman D., 242, 247–49, 273*n*64
notoungulates, 25, 27, 113–14

Olson, Everett C., 34, 273*n*64, 279*n*2, 288*n*62, 289*n*77
Opdyke, Neil, 214
orthogenesis: in evolution, 2; Goldschmidt on, 120, 286*n*28; Lull on, 123, 148; versus orthoselection, 170
orthoselection, 170

Osborn, Henry Fairfield: American Museum of Natural History, 120; on aristogenesis, 121–22; on biology in paleontology, 17; as fundraiser, 299*n*17; *Hipparion* fossils, 200; on Lamarckian inheritance, 122; on mutation and evolution, 122, 123, 286*n*37; principles of paleontology, 121; publications, 120–21
— and Simpson: as colleague, 242; correspondence with Richard Swann Lull, 44; first meeting, 286*n*33; influence on, 22, 24–25; reservations about hiring, 55
Owen, Richard, 113

paleoanthropology: biases in, 292*n*8; changing focus, 189; classification versus phylogenetic interpretation, 183; Cold Spring Harbor symposia, 178–79, 179–80; evolutionary synthesis, 175–76, 182, 193, 291*n*2; influence of other disciplines, 174, 193; naked ape, 191
— Simpson and: as apologist, 189–94; credentials, 176–78, 292*n*7; as critic, 182–89; influence on, 174–94; as mentor, 178–82, 193; opposition to typological thinking, 27, 185, 271*n*43
paleobiogeography: and continental drift, 195–220; Gondwana and, 215–16; migration of species, 109. *See also* biogeography, historical
paleobiology: Lull on, 20; multituberculates, 21; terminology, 34, 172
Paleocene mammals of Montana: collections, 87; ecological in-

compatibility, 93, 106; first field season (1932), 83–84; Gilmore work, 81; illustrations causing publication delay, 85; preservational bias, 93–94; quarries, 86, 280*n*31; second field season (1935), 86–87; Silberling discovery of, 81–82; Simpson work, 25, 80–95, 244, 283*n*24; specimen lots, 92–93; stratigraphy, 92–93; Sydney Prentice as illustrator, 84

paleoecology, 25, 80–95

Paleontological Society, 38

paleontology: in biological sciences, 17–18, 40, 269*n*2; comparative morphology, 17; conflicting approaches, 171; disciplinary independence, 37–40; evolutionary synthesis, 2, 34–35; hypodigm in, 35; Osborn principles, 121; professional societies, 38–39; shift from natural history to natural philosophy, 172; Simpson's impact on, 2; theory in 1930s, 120–26; typology (*see* typology)

— and genetics: conflict with, 126; consistency with new genetics, 131, 287*n*54; phenogenetics, 127; relations with geneticists, 286*n*29

Park, O., 168

Park, T., 168

Parr, Albert: relationship with Simpson, 300*n*33; reorganization of American Museum of Natural History, 38, 245–46; on Simpson's resignation, 250–51, 301*n*48

Patagonia expeditions, 7, 85, 244, 299*n*16

Patterson, Bryan, 259, 302*n*9

Peabody Museum of Natural History, 6, 44, 121

Pedroja, Lydia. *See* Simpson, Lydia Pedroja

phenogenetics. *See* genetics

phyletic evolution, 158–59, 283*n*26, 284*n*2

phylogenetic patterns, 227–30, 297*n*12

physical anthropology, 175–76, 177–78, 291*n*2, 292*n*13

plate tectonics: fossils irrelevant to, 214–15; kinematic versus dynamic concept, 283*n*35; revolution in, 213–15; sea floor spreading and continental drift, 215; Simpson acceptance of, 11, 109, 215–17. *See also* continental drift

population: and fossil record, 107; genetics, 116, 119, 177; and mutation, 119; species as, 26–27, 34–37, 88–89; in taxonomy, 185; variation in, 132

positivism, 77, 279*n*36

preadaptation: adaptive zones, 153; defined, 152; early theorists, 288*n*60; environment, 234; tetrapod limb, 152, 234

Prentice, Sydney, 84, 86

preservational bias, 93–94

primates: classification, 176, 177–78, 183–84, 292*n*14; extinctions, 184; grade concept, 180–81, **182**; specialization and diversity, 179–80

— evolution: adaptive radiations, 178, **181**; evolutionary synthesis, 180–81; fossil evidence, 184; increase in evolutionary grade, **182**; rates of, 145–46

Princeton conference, 165–68

Provine, William, 68, 281*n*2, 297*n*20

quantum evolution: defined, 298n25; explosive phase, **161**; integration with other modes, 160; Simpson on, 107, 159–62, 289n69; systemic discontinuities in, 160–61. *See also* tachytely

race, 190, 192, 294n42
racial senescence, evolutionary, 2
Rainger, R., 269n2, 276n43, 299n17
Rak, Joseph, 57
ratio diagrams, 223–24, **225**
rectilinear evolution, 148–50, 159
religion: Lull on, 286n41; Simpson's attitude, 4, 70, 75
reptiles, Triassic, 208
Roe, Anne: correspondence with Simpson, 71–72; death, 16; first marriage, 8; fundamental evangelicism, 8; illnesses, 14, 253, 294n44; marriage to Simpson, 9, 245; open-heart surgery, 258; professorship at Harvard Graduate School of Education, 13; publication of *Tempo and Mode in Evolution*, 126–27; *Quantitative Zoology* (with Simpson), 8, 25, 100, 112, 282n11; relationship with Simpson prior to marriage, 4, 8–9, 85, 302n20
Roe, Bob, 4, **5**
Romer, Alfred S.: on continental drift, 218; on evolutionary synthesis, 34; offer of Harvard appointment to Simpson, 251, 273n69; president of Society for Vertebrate Paleontology, 38–39
Rudwick, Martin, 239
Runcorn, Keith, 214
Ruthven, Alexander, 245

saltation, 116, 119, 143, 167
Santa Fe formation, 43–44, 50–57, 60–64, 276n44
Scarritt, Horace, 85, 86, 243, 299n15
Schaeffer, Bobb, 242, 248, 252
Schindewolf, Otto, 116, 125–26, 143, 287n46
Schmidt, K. P., 168
scientific method, 70
Shapere, Dudley, 172–73
sibling species, 134
Silberling, Albert C., 82, 83
Simpson, Anne Roe. *See* Roe, Anne
Simpson, Elizabeth (daughter), 6, 258, 294n44
Simpson, George Gaylord: achievements, generally, 1; awareness of personal mortality, 259; biographical sources, xii, 267n1; on biology, 25–28, 29–30, 40; on classification of mammals, 25–26, 176–77; on continental drift (*see* continental drift); and Dobzhansky (*see* Dobzhansky, Theodosius); and *Evolution* (journal), 34; health problems, 12, 14, 249, 253, 273n80; and Henry Fairfield Osborn (*see* Osborn, Henry Fairfield); historical biogeography (*see* biogeography, historical); honors and awards, 14, 40, 127; impact of work, 18; influences on, 270n37; mathematical facility, 100, 282n12; military service, 9–10, 126, 246; musical interests, 264; obituaries and memorials, 267n1; and paleoanthropology (*see* paleoanthropology); on Paleocene mammals of Montana (*see* Paleocene mammals of Montana);

peer recognition, 3, 247–49, 254–55; personal relationships, 3–4; personality, 3, 265–66; philosophical and theoretical viewpoint, 67–68; plate tectonics (*see* plate tectonics); positivistic philosophy, 77, 176, 256, 279*n*36; professional grievances, 64–66, 240, 254–55; public awareness of, 2–3, 11, 244, 262, 299*n*17; quantitative insights, 205–11; on race, 190, 192, 294*n*42; recognition at Princeton conference, 166–68; on religion, 4, 70, 75; and Richard Swann Lull (*see* Lull, Richard Swann); as Sam Magruder, 264–66; on species (*see* species and speciation); spelling experimentation, 275*n*35; on taxonomy, 28, 183–89, 192; on Teilhard de Chardin, 191; travels, 13; visual orientation (*see* visual language of Simpson); and William Diller Matthew, 43 (*see* Matthew, William Diller); and William King Gregory, 23–24, 292*n*9 (*see* Gregory, William King); work on Marsh Collection of Mesozoic Mammals, 6, 43, 44, 59, 66; writings on descriptive morphology, taxonomy, and systematics, 39

— appointments: American Museum of Natural History (*see* American Museum of Natural History); Columbia University, 10, 247; Museum of Comparative Zoology, Harvard University, 12, 240, 251–54; University of Arizona, 14, 254

— childhood, 3–4; apostasy, 68–69; friendships, 4; illnesses, 4; purchase of *Encyclopedia Britannica*, 4, 70

— and Darwin: commentaries, 78–79; Darwinian themes, 69–70, 73–78; influence of, 13, 67–79

— education: academic achievements, 44; courses at Yale, 19–20; cross-disciplinary studies, 18–19, 18–21; doctoral dissertation, 6; high school, 4; study at British Museum of Natural History, 6, 241; undergraduate, 4–5; University of Colorado at Boulder, 4; Yale University, 5, 6, 19–21

— evolution: attitude toward, 68; evolutionary ecology (*see* ecology, evolutionary); evolutionary philosophy, 278*n*18; evolutionary synthesis (*see* evolutionary synthesis); formal introduction to, 72–73; human, 292*n*7, 292*n*9, 292*n*12; as theorist, 13, 107, 168–69

— expeditions: 1924 field expedition (*see below*, field expedition [1924]); Patagonia, 7, 85, 244, 299*n*16; upper headwaters of the Amazon River, 12, 249; Venezuela, 245

— family: daughters, 6, 9, 12; parents, 3–4; relationship with father, 264, 302*n*21

— field expedition (1924): collecting other men's prospects, 57, 59–61; discovery of *Hemicyon*, 60; as field assistant to Matthew, 7, 22, 41–66; first fossil discovery, 47–50;

Simpson, George Gaylord
— field expedition (1924) (*continued*): Frick suppression of paper on Santa Fe formation, 43–44, 62–64, **63**, **64**, 276*n*44; horse skull discovered by Simpson, **48**; joined by Charles Falkenbach, 50; joined by Joseph Rak, 57; map with location of horse bones, **51**; New Mexico field diary, 56–57; New Mexico field map, **56**; objectives, 45–46; paper on Santa Fe formation, 60–62; rock strata with fossil horse skull, **52**; Simpson's notes of discovery, 48–50; sketch by Simpson, **46**; work on Santa Fe beds, 50–57
— marriages: to Anne Roe (*see* Roe, Anne); to Lydia Pedroja (*see* Simpson, Lydia Pedroja); marital difficulties affecting work, 6, 241, 287*n*48, 299*n*16
— photographs: in Army uniform, **10**; with Bob Roe, **5**; late 1940s, **xviii**; Patagonia expedition, 1933–34, **7**; at home in Arizona, **15**; upon graduation from Yale, **44**
— professional affiliations: American Philosophical Society, 112; American Society of Mammalogists, 40; American Society of Zoologists, 40; National Academy of Sciences, 112; Society for Systematic Zoology, 40; Society for the Study of Evolution, 11–12, 34, 247; Society of Vertebrate Paleontology, 11, 38–39, 247
— publications: 1939–1944, generally, 127; *Attending Marvels*, 7, 112, 244; *Concession to the Improbable*, xi, 236, 259; *The Dechronization of Sam Magruder*, 256–66, 301*n*2; *Evolution and Geography*, 11, 198, **206**, 211, 216; fictional works, 256–66, 301*n*3; *Fossils and the History of Life*, 302*n*7, 302*n*9; gap following graduate school, 270*n*33; *Major Features of Evolution*, 11, 235, 289*n*77; *The Meaning of Evolution*, 11, 170, 177–78, 292*n*13; *Principles of Animal Taxonomy*, 110, 180; *Quantitative Zoology* (with Anne Roe), 8, 95, 100, 112, 282*n*11; *Tempo and Mode in Evolution* (*see Tempo and Mode in Evolution*); *United States National Museum Bulletin 169*, 87–88

Simpson, Helen (daughter), 6
Simpson, Helen J. (nee Kinney) (mother), 3, 70
Simpson, Joan (daughter), 6, 257, 298*n*1
Simpson, Joseph A. (father), 3–4, 70, 302*n*21
Simpson, Lydia Pedroja (first wife): complaints affecting Simpson's work, 6, 241, 287*n*14, 299*n*16; daughters, 6; divorce, 9, 245; marriage, 5–6; mental instability, 5–6, 8, 85; separation, 8, 85
Simpson, Margaret (sister), 3, 4
Simpson, Martha (sister), 3; bookplate design, 237, **238**; correspondence, 236–37, 260; living with Simpson and Anne Roe, 9, 85
Simpson, Patricia Gaylord (daughter), 6, 12
Smith, Grafton Elliot, 292*n*7
Smocovitis, V. B., 111
Society for Systematic Zoology, 40

Society for the Study of Evolution: founding of, 166; Simpson's role, 11–12, 34, 247
Society for the Study of Speciation, 166
Society of Vertebrate Paleontology: initial membership, 277*n*62; Simpson's role, 11, 38–39, 247
species and speciation: biological, 36–37, 110–11; cladogenesis, 156; contouring of density, 224, **226**; defined, 37, 98, 106, 110, 284*n*36; descriptive statistics, 98; Dobzhansky on formation of, 117–18; in evolution, 36–37, 108–11, 110, 156; formation by branching, **108**, 108–109; historical biogeography, 109; hypodigm, 35; inferential statistics, 98–99; kinematic versus dynamic concept, 110; naming of, 275*n*18; sibling, 134; Simpson on, 96–111; type specimens, 104
— population concept: adoption of, 34–37; introduction of, 99; versus typology, 26–27, 35, 88–89, 99, 113, 271*n*43, 282*n*19
— variation: inherent, 105, 107; statistical, 101; supraspecific, 108, 113–115, 283*n*25
statistics: in fossil analysis, 27–28, 99–101, **102**, 103–104, 282*n*16; hypothetical datasets, 113; in mammalian classification, 25–26; in species classification, 98–101; in taxonomy, 89–92
Stebbins, G. Ledyard, 167, 169–70
stereophotography, 222, 223, 296*n*6
survivorship, 130–31, 145–46, 151, 287*n*53
sweepstakes routes, 204–205

Swingle, W. W., 20
systematics, 109, 273*n*66

tachytely: defined, 146, 298*n*25; as evolutionary mode, 153, 235. *See also* quantum evolution
taphonomy, 25
taxonomy: classification versus phylogenetic interpretation, 183; defined, 273*n*66; evolutionary rate, 130; inferential statistics, 98–99; levels, 96–97; multituberculates, 88–89; populations in, 185; procedures, **186**; Simpson on, 183–89, 192; statistical analysis, 89–92; type, defined, 282*n*10
Teilhard de Chardin, Pierre, 191
Tempo and Mode in Evolution, 8, 112–73; arguments, 96, **162**; assumptions, 127; basis of title, 106; citations of, 165–71; contents, 29–31; defining Simpson's career, 112; genesis of, 29, 95, 113–15; highlighting mathematical genetics, 116–17; inertia, trend, and momentum, 147–51; lectures on, 267*n*1; morphological discontinuities, 139–44; organism and environment, 151–55; as stipulatory disciplinary change, 171–72; works influencing, 115–17, 284*n*8, 286*n*42; writing of, 126
— evolution: determinants, 131–39; length of generations, 134–35; modes of, 155–62; mutation, 133–34; natural selection, 136–39; and population size, 135–36, **144**, 288*n*59; rates of, 128–31, 144–47; as treatise on, 170; variability, 132–33

— response to: by biologists, 31–32; by paleontologists, 32–33
— reviews, 31–32, 162–65; by geneticists, 162–63; by naturalists, 164; by paleontologists, 164–65
— significance of, 171–73; in evolutionary synthesis, 1–2, 29, 34–35; evolutionary theory, 170–71; paleontology, 112–13
Thompson, Albert, 86
Tieje, Arthur, 4
time, 257, 259–61
Triconodonta, 97, 282*n*5
typology: hypodigm versus, 35–36; population versus, 26–27, 88–89, 113, 271*n*43; as species concept, 35, 99, 282*n*19

ultraviolet fluorescence, 222
University of Arizona, 14, 254
University of California, Berkeley, 6, 241–42, 277*n*61
University of Colorado at Boulder, 4
U.S. National Museum, 81, 82

variability and variation: defined, 132; in evolution, 2, 27, 30–31, 132–33
— species and speciation: inherent, 105, 107; statistical, 101; supraspecific, 108, 113–15, 283*n*25
visual language of Simpson: adaptive grid, 234–36; adaptive landscape, 230–31; bookplate, 237, **238**; contouring of species density, 224, **226**; in deduction, 227–36; diagrams in learning, 221–22; effect versus affect, 222, 236–39; human monuments, 237; images, **222**, 237, 239; organism-environment interactions, 230–36; organization-environment interactions, 221; in personal correspondence, 222, 236–37; personality and, 236; phylogenetic patterns, 227–30, 297*n*12; ratio diagrams, 223–24, **225**; for simplification of data, 223–24; standard morphotype, 224; stereophotography, 222, 223, 296*n*6; technical observations, 223; techniques, 222; Venn diagrams, 231–34, **233**; as visual induction, 223–24; writing as visual process, 221
Vitalist-Inherent Tendency, 148
von Ihering, Hermann, 198

Washburn, Sherwood, 178–79, 184, 193, 292*n*14
Wegener, Alfred: continental drift theory, 196, 197–98; fossil evidence for continental drift, 197–98, 201–202; on Laurasia, 208; Simpson's theories contrasted with, 11
Weidenreich, Franz, 184
Welch, William H., 24
Wenner-Gren Anthropological Foundation, 184
Westoll, T. Stanley, 218–19
Wetmore, F. A., 81, 82, 83, 84
Whitaker, George, 274*n*13
Williams, Ernest, 253
Woodruff, Lorande L., 19
Woodward, Arthur, 292*n*7
Wright, Sewall: on adaptive landscape, 230, 290*n*91, 297*n*20; influencing Simpson's work, 115–16, 289*n*66; misquotation by Washburn, 181–82; on rate of evolution, 167–68; review of *Tempo and Mode in Evolution*, 32, 163, 288*n*58, 288*n*63, 289*n*74